Beyond Planar Graphs

Seok-Hee Hong · Takeshi Tokuyama
Editors

Beyond Planar Graphs

Communications of NII Shonan Meetings

 Springer

Editors
Seok-Hee Hong
School of Computer Science
University of Sydney
Sydney, NSW, Australia

Takeshi Tokuyama
School of Science and Technology
Kwansei Gakuin University
Sanda, Hyogo, Japan

ISBN 978-981-15-6535-9   ISBN 978-981-15-6533-5   (eBook)
https://doi.org/10.1007/978-981-15-6533-5

This Springer imprint is published by the registered company Springer Nature Singapore Pte Ltd.
The registered company address is: 152 Beach Road, #21-01/04 Gateway East, Singapore 189721, Singapore

Preface

Most real-world data sets are relational, which can be modeled as graphs, consisting of vertices and edges. Planar graphs are fundamental for both Graph Theory and Graph Algorithms, and extensively studied: structural properties and fundamental algorithms for planar graphs have been discovered. However, most real-world graphs, such as social networks and biological networks, are *non-planar*. To analyze and visualize such real-world networks, we need to solve fundamental mathematical and algorithmic research questions on sparse non-planar graphs, called *beyond planar graphs*.

Recently, research topics in topological graph theory generalize the notion of planarity to beyond-planar graphs, i.e., non-planar graphs with topological constraints such as specific types of crossings, or with forbidden crossing patterns. Examples include:

- *k-planar graphs*, which can be embedded with at most k crossings per edge;
- *k-quasi-planar graphs*, which can be embedded without k mutually crossing edges.

Consequently, combinatorics (such as edge density), algorithmics (such as testing/embedding algorithms), and geometric representations (such as straight-line drawings) of beyond-planar graphs have emerged as new research directions.

The NII (National Institute of Informatics) Shonan Meeting No-089 *Algorithmics on Beyond Planar Graphs* was held on November 27–December 1, 2016 in Shonan, Japan, to bring world-renowned researchers on Graph Algorithm, Graph Drawing, Computational Geometry, Graph Theory, and Combinatorial Optimization.

The main aim of the workshop was to identify research opportunities on Beyond Planar Graphs and collaboratively develop innovative theory and algorithms for sparse non-planar topological graphs with specific applications to large and complex network visualization.

The workshop had 26 participants from 7 countries, and consisted of 7 invited talks, open problem sessions, discussion sessions, and report sessions from each working group. Outcomes of the workshop include the Shonan Meeting Report, research articles as well as invited contributions to the book chapters from the participants.

This book contains a selection of book chapters initiated from the Shonan Workshop No-089 on Beyond Planar Graphs. More specifically, it consists of 13 chapters that represent recent advances in various areas of beyond planar graph research. Each book chapter was peer-reviewed according to the book standards.

The main aims and objectives of this book include:

- timely provide the state-of-the-art survey and a bibliography on beyond planar graphs;
- set the research agenda on beyond planar graphs by identifying fundamental research questions and new research directions;
- foster cross-disciplinary research collaboration between Computer Science (Graph Drawing and Computational Geometry) and Mathematics (Graph Theory and Combinatorics).

This book is the first general and extensive review of the algorithmic and mathematical results of beyond planar graphs. New algorithms for beyond planar graphs will be in high demand by practitioners in various application domains to solve complex visualization problems. As such, this book will be a valuable resource for researchers in Graph theory, Algorithms, and Theoretical Computer Science, and will stimulate further deep scientific investigations into many areas of beyond planar graphs.

We wish to thank all the authors for contributing their chapters to this book. We also thank all the participants of the Shonan Workshop No-089 for their valuable contribution and participation during the workshop, which greatly helped to improve many aspects of the chapters published in this book.

Finally, we would like to thank NII for the opportunity to organize a successful meeting to enable these exciting initiatives, and Springer for the opportunity to edit this book, with dedicated assistance and support to make this book possible.

Sydney, Australia   Seok-Hee Hong
Sanda, Japan   Takeshi Tokuyama
March, 2020

Contents

Chapter 1
Beyond Planar Graphs: Introduction

Seok-Hee Hong

Abstract Recent research topics in topological graph theory and graph drawing generalize the notion of planarity to sparse non-planar graphs called *beyond planar graphs* with forbidden crossing patterns. In this chapter, we introduce various types of beyond planar graphs and briefly review known results on the edge density, computational complexity, and algorithms for testing beyond planar graphs.

1.1 Beyond Planar Graphs: Edge Density

Recent research topics in topological graph theory and graph drawing generalize the notion of planarity to sparse non-planar graphs, called *beyond planar graphs*, either with forbidden edge crossing patterns or with specific types of edge crossings. Examples include:

- *k-planar graphs*: graphs which can be embedded with at most k crossings per edge [40].
- *k-quasi-planar graphs*: graphs which can be embedded without k mutually crossing edges [2].
- *RAC graphs*: graphs which can be embedded with right angle crossings [19].
- *fan-crossing-free graphs*: graphs which can be embedded without fan-crossings [17].
- *fan-planar graphs*: graphs which can be embedded such that each edge is crossed by a bundle of edges incident to a common vertex [35].
- *k-gap-planar graphs*: graphs which can be embedded such that each crossing is assigned to one of the two involved edges and each edge is assigned at most k of its crossings.

Figure 1.1 shows examples of forbidden crossing patterns for beyond planar graphs.

S.-H. Hong (✉)
University of Sydney, Sydney, Australia
e-mail: seokhee.hong@sydney.edu.au

© Springer Nature Singapore Pte Ltd. 2020
S.-H. Hong and T. Tokuyama (eds.), *Beyond Planar Graphs*,
https://doi.org/10.1007/978-981-15-6533-5_1

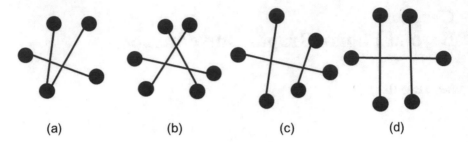

Fig. 1.1 Examples of crossing patterns: **a** fan-crossing (fan-planar and 2-planar graph, but not fan-crossing-free graph); **b** 3 mutually crossing edges (not quasi-planar graph); **c** fan-crossing-free and 2-planar graph (but not fan-planar); **d** RAC and 2-planar graph

Combinatorial aspects of beyond planar graphs are well studied, for example, the maximum number of edges of beyond planar graphs:

- k-planar graphs: Pach and Toth [40] proved that 1-planar graphs with n vertices have at most $4n - 8$ edges.
- k-quasi-planar graphs: Agarwal et al. [2] (respectively, Ackerman [1]) proved that 3 (respectively, 4)-quasi-planar graphs have linear number of edges. Fox et al. [26] showed that k-quasi-planar graphs have at most $O(n \log^{1+o(1)} n)$ edges.
- RAC graphs: Didimo et al. [19] proved that RAC graphs have at most $4n - 10$ edges.
- fan-crossing-free graphs: Cheong et al. [17] proved a tight bound of $4n - 8$ on the maximum number of edges for a 2-fan-crossing-free graph, and an upper bound of $3(k - 1)(n - 2)$ edges for $k \geq 3$.
- fan-planar graphs: Kaufmann and Ueckerdt [35] showed that fan-planar graphs have at most $5n - 10$ edges.
- k-gap-planar graphs: Bae et al. [7] proved that every k-gap-planar graph has $O(\sqrt{k}n)$ edges (for $k = 1$, an upper bound is $5n - 10$). They also study relationships to other classes of beyond planar graphs.

We now briefly review latest results on beyond planar graphs, mainly focusing on the computational complexity and algorithmic aspects.

1.2 Computational Complexity: NP-Hardness

Recently, computational complexity for testing beyond planarity has been studied. More specifically:

- 1-planar graphs: Grigoriev and Bodlaender [29], and Korzhik and Mohar [37] independently proved that testing 1-planarity of a graph is NP-complete. Auer et al. [6]. showed that it remains NP-hard, even if a *rotation system* (i.e., the circular ordering of edges for each vertex) is given.

Furthermore, Cabello and Mohar [15] showed that NP-hardness holds even if the input graph is an *almost planar graph* (i.e., deletion of an edge makes the resulting graph planar). More recently, Bannister et al. [8] studied the fixed parameter complexity of 1-planarity.

- RAC graphs: Argyriou et al. [4] proved that testing whether a given graph admits a straight-line RAC drawing is NP-hard, by presenting an infinite class of graphs with unique RAC embedding.
- fan-planar graphs: Binucci et al. [12] proved that testing fan-planarity of graphs is NP-complete; Bekos et al. [10] showed that it remains NP-hard, even if a rotation system is given.
- gap-planar graphs: Bae et al. [7] proved that testing k-gap-planarity of graphs is NP-complete.

1.3   Polynomial-Time Testing Algorithm

On the positive side, polynomial-time algorithms are available for testing restricted subclasses of beyond planar graphs with additional constraints, as well as computing such an embedding, if it exists.

For example, algorithms for testing special subclasses of 1-planar graphs are well studied:

- *Maximal-1-planar graphs*: Eades et al. [21] showed that the problem of testing the maximal 1-planarity (i.e., addition of an edge destroys 1-planarity) of a graph can be solved in linear time, if a rotation system is given. The embedding is unique, if it exists, and the algorithm also produces the embedding.
- *Outer-1-planar graphs*: Hong et al. [30] and Auer et al. [5] independently presented a linear-time algorithm for testing outer-1-planarity (i.e., 1-planar embedding with each vertex lies on the outer face) of a graph. The algorithm also computes such an embedding, if it exists.
- *Optimal 1-planar graphs*: Optimal-1-planar graph is a special subclass of 1-planar graphs with the maximum of $4n - 8$ edges [41]. A linear-time algorithm was given for testing optimal 1-planarity by Brandenburg [14], using a reduction from optimal 1-planar graphs to irreducible extended wheel graphs.

Figure 1.2 shows examples of maximal 1-planar graphs and outer-1-planar graphs.

For other types of beyond planar graphs, polynomial-time algorithms are also available for testing restricted subclasses of beyond planar graphs with additional constraints. Examples include:

- *Outer-2-planar graphs*: A graph is outer-2-planar, if it admits a drawing where each vertex is placed on the outer boundary and no edge has more than two crossings. A graph is *fully outer-2-planar*, if it admits an outer-2-planar embedding such that no crossing appears along the outer boundary.

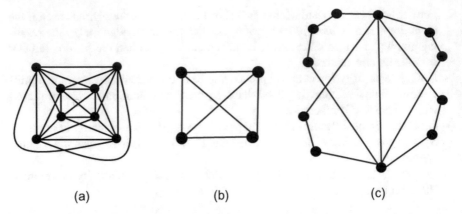

Fig. 1.2 Examples of: **a** maximal 1-planar graphs; **b** triconnected outer-1-planar graphs; **c** biconnected outer-1-planar graphs

Hong and Nagamochi [32] showed that every triconnected full-outer-2-planar graph has a constant number of full-outer-2-planar embeddings. Based on these properties, linear-time algorithms for testing full-outer-2-planarity of a connected, biconnected, and triconnected graph were presented. The algorithms also produce a full-outer-2-planar embedding of a graph, if it exists.

- *Outer k-planar graphs*: Chaplick et al. [16] showed that every outer k-planar graph has a small balanced separator of size at most $2k + 3$, which allow testing outer k-planarity in quasi-polynomial time.

  It was also shown that *closed outer k-planarity* (i.e., the vertex sequence on the boundary is a cycle in the graph) is linear time testable, since outer k-planar graphs have bounded treewidth.
- *Circular-RAC graphs*: Circular-RAC drawing is a circular layout where each vertex lies on the circle and all crossings are with right angles. Dehkordi et al. [18] presented a characterization for circular-RAC graphs, and a linear-time algorithm for testing and constructing such a drawing, if it exists.
- *2-layer RAC graphs*: A 2-layer RAC drawing of a bipartite graph is a straight-line drawing, where each vertex is placed on one of two parallel lines such that no two vertices on the same line are adjacent, and each crossing angle is a right angle. Di Giacomo et al. [27] characterized 2-layer RAC graphs, and presented linear-time testing and embedding algorithms.
- *Maximal outer-fan-planar graphs*: A graph is maximal outer-fan-planar if it has a fan-planar embedding, where every vertex is on the outer face, and insertion of an edge destroys its outer-fan-planarity. Bekos et al. [10] presented a linear-time algorithm for testing whether a graph is maximal outer-fan-planar. The algorithm also computes such an embedding, if it exists.

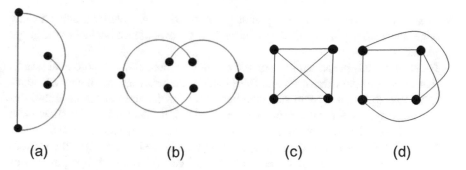

(a)	(b)	(c)	(d)

Fig. 1.3 Examples of: **a** B graph; **b** W graph; **c** straight-line 1-planar drawing of K_4; **d** 1-planar embedding of K_4 containing the B subgraph

1.4 Straight-Line Drawing

The classical *Fáry's Theorem* [25] showed that every *plane graph* (i.e., a planar graph with a given planar embedding) admits a planar straight-line drawing. Indeed, planar straight-line drawing is one of the most popular drawing conventions in Graph Drawing; consequently many straight-line drawing algorithms are available for planar graphs [9, 39].

On the other hand, Thomassen [42] showed that 1-plane graphs (i.e., 1-planar graphs with a given 1-planar embedding) have two forbidden subgraphs, called B graph and W graph, to admit a straight-line drawing. Figure 1.3 shows two forbidden subgraphs of 1-planar graphs.

As such, it opened the way for the investigation for straight-line drawings of beyond planar graphs:

- *1-plane graphs*: Based on the forbidden subgraph characterization by Thomassen [42], Hong et al. [33] presented a linear-time testing and drawing algorithm to construct a straight-line drawing of 1-plane graphs, if it exists. It was also shown that some 1-planar graphs require exponential area for any straight-line drawing.
- *Re-embedding 1-plane graphs*: Re-embedding a 1-plane graph is to change the rotation system or the outer face of the given 1-planar embedding of the 1-plane graph, while preserving the same set of pairs of crossing edges.

  Hong and Nagamochi [31] considered the problem of re-embedding a 1-plane graph, which contains the forbidden subgraphs (i.e., B graph or W graph), to a new 1-planar embedding which admits a straight-line drawing (i.e., 1-planar embedding without B graph or W graph). They presented a characterization of forbidden configuration.

  Based on the characterization, a linear-time algorithm for finding a straight-line drawable 1-planar embedding or the forbidden configuration was presented.

- *Almost planar graphs*: Almost planar graph consists of a planar graph plus one edge, also called graphs with *1-skewness* (i.e., removal of an edge makes the graph planar).

  Eades et al. [22] presented a characterization of almost planar topological graphs that admit a straight-line drawing. Based on the characterization, linear-time algorithms were presented for testing whether an almost planar graph admits a straight-line drawing, and for constructing such a drawing if it exists. It was also shown that some almost planar graphs require exponential area for any straight-line drawing.

- *General embedded non-planar graphs*: Nagamochi [38] investigated the stretchability problem (i.e., straight-line drawings) of general embedded graphs. It was shown that there is a 3-planar embedding and quasi-planar embedding that admits no straight-line drawing, which cannot be characterized by forbidden configuration.

  He also considered a problem of whether a given embedded graph G admits a straight-line drawing under the same *frame*, which is defined by a fixed biconnected planar spanning subgraph of G, and presented forbidden configurations (i.e., a given embedding admits a straight-line drawing under the same frame if and only if it contains no forbidden configuration).

  If a given embedding is quasi-planar (i.e. no pairwise crossing edges) and its crossing-free edges induce a biconnected spanning subgraph, then the stretchability can be tested in polynomial time using forbidden configurations.

1.5   Outlook and Open Problem

This chapter introduces beyond planar graphs and briefly reviews known results on the edge density, computational complexity and algorithmic results on testing and drawing beyond planar graphs.

Many combinatorial results are also studied for beyond planar graphs, including structural properties, various geometric representations, as well as the relationships between beyond planar graphs. Examples include:

- *Structural properties*: Structures of 1-planar graphs are well studied.

  For example, Borodin [13] studied the coloring problem of 1-planar graphs. Fabrici and Madaras [24] presented structural results on 1-planar graphs, while Hudak et al. [34] studied structural properties of maximal 1-planar graphs. Suzuki [41] investigated structural properties of optimal 1-planar graphs.

- *Geometric representation*: Various geometric representations of beyond planar graphs, such as orthogonal drawings, polyline drawings, visibility representations, and book embeddings are also studied.

For example, Biedl et al. [11] studied RVR (Rectangle Visibility Representation) of embedded graphs. Di Giacomo et al. [28] studied polyline drawings of topological graphs with few bends per edge.

- *Relationships between beyond planar graphs*: Relationships between k-planar graphs, RAC graphs, k-quasi-planar graphs, fan-planar graphs and gap-planar graphs are well studied.
  For example, Eades and Liotta [23] studied the relationship between RAC and 1-planar graphs. Angelini et al. [3] showed that every simple k-planar topological graph can be transformed into a simple k-quasi-planar topological graph.

For more details, we refer to corresponding chapters in this book and a recent survey on 1-planar graphs [36] and beyond planar graphs [20].

Finally, we conclude with open problems related to the topics covered in this chapter.

- *Computational complexity*: For most beyond planar graphs, testing problem is known to be NP-complete. However, it is still open for some classes of beyond planar graphs.

  - **Open Problem 1**: Is it NP-complete to test quasi-planarity?
  - **Open Problem 2**: Is it NP-complete to test whether a given graph is a fan-crossing-free graph?

- *Testing algorithm*: Polynomial-time algorithms are available for testing restricted subclass of beyond planar graphs. For example, testing problem becomes tractable when further restrictions such as a rotation system, maximality/optimality, or outer-beyond planarity are assumed.

  - **Open Problem 3**: Is it polynomial time solvable to test maximal quasi-planarity?
  - **Open Problem 4**: Is it polynomial time solvable to test whether a given graph is a maximal fan-crossing-free graph?

- *Straight-line drawability*: Forbidden subgraph characterization to admit a straight-line drawing and linear-time algorithm to construct straight-line drawing if it exists are known for 1-planar graphs and almost planar graphs. For other beyond planar graphs, straight-line drawability problem need further investigation.

  - **Open Problem 5**: Characterize forbidden configuration of RAC graphs to admit a straight-line drawing. Is there an efficient algorithm to construct a straight-line drawing of a RAC graph?
  - **Open Problem 6**: Characterize forbidden configuration of 2-skewness graphs (i.e., removal of two edges makes the resulting graph planar) to admit a straight-line drawing. Is there an efficient algorithm to construct a straight-line drawing of a 2-skewness graph?

Acknowledgements This work is supported by ARC (Australian Research Council) Discovery Project grant.

References

1. Ackerman, E.: On the maximum number of edges in topological graphs with no four pairwise crossing edges. Discret. Comput. Geom. **41**(3), 365–375 (2009). https://doi.org/10.1007/s00454-009-9143-9
2. Agarwal, P.K., Aronov, B., Pach, J., Pollack, R., Sharir, M.: Quasi-planar graphs have a linear number of edges. Combinatorica **17**(1), 1–9 (1997). https://doi.org/10.1007/BF01196127
3. Angelini, P., Bekos, M.A., Brandenburg, F.J., Da Lozzo, G., Di Battista, G., Didimo, W., Hoffmann, M., Liotta, G., Montecchiani, F., Rutter, I., Tóth, C.D.: Simple k-planar graphs are simple (k839+82391)-quasiplanar. J. Comb. Theory, Ser. B **142**, 1–35 (2020). https://doi.org/10.1016/j.jctb.2019.08.006
4. Argyriou, E.N., Bekos, M.A., Symvonis, A.: The straight-line RAC drawing problem is np-hard. J. Graph Algorithms Appl. **16**(2), 569–597 (2012). https://doi.org/10.7155/jgaa.00274
5. Auer, C., Bachmaier, C., Brandenburg, F.J., Gleißner, A., Hanauer, K., Neuwirth, D., Reislhuber, J.: Outer 1-planar graphs. Algorithmica **74**(4), 1293–1320 (2016). https://doi.org/10.1007/s00453-015-0002-1
6. Auer, C., Brandenburg, F.J., Gleißner, A., Reislhuber, J.: 1-planarity of graphs with a rotation system. J. Graph Algorithms Appl. **19**(1), 67–86 (2015). https://doi.org/10.7155/jgaa.00347
7. Bae, S.W., Baffier, J., Chun, J., Eades, P., Eickmeyer, K., Grilli, L., Hong, S., Korman, M., Montecchiani, F., Rutter, I., Tóth, C.D.: Gap-planar graphs. Theor. Comput. Sci. **745**, 36–52 (2018). https://doi.org/10.1016/j.tcs.2018.05.029
8. Bannister, M.J., Cabello, S., Eppstein, D.: Parameterized complexity of 1-planarity. J. Graph Algorithms Appl. **22**(1), 23–49 (2018). https://doi.org/10.7155/jgaa.00457
9. Battista, G.D., Eades, P., Tamassia, R., Tollis, I.G.: Graph Drawing: Algorithms for the Visualization of Graphs. Prentice-Hall (1999)
10. Bekos, M.A., Cornelsen, S., Grilli, L., Hong, S., Kaufmann, M.: On the recognition of fan-planar and maximal outer-fan-planar graphs. Algorithmica **79**(2), 401–427 (2017). https://doi.org/10.1007/s00453-016-0200-5
11. Biedl, T.C., Liotta, G., Montecchiani, F.: Embedding-preserving rectangle visibility representations of nonplanar graphs. Discret. Comput. Geom. **60**(2), 345–380 (2018). https://doi.org/10.1007/s00454-017-9939-y
12. Binucci, C., Giacomo, E.D., Didimo, W., Montecchiani, F., Patrignani, M., Symvonis, A., Tollis, I.G.: Fan-planarity: properties and complexity. Theor. Comput. Sci. **589**, 76–86 (2015). https://doi.org/10.1016/j.tcs.2015.04.020
13. Borodin, O.V.: Solution of the Ringel problem on vertex-face coloring of planar graphs and coloring of 1-planar graphs. Metody Diskret. Analiz. **41**, 12–26, 108 (1984)
14. Brandenburg, F.J.: Recognizing optimal 1-planar graphs in linear time. Algorithmica **80**(1), 1–28 (2018). https://doi.org/10.1007/s00453-016-0226-8
15. Cabello, S., Mohar, B.: Adding one edge to planar graphs makes crossing number and 1-planarity hard. SIAM J. Comput. **42**(5), 1803–1829 (2013). https://doi.org/10.1137/120872310
16. Chaplick, S., Kryven, M., Liotta, G., Löffler, A., Wolff, A.: Beyond outerplanarity. In: F. Frati, K. Ma (eds.) Graph Drawing and Network Visualization - 25th International Symposium, GD 2017, Boston, MA, USA, September 25-27, 2017, Revised Selected Papers, Lecture Notes in Computer Science, vol. 10692, pp. 546–559. Springer (2017). https://doi.org/10.1007/978-3-319-73915-1_42
17. Cheong, O., Har-Peled, S., Kim, H., Kim, H.: On the number of edges of fan-crossing free graphs. Algorithmica **73**(4), 673–695 (2015). https://doi.org/10.1007/s00453-014-9935-z
18. Dehkordi, H.R., Eades, P., Hong, S., Nguyen, Q.H.: Circular right-angle crossing drawings in linear time. Theor. Comput. Sci. **639**, 26–41 (2016). https://doi.org/10.1016/j.tcs.2016.05.017
19. Didimo, W., Eades, P., Liotta, G.: Drawing graphs with right angle crossings. Theor. Comput. Sci. **412**(39), 5156–5166 (2011). https://doi.org/10.1016/j.tcs.2011.05.025
20. Didimo, W., Liotta, G., Montecchiani, F.: A survey on graph drawing beyond planarity. ACM Comput. Surv. **52**(1), 4:1–4:37 (2019). https://doi.org/10.1145/3301281

21. Eades, P., Hong, S., Katoh, N., Liotta, G., Schweitzer, P., Suzuki, Y.: A linear time algorithm for testing maximal 1-planarity of graphs with a rotation system. Theor. Comput. Sci. **513**, 65–76 (2013). https://doi.org/10.1016/j.tcs.2013.09.029
22. Eades, P., Hong, S., Liotta, G., Katoh, N., Poon, S.: Straight-line drawability of a planar graph plus an edge. In: F. Dehne, J. Sack, U. Stege (eds.) Algorithms and Data Structures - 14th International Symposium, WADS 2015, Victoria, BC, Canada, August 5-7, 2015. Proceedings, Lecture Notes in Computer Science, vol. 9214, pp. 301–313. Springer (2015). https://doi.org/10.1007/978-3-319-21840-3_25
23. Eades, P., Liotta, G.: Right angle crossing graphs and 1-planarity. Discret. Appl. Math. **161**(7–8), 961–969 (2013). https://doi.org/10.1016/j.dam.2012.11.019
24. Fabrici, I., Madaras, T.: The structure of 1-planar graphs. Discrete Mathematics **307**(7–8), 854–865 (2007)
25. Fáry, I.: On straight line representations of planar graphs. Acta Sci. Math. Szeged **11**, 229–233 (1948)
26. Fox, J., Pach, J., Suk, A.: The number of edges in k-quasi-planar graphs. SIAM J. Discrete Math. **27**(1), 550–561 (2013). https://doi.org/10.1137/110858586
27. Giacomo, E.D., Didimo, W., Eades, P., Liotta, G.: 2-layer right angle crossing drawings. Algorithmica **68**(4), 954–997 (2014). https://doi.org/10.1007/s00453-012-9706-7
28. Giacomo, E.D., Eades, P., Liotta, G., Meijer, H., Montecchiani, F.: Polyline drawings with topological constraints. Theor. Comput. Sci. **809**, 250–264 (2020). https://doi.org/10.1016/j.tcs.2019.12.016
29. Grigoriev, A., Bodlaender, H.L.: Algorithms for graphs embeddable with few crossings per edge. Algorithmica **49**(1), 1–11 (2007). https://doi.org/10.1007/s00453-007-0010-x
30. Hong, S., Eades, P., Katoh, N., Liotta, G., Schweitzer, P., Suzuki, Y.: A linear-time algorithm for testing outer-1-planarity. Algorithmica **72**(4), 1033–1054 (2015). https://doi.org/10.1007/s00453-014-9890-8
31. Hong, S., Nagamochi, H.: Re-embedding a 1-plane graph into a straight-line drawing in linear time. In: Y. Hu, M. Nöllenburg (eds.) Graph Drawing and Network Visualization - 24th International Symposium, GD 2016, Athens, Greece, September 19-21, 2016, Revised Selected Papers, Lecture Notes in Computer Science, vol. 9801, pp. 321–334. Springer (2016). https://doi.org/10.1007/978-3-319-50106-2_25
32. Hong, S., Nagamochi, H.: A linear-time algorithm for testing full outer-2-planarity. Discret. Appl. Math. **255**, 234–257 (2019). https://doi.org/10.1016/j.dam.2018.08.018
33. Hong, S.H., Eades, P., Liotta, G., Poon, S.H.: Fáry's theorem for 1-planar graphs. In: J. Gudmundsson, J. Mestre, T. Viglas (eds.) Proceedings of COCOON 2012, Lecture Notes in Computer Science, vol. 7434, pp. 335–346. Springer (2012)
34. Hudák, D., Madaras, T., Suzuki, Y.: On properties of maximal 1-planar graphs. Discussiones Mathematicae Graph Theory **32**(4), 737–747 (2012). https://doi.org/10.7151/dmgt.1639
35. Kaufmann, M., Ueckerdt, T.: The density of fan-planar graphs. CoRR (2014). http://arxiv.org/abs/1403.6184
36. Kobourov, S.G., Liotta, G., Montecchiani, F.: An annotated bibliography on 1-planarity. Comput. Sci. Rev. **25**, 49–67 (2017). https://doi.org/10.1016/j.cosrev.2017.06.002
37. Korzhik, V.P., Mohar, B.: Minimal obstructions for 1-immersions and hardness of 1-planarity testing. J. Graph Theory **72**(1), 30–71 (2013)
38. Nagamochi, H.: Straight-line drawability of embedded graph. Technical Report 2013–005, Department of Applied Mathematics and Physics, Kyoto University, Japan (2013)
39. Nishizeki, T., Rahman, M.S.: Planar Graph Drawing, *Lecture Notes Series on Computing*, vol. 12. World Scientific (2004). https://doi.org/10.1142/5648
40. Pach, J., Tóth, G.: Graphs drawn with few crossings per edge. Combinatorica **17**(3), 427–439 (1997)
41. Suzuki, Y.: Optimal 1-planar graphs which triangulate other surfaces. Discret. Math. **310**(1), 6–11 (2010). https://doi.org/10.1016/j.disc.2009.07.016
42. Thomassen, C.: Rectilinear drawings of graphs. J. Graph Theory **12**(3), 335–341 (1988)

Chapter 2
Quantitative Restrictions on Crossing Patterns

Csaba D. Tóth

Abstract This chapter is dedicated to beyond-planar graphs defined in terms of quantitative restrictions on the intersection pattern of edges. These classes include k-planar graph, k-quasiplanar graphs, k-gap-planar graphs, and k-locally planar graphs. The chapter reviews typical proof techniques, upper and lower bounds on the number of edges in these classes, as well as recent results on containment relations between these classes, and concludes with a collection of open problems.

2.1 Introduction

A graph is *planar* if it can be drawn in the plane such that no two edges cross. By relaxing this condition on edge crossings in a drawing, we arrive at a thriving family of graph classes that go beyond planarity. This chapter surveys graphs that can be defined in terms of quantitative bounds on the crossing pattern of the edges in a drawing. It reviews typical proof techniques, upper and lower bounds on various graph parameters for these graph classes, as well as recent results on containment relations between these classes. One of the most basic restrictions on a drawing, which was initiated by Ringel [63] in the 1960s, requires that every edge is involved in at most k crossings, for some constant $k \in \mathbb{N}$. The drawings satisfying this condition are called k-*planar* (see Sect. 2.3), several variants of this concept have been studied over the last few decades.

Typical proof techniques for k-planar drawings may modify some of the edges (by truncating or rerouting the drawing of an edge) locally or globally. Some proofs crucially depend on the assumption that two edges cross at most a constant number of times, hence the maximum number of crossings between two edges becomes an important parameter in the corresponding quantitative results. Section 2.2 briefly reviews the framework developed for handling multigraphs and drawings with multiple crossings between a pair of edges.

C. D. Tóth (✉)
California State University Northridge, Los Angeles, CA, USA
e-mail: csaba.toth@csun.edu

© Springer Nature Singapore Pte Ltd. 2020
S.-H. Hong and T. Tokuyama (eds.), *Beyond Planar Graphs*,
https://doi.org/10.1007/978-981-15-6533-5_2

Section 2.3 showcases the graph classes defined in terms of quantitative bounds on crossing patterns in a drawing (k-planar graphs, k-gap-planar graphs, k-quasiplanar graphs, and k-locally planar graphs). Section 2.4 reviews upper and lower bounds on the density of graphs in these classes, Sect. 2.5 is devoted to containment relations between these classes, and Sect. 2.6 to the computational complexity of the corresponding recognition problems, i.e., deciding whether a given graph belongs to a given class of graphs. We conclude with a selection of open problems in Sect. 2.7.

Graphs defined by *non*quantitative restrictions on the crossing pattern in drawing are beyond the scope of this chapter. We do not discuss parity conditions or any other number theoretic or algebraic constraints. For example, a common generalization of the strong and the weak Hanani–Tutte theorems states that if a graph G admits a drawing D in which any two edges cross an even number of times, then G is planar and has a crossing-free drawing in which the rotation of every vertex is the same as in D [37]. We refer to a recent book by Schaefer [66] for other nonquantitative results. One can also impose topological constraints on the crossing edges (such as *fan-planar* graphs discussed in Chap. 7, *near-independent crossing planar* graphs [12], or *planarly connected crossings* [7]). These are not discussed in this chapter.

2.2 Simple Graphs and Simple Topological Graphs

2.2.1 Simple Graphs Versus Multigraphs

Most results in graph theory concern simple graphs (i.e., graphs without loops and multiedges). Some of the proof techniques developed for topological graphs, however, involve local operations that successively reroute edges, possibly modifying the neighborhood of one endpoint regardless of the location of other endpoint. These techniques inevitably produce multigraphs. A *graph* is a pair $G = (V, E)$, where V is a set (*vertices*) and E is a set of 2-element subsets of V (*edges*). In a *multigraph* $G = (V, E)$, E is a multiset of 1- or 2-element subsets of V (the 1-element subsets are *loops*).

Without additional constraints, a multigraph on n vertices may have arbitrarily many edges. For topological graphs, the most helpful constraints are defined in terms of homotopies (i.e., continuous deformations), requiring that no two parallel edges are homotopic with respect to V, and no loop is null-homotopic. Intuitively, this means that no edge can be continuously deformed into a parallel edge, and no loop can be contracted to its endpoint, without passing through any other vertex (i.e., the deformation maintains a valid topological graph at all times). While the choice of the outer face (e.g., a projective transformation) has no impact on the combinatorial properties of a drawing, it does make a difference for continuous deformations. See Fig. 2.2 for an illustration. For this reason, homotopies are defined on a sphere $\mathbb{S}^2$ (which is the one-point compactification of the plane).

Fig. 2.1 Left: Three Jordan arcs between u and v that are pairwise nonhomotopic in $\mathbb{R}^2 \setminus V$; but two of them are homotopic in $\mathbb{S}^2 \setminus V$. Middle: Three Jordan arcs between u and v, two of which are homotopic. Right: a null-homotopic loop incident to w, and two loops incident to u that are homotopic in $\mathbb{S}^2 \setminus V$

A *topological multigraph* is a pair $G = (V, E)$, where V is a set of points in the sphere $\mathbb{S}^2$, and E is a set of Jordan arcs and Jordan curves in $\mathbb{S}^2$ such that each Jordan arc connects two points in V without passing through any point in V, and each Jordan curve (i.e., loop) is incident to a unique point in V. A topological multigraph naturally defines an abstract multigraph on V.

Homotopy equivalence. An *arc* in a manifold M is the image of a continuous function $\gamma : [0, 1] \to M$. Intuitively, two arcs between the same pair of points are homotopy equivalent (for short, homotopic) with respect to a set of "obstacles" if we can continuously deform one arc into the other such that their endpoints remain fixed and they avoid the obstacles during the deformation. Refer to Fig. 2.1 for examples. Formally, a *homotopy* between two arcs, $\gamma_0 : [0, 1] \to M$ and $\gamma_1 : [0, 1] \to M$, between $u = \gamma_0(0) = \gamma_1(0)$ and $v = \gamma_0(1) = \gamma_1(1)$ is a continuous function $H : [0, 1]^2 \to M$ such that its boundary values are given by $H(0, t) = \gamma_0(t)$, $H(1, t) = \gamma_1(t)$, $H(t, 0) = u$, and $H(t, 1) = v$ for all $t \in [0, 1]$. A closed arc $\gamma_0 : [0, 1] \to M$, where $w = \gamma_0(0) = \gamma_0(1)$, is *null-homotopic* if γ is homotopic to the constant arc $\gamma_1 : [0, 1] \to \{w\}$; see Fig. 2.1 for examples.

Homotopic edges in a topological graph. Let $G = (V, E)$ be a topological multigraph on a sphere. Two parallel edges, $e_0 = uv$ and $e_1 = uv$, represented by $\gamma_0 : [0, 1] \to \mathbb{S}^2$ and $\gamma_1 : [0, 1] \to \mathbb{S}^2$, respectively, are *homotopy equivalent* (for short, *homotopic*) if γ_0 and γ_1 are homotopic in the punctured sphere $\mathbb{S}^2 \setminus (V \setminus \{u, v\})$. Similarly, a loop e_0 incident to $w \in V$, represented by the arc $\gamma_0 : [0, 1] \to \mathbb{S}^2$, is *null-homotopic* if γ_0 is null-homotopic in the punctured sphere $\mathbb{S}^2 \setminus (V \setminus \{w\})$.

We consider topological multigraphs without homotopic parallel edges and null-homotopic loops; we call them *homotopy-free topological multigraphs*, for short. They are also known as *generalized topological graphs* [8].

Applications. Homotopy-free topological multigraphs have been useful for handling local rerouting operations that modify part of an edge in the neighborhood of an endpoint, and may inadvertently create parallel edges or loops. This technique has been used for bounding the maximum number of edges in n-vertex 2- and 3-planar graphs by Pach and Tóth [60] and by Bekos et al. [18]; for quasiplanar graphs by Ackerman and Tardos [8]; and for 1-gap-planar graphs by Bae et al. [14].

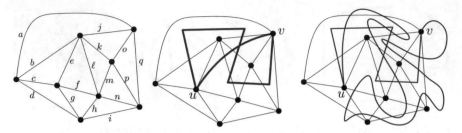

Fig. 2.2 Left: A triangulation. Edges are directed from left-to-right. Middle: Two parallel edges that are not homotopic; they are encoded by $\ell^{-1}k^{-1}$ and $b^{-1}jk\ell\ell^{-1}mp^{-1} = b^{-1}jkmp^{-1}$. Right: Two homotopic parallel edges: $b^{-1}jk\ell\ell^{-1}mp^{-1} = b^{-1}jkmp^{-1}$ and $b^{-1}a^{-1}ajj^{-1}jkmnii^{-1}h^{-1}f^{-1}fgdd^{-1}g^{-1}hii^{-1}n^{-1}p^{-1}q^{-1}qo^{-1}oq^{-1} = b^{-1}jkmp^{-1}$

If we perform local operations over homotopy-free topological multigraphs, then we are allowed to create parallel edges and loops, but we still need to ensure that no two parallel edges are homotopic and no loop is null-homotopic. Homotopy between the edges can be discretized and detected efficiently by a so-called *cross-metric representation* [29, 30]. Suppose that we wish to encode the edges of a topological graph $G = (V, E)$. We first find a cellular plane graph $G_0 = (V, E_0)$, a so-called *cut graph*, which is a topological graph where no two edges cross and every face is homeomorphic to a disk. The graphs G and G_0 have the same vertex set, and we assume that each edge in E_0 is either in E or intersects every edge in E finitely many times. For example, G_0 can be taken to be a triangulation or an edge-maximal plane (multi-)graph. We direct the edges in E_0 arbitrarily and label them by $e_1, e_2, \ldots, e_m$, where $m = |E_0|$. Then the homotopy type of an edge $e \in E \setminus E_0$ can be encoded as a word $w(e)$ over the alphabet $\{e_i, e_i^{-1} : i = 1, \ldots, m\}$. Specifically, the ith symbol in $w(e)$ corresponds to the ith edge crossed by e, say $e_j \in E_0$; the ith symbol is e_j or e_j^{-1} depending on whether e crosses e_j from left-to-right or right-to-left. Two parallel edges $e_1, e_2 \in E \setminus E_0$ are homotopic if and only if $w(e_1) = w(e_2)$ or if $w(e_1)$ and $w(e_2)$ reduce to the same word after successively performing all cancelations of the type $e_i e_i^{-1}$ or $e_i^{-1} e_i$; and the omission of all edges incident to any common endpoint of e_1 and e_2. Figure 2.2 depicts two examples.

Triangulations. A cut graph is typically a triangulation, which can be defined in a broad sense over topological multigraphs [29]. A crossing-free topological multigraph $G = (V, E)$ is a *triangulation* if it is cellular and every face is incident to either precisely three distinct edges or a bridge and a loop (Fig. 2.3). Alternatively, a triangulation can be defined as an edge-maximal crossing-free and homotopy-free topological multigraph.

A triangulation is known to be 3-connected if the underlying abstract graph is simple. By Whitney's theorem [75] every 3-connected planar graph has a combinatorially unique embedding in the sphere (hence a combinatorially unique embedding in the plane up to the choice of the outer face), which means that the cyclic order of the edges incident to every vertex is determined up to a reflection. These properties do not extend to homotopy-free multigraph triangulations. Every multigraph trian-

Fig. 2.3 Left: A simple triangulation. Right: a homotopy-free multigraph triangulation. For $n = 6$ vertices, both have $3n - 6 = 12$ edges. The outer face is bounded by three edges in both

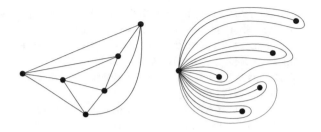

gulation is connected but need not be 3-connected, and an (abstract) multigraph may correspond to combinatorially different embeddings.

For every $n \in \mathbb{N}$, let $\mathscr{T}_n$ be the set of edge-maximal planar simple graphs on n vertices; every graph in $\mathscr{T}_n$ admits an embedding in $\mathbb{S}^2$ as a triangulation. Analogously, let $\mathscr{T}_n^*$ be the set of edge-maximal (abstract) multigraphs on n vertices that admit an embedding as a homotopy-free topological multigraph. Clearly, $\mathscr{T}_n \subseteq \mathscr{T}_n^*$. Euler's polyhedron theorem applies to homotopy-free topological multigraphs, hence every graph in $\mathscr{T}_n^*$ has at most $3n - 6$ edges for $n \geq 3$. In a simple triangulation in $\mathscr{T}_n$, the maximum vertex-degree is $n - 1$. However, every edge in a multigraph triangulation in $\mathscr{T}_n^*$ may be incident to a common vertex; in fact, a triangulation could consist of a spanning star of $n - 1$ edges, and $2n - 5$ loops incident to the center of the star (see Fig. 2.3).

In applications [14, 18, 60], we are given an edge-maximal simple topological graph $G = (V, E)$ with some forbidden crossing pattern. One can show that G contains a crossing-free triangulation, which is a simple triangulation. With respect to such a triangulation, additional properties can be deduced: Either a property holds, or one can insert a new edge by locally rerouting existing edges, contradicting maximality. For certifying whether two edges are homotopic (or an edge is null-homotopic), multigraph triangulations and simple triangulations are equally useful. However, it is easier to work with a simple triangulation, as parallel edges and loops would lead to unnecessary special cases.

Flips in Multigraph Triangulations. A flip graph is defined on the n-vertex triangulations, where two triangulations are adjacent if one can be obtained from the other by deleting an edge and inserting another edge. Let $\mathscr{G}_n$ be the flip graph of simple triangulations $\mathscr{T}_n$, and $\mathscr{G}_n^*$ the flip graph of multigraph triangulations. The diameter of $\mathscr{G}_n$ is $\Theta(n)$, the current best lower and upper bounds are between $\frac{7}{3}n + \Theta(1)$ [36] and $5n - 23$ for $n \geq 6$ [27].

It is not difficult to see that $\mathscr{G}_n^*$ is also connected. Indeed, it is enough to show that a sequence of flips can carry any multigraph triangulations $G \in \mathscr{G}_n^*$ into some common triangulation in $\mathscr{G}_n^*$. We define the canonical triangulation $G_0 \in \mathscr{G}_n^*$ as a spanning star centered at a vertex v_0 and $2n - 5$ loops incident to v_0; see Fig. 2.3. Given an arbitrary triangulation $G \in \mathscr{G}_n^*$, let v_0 be a vertex of maximum degree. We transform G into canonical form in two phases: (1) While there is any face not incident to v_0, let e be an edge separating a face incident to v_0 and a face nonincident to v_0 (note that e is not incident to v_0); a flip replaces e with an edge incident to v_0, and increases the number of faces incident to v_0 by one. (2) While there is an edge e not incident to v_0 (separating two faces that are each incident to v_0), then a flip replaces e with a loop incident to v_0. The first phase creates a star in at most $n - 1 - \deg(v_0) = n - 1 - \Delta(G)$ steps, and the second phase creates $(3n - 6) - (n - 1) = 2n - 5$ loops. So the number of operations is $3n - 6 - \Delta(G)$. Consequently, $\mathscr{G}_n^*$ is connected, and its diameter is at most $6n - \Theta(1)$ for $n \geq 3$.

For the diameter of $\mathscr{G}_n^*$, one can easily establish a lower bound of $3n - 12$, for $n \geq 3$. Consider the distance between the triangulation $G_0 \in \mathscr{G}_n^*$ defined above, and a triangulation $G_1 \in \mathscr{G}_n \subseteq \mathscr{G}_n^*$ with maximum degree at most 6. Since an edge flip affects only one edge at a time, it can decrease the maximum number of edges incident to a vertex by at most one. Consequently, G_0 and G_1 are at distance at least $(3n - 6) - 6$ in the flip graph $\mathscr{G}_n^*$.

2.2.2   t-Simple Topological Graphs

In a *geometric graph* (i.e., a straight-line drawing of a graph), every pair of edges intersect at most once: either at a common endpoint or at a common interior point where the two edges cross transversely. A topological graph that satisfies the same condition is called *simple*; otherwise it is *nonsimple*. This binary notion can be refined by quantifying the number of crossings between pairs of edges. A topological graph is *t-simple*, for an integer $t \geq 0$, if any two edges have at most t points in common, each of which is either a common endpoint or a common interior point where the two edges cross transversely. In particular, plane graphs and simple topological graphs are 1-simple. Note that every graph admits a 1-simple drawing (e.g., a straight-line drawing). Some of the results in the literature are stated for simple topological graphs even if this restriction is not needed for their proof. Research in the early days of topological graph theory focused on the crossing number of a graph, and it is well known that every graph admits a simple topological drawing that minimizes the number of crossings (see, e.g., [73]).

The simplicity parameter t plays an important role in combination with other constraints on topological graphs.

- Pach et al. [57, Lemma 1.1] proved that for $k \in \{0, 1, 2, 3\}$ every k-planar graph (which can be drawn with at most k crossings per edge) admits a *simple* topological drawing with at most k crossing per edge. These results, however, do not extend to $k \geq 4$. There exits a 4-planar graph such that some pair of adjacent edges cross in every drawing in which there are at most 4 crossings per edge, and there exists a 5-planar graph such that some pair of independent edges cross more than once in every drawing in which there are at most 5 crossings per edge [66, Chap. 7].
- Ackerman and Tardos [8] proved that every n-vertex simple topological quasiplanar graph has at most $6.5n - O(1)$ edges, but there exist (nonsimple) topological quasiplanar graphs with $7n - O(1)$ edges (see Sect. 2.4).

- Kaufmann et al. [43] generalized the classic Crossing Lemma to homotopy-free topological multigraphs in which adjacent edges (including parallel edges) do not cross each other. They show that if the number of edges is $m \geq 4n$, there are $\Omega(m^3/n^2)$ crossings. The conditions that adjacent edges do not cross is essential: Pach and Tóth [61] construct 2-simple topological multigraphs with $m = (n/3)^3$ edges and less than $2\binom{m}{2} = O(n^6)$ crossings; significantly fewer than $\Omega(m^3/n^2) = \Omega(n^7)$.

2.3 Overview of Graph Classes

The graph classes considered in this chapter are characterized by having a drawing in the plane (or other manifolds) without certain crossing patterns. These are hereditary graph classes, as the defining properties are invariant under edge deletion (both from the graph and from a drawing). However, these classes are not minor closed. For example, every complete graph is the minor of a 1-planar graph: in an arbitrary drawing of K_n, one can subdivide the edges such that each edge crosses at most one other edge [38].

k-**Planar Graphs** A topological graph is k-*planar*, for $k \geq 0$, if every edge is involved in at most k crossings (with k or fewer other edges). An (abstract) graph is k-*planar* if it can be drawn in the plane as a k-planar topological graph. Clearly, planar graphs are precisely the 0-planar graphs. Restrictions to geometric graphs (i.e., straight-line drawings) and t-simple topological graphs are of interest, as well.

Every graph is k-planar for a sufficiently large integer k. The *local crossing number* of a graph G, denoted $\text{lcn}(G)$, is the minimum integer $k \geq 0$ such that G is k-planar. Restrictions to geometric graphs and simple topological graphs, respectively, lead to the *rectilinear local crossing number* $\overline{\text{lcr}}(G)$ and the *simple local crossing number* $\text{lcr}^*(G)$.

Motivated by a map-coloring problem, 1-planar graphs were introduced by Ringel [63] in 1965. He proved that 1-planar graphs are 7-colorable. Later Borodin [21, 22] showed that the chromatic number of every 1-planar graph is at most 6, which is optimal since K_6 is 1-planar; Fig. 2.4(right). The density of k-*planar* graphs plays a crucial role in proving the current best constants for the classic *Crossing Lemma* [5, 57, 60]. Székely's probabilistic method [69] derives a lower bound for the crossing number $\text{cr}(G)$ of a graph $G = (V, E)$ by choosing a Bernoulli sample $V' \subset V$ with probability $p = \Theta(|V|/|E|)$, where the induced graph $G' = (V', E')$ has comparable number of edges and vertices in expectation. A trivial lower bound $\text{cr}(G') \geq |E'| - 3(|V'| - 2)$ follows from the density of planar graphs. This bootstrap inequality can be improved to $\text{cr}(G') \geq 5|E'| - \frac{139}{6}(|V'| - 2)$ by using 4-planar graphs and an upper bound of $6(n - 2)$ on the number of edges in 4-planar graphs with n vertices (Sect. 2.4).

Besides density, several other graph parameters of k-planar graphs are close to planar graphs for constant $k \in \mathbb{N}$. Dujmović et al. [31] proved that the treewidth of a k-planar n-vertex graph is $O(\sqrt{k+1} \cdot n)$, and this bound is the best possible. The result generalizes to k-planar topological graphs on surfaces of genus g, where the treewidth is $O(\sqrt{(g+1)(k+1)} \cdot n)$. Bekos et al. [17] proved that the book thickness of 1-planar graphs is $O(1)$; it remains open to bound the book thickness of k-planar graphs by a function of k for all $k \geq 2$. Structural results are also available. For example, Ackerman [4] showed that the edge set of every 1-planar graph can be decomposed into a planar graph and a forest. A large number of recent results are available for 1-planar graphs; refer to an annotated bibliography by Kobourov, Liotta, and Montecchiani [44].

k-Gap-Planar Graphs A topological graph $G = (V, E)$ is k-*gap-planar*, for $k \geq 0$, if every subset of edges $E' \subseteq E$ is involved in at most $k|E'|$ crossings. A (abstract) graph is k-*gap-planar* if it can be drawn in the plane as a k-gap-planar topological graph. By Hall's theorem, every k-gap-planar topological graph admits an assignment of its crossings to edges such that each crossing is assigned to one of the two crossing edges, and every edge is responsible for at most k crossings.

The assignment of a crossings to edges can be interpreted as an antisymmetric binary relation between the edges: When edges e_1 and e_2 cross at a point c and the crossing is assigned to e_1, we can say that e_1 *crosses* e_2 (or e_2 is crossed by e_1). In a k-gap-planar simple topological graph, every edge crosses at most k other edges. An antisymmetric crossing relation is motivated by *edge casing*, a common technique in visualization that alleviates visual clutter generated by intersecting curves in a diagram [10, 34]. The edge casing technique eliminates crossings by locally interrupting one of the two crossing edges; see Fig. 2.4(right).

Eppstein et al. [34] studied several optimization problems related to edge casing. They show that, given a topological graph G, one can find the minimum integer $k \geq 0$ such that G is k-gap-planar in time polynomial in the description complexity of G. However, Bae et al. [14] show that it is NP-complete to decide whether a given (abstract) graph G is 1-gap-planar.

Eppstein and Gupta [33] introduced a closely related concept, which further requires the antisymmetric crossing relation to be acyclic. A topological graph has k-*degenerate crossings* if the edges admit a total order in which each edge crosses at most k previous edges. For simple topological graphs, this is equivalent to the condition that the intersection graph of the (open) edges is k-degenerate. It is clear from the definition that every k-degenerate crossing graph is a k-gap-planar graph for all $k \in \mathbb{N}$. The converse is false already for $k = 1$ [14], but it is easy to see that every k-gap-planar graph is a $2k$-degenerate crossing graph for all $k \in \mathbb{N}$.

Recently, Ossona de Mendez et al. [53] introduced a similar concept: A graph G is k-*close-to-planar* if every subgraph G' of G with m' edges satisfies $\mathrm{cr}(G') \leq km'$. It is clear that every k-gap-planar graph is k-close-to-planar. The converse is already false for $K_{6,6}$, which is 1-gap-planar [13] but not 1-close-to-planar.

k-Quasiplanar Graphs A topological graph $G = (V, E)$ is k-*quasiplanar*, for $k \geq 2$, if E does not contain k edges that pairwise cross. A (abstract) graph is k-*quasiplanar* if it admits a drawing as a k-quasiplanar topological graph. Clearly, 2-quasiplanar graphs are precisely the planar graphs, so $k = 3$ is the first interesting value. For brevity, 3-quasiplanar graphs are often called just *quasiplanar*.

Fig. 2.4 Two realizations of
K_6. Left: A 1-planar
topological graph with a total
of $\mathrm{cr}(K_6) = 3$ crossings.
Right: A 3-planar geometric
graph, which is 1-gap-planar
using suitable edge casing.
Both topological graphs are
quasiplanar

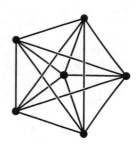

For simple topological graphs, k-quasiplanarity is equivalent to the condition that the intersection graph of the (open) edges is K_k-free. Many other Turán-type problems on topological graphs excluding certain crossing patterns have been considered in the literature: for example, excluding a complete bipartite graph [6, 54], or a join of two paths [56, 71]. As noted above, there exist 3-quasiplanar graphs for which every 3-quasiplanar drawing is nonsimple [8].

k-**Locally Planar Graphs** A topological graph G is k-*locally planar* if no path of length at most k has any self-crossing. An abstract graph is k-*locally planar* if it has a drawing as a k-locally planar topological graph. Clearly, an n-vertex graph G is planar if and only if it is k-locally planar for $k = \mathrm{diam}(G)$. As opposed to other notions of beyond-planar graphs, a graph is closer to planarity for larger values of k.

For even $k \geq 0$, the $(k/2)$-neighborhood of every vertex is crossing-free in a k-locally planar topological graph. Tardos [70] points out that this condition is much stronger than the similar condition for abstract graphs requiring the $(k/2)$-neighborhood of every vertex to be planar. For every $k \in \mathbb{N}$, there exist graphs with $\Omega(n^{k/(k-1)})$ edges and girth larger than k, which meet the latter condition, and yet such a graph does not admit 3-locally planar straight-line drawing by the density result of Pach et al. [55].

2.4   Density

The current best upper bounds for the number of edges in a graph with n-vertex for the families defined in Sect. 2.3 are listed in Table 2.1. We mention a few additional results.

A 1-planar graph with $n \geq 3$ vertices has at most $4n - 8$ edges [19, 60], and this bound is the best possible for every $n \geq 12$. However, for geometric graphs, Didimo [4] proved a slightly stronger upper bound of $4n - 9$. For $k = 1, 2, 3$, the lower bound constructions are based on tiling of the sphere with convex quadrilaterals, pentagons, and hexagons, respectively, and all diagonals in each tile [57, 60]. These constructions can be realized as a geometric graph in the plane apart from the outer face and its neighbors, so these bounds are tight for geometric graphs up to some additive constant. The lower bound construction for 2-planar graphs can be adapted to

Table 2.1 Upper bounds for the number of edges in an n-vertex graph for $n \geq 3$. $\alpha(n)$ denotes the inverse of the Ackermann function; $f(k)$ and $f(k, t)$ denote some function of k and of k and t, respectively. A bound is *tight* if it is attained for infinitely many graphs. Multigraphs refer to homotopy-free topological multigraphs

Graph class	Bound	Tight	Multigraphs	References
Planar	$3(n-2)$	Yes	Yes	Euler's formula
1-planar	$4(n-2)$	Yes	Yes	Bodendiek et al. [19, 60]
2-planar	$5(n-2)$	Yes	Yes	Pach and Tóth [18, 60]
3-planar	$5.5(n-2)$	Yes[a]	Yes	Pach and Tóth [18, 57]
4-planar, 1-simple	$6(n-2)$	–	–	Ackerman [5]
k-planar	$\Theta(\sqrt{k}n)$	Yes	Yes[b]	Pach and Tóth [60, 61]
1-gap-planar	$5(n-2)$	Yes	Yes	Bae et al. [14]
k-gap-planar	$\Theta(\sqrt{k}n)$	Yes	Yes[b]	Bae et al. [14]
3-quasiplanar, 1-simple	$6.5(n-2)$	Yes	–	Ackerman and Tardos [8]
3-quasiplanar	$8n-20$	Yes[a]	Yes	Ackerman and Tardos [8]
4-quasiplanar	$72(n-2)$	No	–	Ackerman [3]
k-quasiplanar, 1-simple	$f(k)n\log n$	–	–	Suk and Walczak [68]
k-quasiplanar, t-simple	$f(k, t)n\log n$	–	–	Rok and Walczak [64]
k-quasiplanar	$n(\log n)^{O(\log k)}$	–	–	Fox and Pach [35]
3-locally planar, straight	$\Theta(n\log n)$	Yes	–	Pach et al. [55]
k-locally planar, straight	$O(n\log^{1/\lfloor k/2 \rfloor} n)$	–	–	Pach et al. [55], Tardos [70]
3-locally planar	$O(n^{3/2})$	–	–	Pach et al. [55]

[a]The bound is tight for multigraphs. For (simple) graphs, the current best lower bounds are $5.5n - 15$ for 3-planar graphs [18], and $7.5n - O(1)$ for quasiplanar graphs [8]
[b]The bound holds for topological multigraphs where no two adjacent edges cross each other [43, 61]

1-gap-planar graphs, so the upper bound $5(n-2)$ is tight for 1-gap-planar geometric graphs, as well, apart from an additive constant.

A k-planar graph $G = (V, E)$ admits a drawing with at most $k|E|/2$ crossings, hence $\mathrm{cr}(G) \leq k|E|/2$. Combined with the lower bound $\mathrm{cr}(G) \geq \Omega(|E|^3/n^2 - |E|)$ from the Crossing Lemma, this yields $|E| \leq O(\sqrt{k}n)$, which is tight apart from the constant factor. The current best constants in the Crossing Lemma [5] yield an upper bound of $3.81\sqrt{k}n$. This bound holds for nonsimple drawings, as well. The Crossing

Lemma has recently been extended to some homotopy-free topological multigraphs [43, 61]: The bound $\mathrm{cr}(G) \geq \Omega(|E|^3/n^2 - |E|)$ carries over, albeit with weaker constant coefficients, and restricted to topological drawings where adjacent edges (including parallel edges) do not cross each other. If adjacent nonparallel edges are allowed to cross any number of times, the lower bound degrades to $\Omega(|E|^{5/2}/n^{3/2} - |E|)$ [43], which yields $|E| \leq O(k^{2/3}n)$.

For k-quasiplanar graphs, $k \geq 2$, the current best lower bound for the maximum number of edges is $\Omega(kn)$. Pach, Shahrokhi, and Szegedy [59] conjectured that there exists a constant c_k for every $k \geq 3$ such that every n-vertex k-quasiplanar graph has at most $c_k n$ edges.

Maximality. Every edge-maximal planar graph (i.e., triangulation) on $n \geq 3$ vertices has $3n - 6$ edges. The same tight bound holds for edge-maximal crossing-free topological graphs (if an embedded graph has fewer edges, then one of the faces is not a triangle, and the graph can be augmented with one of the diagonals). For edge-maximal beyond-planar graphs, these concepts are no longer equivalent.

An abstract graph $G = (V, E)$ in a family $\mathscr{F}$ is *edge-maximal* (or *saturated*) if it is not a proper subgraph of any other graph in $\mathscr{F}$ on the same vertex set V. Similarly, a topological graph $G = (V, E)$ in a family $\mathscr{F}$ is *edge-maximal* (*saturated*) if there is no topological graph $G' = (V, E')$ such that $E \subset E'$ but $E \neq E'$ (where both E and E' are sets of Jordan arcs). All bounds in Table 2.1 are for abstract graphs.

Improving earlier results by Brandenburg et al. [24], Barát and Tóth [16] showed that every saturated 1-planar graph or 1-planar topological graph has at least $\frac{20}{9}n - \frac{10}{3} \approx 2.22n$ 3.33 edges for $n \geq 4$. In particular, a saturated 1-planar graph may have fewer edges than a plane graph on the same vertex set.

A saturated t-simple topological graph on n vertices may have $O(n)$ edges for any $t \ll n$. For every $t \in \mathbb{N}$, a t-simple topological graph is *saturated* if no further edge can be added to produce a t-simple topological graph. This means that any Jordan arc between two nonadjacent vertices would cross some existing edge at least $t + 1$ times. Denoting by $s_t(n)$ the minimum number of edges in a saturated t-simple topological graph with n vertices, Kynčl et al. [46] and Hajnal et al. [39] proved that $1.5 \leq s_1(n) \leq 7n$ and $s_t(n) \leq 14.5n$ for every $t \geq 2$. It remains an open problem whether the function $S(t) = \liminf_{n\to\infty} \frac{s_t(n)}{n}$ converges as t tends to infinity.

Complete Graphs. A good indicator of how close a family of beyond-planar graphs may be to planarity is the maximum size of a clique it contains. Exact bounds for the local crossing numbers of complete graphs are known in only very few cases. Ábrego and Fernández-Merchant [2] determined the *rectilinear local crossing numbers* $\overline{\mathrm{lcr}}(K_n)$ for all $n \in \mathbb{N}$. They prove

$$\overline{\mathrm{lcr}}(K_n) = \left\lceil \frac{1}{2}\left(n - 3 - \left\lceil \frac{n-3}{3} \right\rceil\right)\left\lceil \frac{n-3}{3} \right\rceil \right\rceil = \Theta(n^2),$$

for $n \in \mathbb{N} \setminus \{8, 14\}$, $\overline{\mathrm{lcr}}(K_8) = 4$ and $\overline{\mathrm{lcr}}(K_{14}) = 15$. This is clearly an upper bound for both $\mathrm{lcn}^*(K_n)$ and $\mathrm{lcn}(K_n)$, but only the trivial lower bound $\mathrm{cr}(K_n)/\binom{n}{2} = \Theta(n^2)$ is known apart from sporadic examples. In contrast, Kynčl and Valtr [49] studied the *minimum* integer $h(n)$ such that every simple topological drawing of K_n contains some edge that crosses at most $h(n)$ others. They show that $\Omega(n^{3/2}) \le h(n) \le O(n^2/\log^{1/4} n)$. Note that there is no direct relation between $\mathrm{lcn}^*(K_n)$ and $h(n)$.

Cliques and complete bipartite graphs are instrumental in hardness reductions: They can form subgraphs that attain the maximum number of crossings in any realization, and hence can serve as a "blocker" gadget that prevents interaction between various other parts of the graph. For example, Grigoriev and Bodlaender [38] showed that in every 1-planar drawing of K_6, between every pair of vertices there exists a path in which every edge is already crossed once. Bae et al. [14] proved that in every 1-gap-planar drawing of $K_{3,12}$, every pair of vertices of degree three are part of a cycle in which every edge crosses some other edge.

Algorithmic questions for complete graphs have also been addressed. Kynčl [48] showed that, given a complete graph K_n together with a binary relation R between the edges, one can decide in polynomial time whether K_n is realizable as a simple topological graph such that two edges cross if and only if they are related in R. In fact, Kynčl proves that such a realization of K_n, $n \ge 6$ is possible if and only if every subgraph induced by 6 vertices is realizable (and the realizability of all K_5 subgraphs does not suffice). Kynčl [47] uses this property for bounding the number of possible intersection patterns of the edges in a simple topological graphs, which is shown to be $2^{n(\log n - O(1))}$. Ábrego et al. [1] enumerated *all* simple topological realizations of K_n for $n = 1, \ldots, 9$.

Colorings. The chromatic number of every 1-planar graph is at most 6 [21, 22]; and the density results in Sect. 2.4 imply that every k-planar and k-gap-planar graph is $O(\sqrt{k})$-degenerate, hence $O(\sqrt{k})$-colorable; and these bounds are tight apart from constant factors. Recently, Ossona De Mendez et al. [53] showed that these every graph in these classes is $(3, O(k^{5/2}))$-choosable, that is, it admits a vertex 3-coloring such that each color class induces a subgraph of maximum degree $O(k^{5/2})$. However, this bound is not known to be tight; and no tight bounds are known for the chromatic number of k-quasiplanar and k-locally planar graphs.

2.5   Inclusions

The definitions of k-planar, k-gap-planar, k-quasiplanar, and k-locally planar graphs imply the obvious inclusions

$$k\text{-PLANAR} \subsetneq (k+1)\text{-PLANAR} \quad \text{for } k \ge 0. \tag{2.1}$$

$$k\text{-GAP-PLANAR} \subsetneq (k+1)\text{-GAP-PLANAR} \quad \text{for } k \ge 0. \tag{2.2}$$

$$k\text{-QUASIPLANAR} \subsetneq (k+1)\text{-QUASIPLANAR} \quad \text{for } k \ge 2. \tag{2.3}$$

$$(k+1)\text{-LOCALLY-PLANAR} \subsetneq k\text{-LOCALLY-PLANAR} \quad \text{for } k \ge 1. \tag{2.4}$$

In each case, it is an easy exercise to show proper containment.

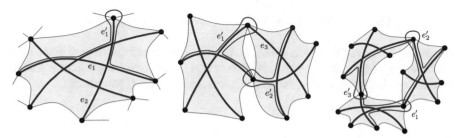

Fig. 2.5 Left: Four pairwise crossing edges. The crossing between e_1 and e_2 can be eliminated by rerouting e_1 around the endpoint of e_2. Middle and Right: If we reroute an arbitrary edge from each pairwise crossing triple, one might create new triples of pairwise crossing edges. A *twin* configuration (middle) and a *swirl* configuration (right)

A restriction to *simple* topological realizations always yields a subclass, but in most cases it is far from obvious whether it is a proper subclass. As noted above, Pach et al. [57, Lemma 1.1] proved that for $k \in \{0, 1, 2, 3\}$ every k-planar graph admits a *simple* topological drawing with at most k crossings per edge, but this result does not extend to $k \geq 4$.

$$\text{SIMPLE } k\text{-PLANAR} = k\text{-PLANAR for } 0 \leq k \leq 3. \tag{2.5}$$

Very few inclusion relations are known between distinct families of beyond-planar graphs. Current proof techniques typically transform a given drawing of a graph into another by rerouting some of the edges. The inclusion relation

$$k\text{-PLANAR} \subsetneq (k + 1)\text{-QUASIPLANAR for } k \geq 2 \tag{2.6}$$

has been proven for $k \geq 2$ by Angelini et al. [9]. It does not extend to $k = 1$, as 2-QUASIPLANAR = PLANAR, but 1-PLANAR contains nonplanar graphs (e.g., K_5 or $K_{3,3}$). The converse of (2.6) is false: For every $k \geq 2$, there exist 3-quasiplanar graphs G_k for which $\mathrm{lcr}(G_k) = k$ [66]. However, it remains an open problem whether (2.6) can be strengthened: Angelini et al. [9] ask whether there exists a sublinear function $f : \mathbb{N} \to \mathbb{N}$ such that every k-planar graph is $f(k)$-quasiplanar.

The proof of (2.6) is constructive: Given a k-planar topological graph that is not $(k + 1)$-quasiplanar, one can identify the $(k + 1)$-tuples of edges that pairwise cross, and successively reroute at least one edge from each until all such $(k + 1)$-tuples are eliminated (Fig. 2.5). The challenging part of the proof is to ensure that the rerouting algorithm does not create any new pairwise crossing $(k + 1)$-tuples. This can be done by careful choices utilizing Hall's theorem in all cases. The case $k = 2$ requires heavier machinery, as the basic rerouting strategy can easily create triples of pairwise crossing edges; see Fig. 2.5 for examples.

Bae et al. [14] show that the class of k-gap-planar graphs is sandwiched between the $(2k)$-planar and $(2k + 2)$-quasiplanar families:

$$(2k)\text{-PLANAR} \subsetneq k\text{-GAP-PLANAR} \subsetneq (2k+2)\text{-QUASIPLANAR}.$$

However, the hierarchy of gap-planar graphs (2.2) does not interleave with the hierarchies (2.1)–(2.3): For every $k \in \mathbb{N}$, Bae et al. [14] construct 1-gap-planar graphs that are not k-planar; and 3-quasiplanar graphs that are not k-gap-planar.

2.6 Recognition Algorithms

There are linear-time planarity testing algorithms [20, 23, 42, 67, 74], and if a graph is planar, one can also find a straight-line embedding in linear time [28]. Unfortunately, the recognition problem for most classes of beyond-planar graphs is NP-hard. In application domains, the input is often given as a topological graph that certifies membership in a beyond-planar class. One can often adapt graph algorithms originally designed for plane graphs to work on topological graphs in beyond-planar families.

For abstract graphs, there are very few positive results, and they are limited to restricted families of 1-planar graphs. Recognizing 1-planar graphs is NP-Complete. Grigoriev and Bodlaender [38] reduce the problem from 3- PARTITION, Korzhik and Mohar [45] reduce the problem from 3- COLORABILITY OF PLANAR GRAPHS OF MAXIMUM DEGREE AT MOST FOUR. The recognition of 1-planar graphs remains NP-hard even for graphs of bounded bandwidth, pathwidth, or treewidth [15], for 3-connected graphs [11] with or without a given rotation system, and for graphs obtained by augmenting a planar graph by one edge [26].

Eades et al. [32] presented an $O(n)$-time algorithm for recognizing *edge-maximal* 1-planar graphs with a given rotation system. They also showed that the rotation system determines a unique realization as a 1-planar topological graph up to homeomorphisms (and the choice of the outer face). Recently, Brandenburg [25] designed an $O(n)$-time algorithm for recognizing 1-planar graphs with n vertices and $4n - 8$ edges, which is the worst-case upper bound for the number of edges in such a graph (cf. Table 2.1).

2.7 Open Problems

We conclude this chapter with a selection of open problems.

1. Closing the gaps between the upper and lower bounds in Table 2.1 for the maximum sizes of n-vertex beyond-planar graphs is an obvious open problem. For example, Pach, Sharokhi, and Szegedy [59] conjectures that for every $k \in \mathbb{N}$, there exists a constant $c_k > 0$ such that every k-quasiplanar graph with n vertices has at most $c_k n$ edges. It has been settled in the affirmative for $k \leq 4$ [3].

2. The current best bound for the number of edges in a k-locally planar graph is $O(n^{3/2})$, for every $k \geq 3$, but a 3-locally planar *geometric* graph has $O(n \log n)$ edges [55]. The lower bound constructions for this problem are designed for geometric graphs [70]. Interestingly, the $O(n^{3/2})$ bound is the best possible for 3-locally planar topological graphs in which all pairs of edges in a path of length 3 cross an *even* number of times [55]. Can the $O(n^{3/2})$ bound be improved for simple topological graphs?

3. For $k \in \{0, 1, 2, 3\}$, every k-planar graph can be realized as a k-planar simple topological graph in which every pair of edges intersect at most once (at a common endpoint or at a common interior point where the two edges cross transversely) [57]. This no longer holds for $k \geq 4$ [66], although crossings between adjacent edges can be avoided for $k = 4$. Answering a question by Schaefer [65], Liu et al. [51] recently showed that the *simple* local crossing number $\mathrm{lcn}^*(G)$ is bounded by a function of the local crossing number $\mathrm{lcn}(G)$ of a graph G; specifically, $\mathrm{lcn}^*(G) \leq O((\mathrm{lcn}(G))^{3/2} \cdot 3^{\mathrm{lcn}(G)})$. It remains an open problem whether $\mathrm{lcn}^*(G)$ is bounded by a polynomial in $\mathrm{lcn}(G)$?

4. Is there a simple characterization for graphs that admit a *straight-line k-planar, k-gap-planar, or k-quasiplanar realization*? For $k = 1$, Thomassen [72] characterized 1-planar geometric graphs in terms of two forbidden configurations, which has lead to a linear-time algorithm [40, 41] that determines whether a 1-planar topological graph admits a homeomorphic straight-line drawing (i.e., whether it is stretchable). Even for a plane graph (with given rotation system) and one extra edge st, it is challenging to find a homeomorphic straight-line embedding that minimizes the number of edges that cross st [62].

5. Pach, Radoičić, and Tóth [58] propose the following strengthening of the linear bound on the size of k-quasiplanar graphs: Is there a constant $c_k \in \mathbb{N}$ for every $k \geq 3$ such that the edges of every k-quasiplanar topological graph have a c_k-coloring so that no two edges of the same color cross each other. Equivalently, is the intersection graph of (open) edges of topological graphs χ-bounded? The answer is in the affirmative for *simple* topological graphs in which a Jordan curve crosses every edge [50, 52].

6. Ackerman et al. [6] propose a bipartite generalization of quasiplanarity: For $k, \ell \in \mathbb{N}$, a (k, ℓ)-*grid* in a topological graph consists of two disjoint sets of edges of size k and ℓ, respectively, such that every edge in the first set crosses every edge in the second. A (k, ℓ)-grid is *natural* if the $k + \ell$ edges involved are pairwise nonadjacent. They conjecture that for every $k, \ell \in \mathbb{N}$ there exists a constant $c_{k,\ell}$ such that every simple topological graph with n vertices with no natural (k, ℓ)-grid has at most $c_{k,\ell} n$ edges. A linear upper bound is known if all (k, ℓ)-grids are excluded [54]; and an $O(n \log^* n)$ bound if only natural (k, ℓ)-grids are excluded [6].

7. The study of 1-planar graphs was motivated by a map-coloring problem [63], and we now have tight bounds for the chromatic number of k-planar and k-gap-planar graphs (Sect. 2.4). No similar tight bounds are available for other families of beyond-planar graphs. Other graph parameters, such as bisection width, path width, treewidth, thickness, book thickness, and crossing number, would be of

interest, as well. For example, the book thickness of a 1-planar graph is at most 39 [17]. Is the book thickness of k-planar graphs bounded by a function of k?

8. It is NP-complete to recognize 1-planar and 1-gap-planar graphs. Is it NP-complete to decide membership in other families of beyond-planar graphs, such as k-quasiplanar graphs and k-locally planar graphs?

Acknowledgements We thank Yusuke Suzuki for many helpful comments on an earlier version of this chapter. Work on this survey was supported in part by the NSF awards CCF-1422311, CCF-1423615, and DMS-1800734.

References

1. Ábrego, B.M., Aichholzer, O., Fernández-Merchant, S., Hackl, T., Pammer, J., Pilz, A., Ramos, P., Salazar, G., Vogtenhuber, B.: All good drawings of small complete graphs. In: Book of Abstracts of the 31st European Workshop on Computational Geometry (EuroCG), pp. 57–60, Ljubljana (2015)
2. Ábrego, B.M., Fernández-Merchant, S.: The rectilinear local crossing number of K_n. J. Comb. Theory Ser. A **151**, 131–145 (2017). https://doi.org/10.1016/j.jcta.2017.04.003
3. Ackerman, E.: On the maximum number of edges in topological graphs with no four pairwise crossing edges. Discret. Comput. Geom. **41**(3), 365–375 (2009). https://doi.org/10.1007/s00454-009-9143-9
4. Ackerman, E.: A note on 1-planar graphs. Discret. Appl. Math. **175**, 104–108 (2014). https://doi.org/10.1016/j.dam.2014.05.025
5. Ackerman, E.: On topological graphs with at most four crossings per edge. Comput. Geom. **85** (2019). https://doi.org/10.1016/j.comgeo.2019.101574
6. Ackerman, E., Fox, J., Pach, J., Suk, A.: On grids in topological graphs. Comput. Geom. **47**(7), 710–723 (2014). https://doi.org/10.1016/j.comgeo.2014.02.003
7. Ackerman, E., Keszegh, B., Vizer, M.: On the size of planarly connected crossing graphs. J. Graph Algorithms Appl. **22**(1), 11–22 (2018). https://doi.org/10.7155/jgaa.00453
8. Ackerman, E., Tardos, G.: On the maximum number of edges in quasi-planar graphs. J. Comb. Theory Ser. A **114**(3), 563–571 (2007). https://doi.org/10.1016/j.jcta.2006.08.002
9. Angelini, P., Bekos, M.A., Brandenburg, F.J., Da Lozzo, G., Di Battista, G., Didimo, W., Hoffmann, M., Liotta, G., Montecchiani, F., Rutter, I., Tóth, C.D.: Simple k-planar graphs are simple $(k + 1)$-quasiplanar. J. Comb. Theory Ser. B **142**, 1–35 (2020). https://doi.org/10.1016/j.jctb.2019.08.006
10. Appel, A., Rohlf, F.J., Stein, A.J.: The haloed line effect for hidden line elimination. SIGGRAPH Comput. Graph. **13**(2), 151–157 (1979). https://doi.org/10.1145/965103.807437
11. Auer, C., Brandenburg, F.J., Gleißner, A., Reislhuber, J.: 1-planarity of graphs with a rotation system. J. Graph Algorithms Appl. **19**(1), 67–86 (2015). https://doi.org/10.7155/jgaa.00347
12. Bachmaier, C., Brandenburg, F.J., Hanauer, K., Neuwirth, D., Reislhuber, J.: NIC-planar graphs. Discret. Appl. Math. **232**, 23–40 (2017). https://doi.org/10.1016/j.dam.2017.08.015
13. Bachmaier, C., Rutter, I., Stumpf, P.: 1-gap planarity of complete bipartite graphs (Poster). In: Biedl, T.C., Kerren, A. (eds.) Proceedings of 26th Symposium on Graph Drawing and Network Visualization. LNCS, vol. 11282. Springer, Cham (2018)
14. Bae, S.W., Baffier, J.F., Chun, J., Eades, P., Eickmeyer, K., Grilli, L., Hong, S.H., Korman, M., Montecchiani, F., Rutter, I., Tóth, C.D.: Gap-planar graphs. Theor. Comput. Sci. (2018). https://doi.org/10.1016/j.tcs.2018.05.029
15. Bannister, M.J., Cabello, S., Eppstein, D.: Parameterized complexity of 1-planarity. J. Graph Algorithms Appl. **22**(1), 23–49 (2018). https://doi.org/10.7155/jgaa.00457

16. Barát, J., Tóth, G.: Improvements on the density of maximal 1-planar graphs. J. Graph Theory **88**(1), 101–109 (2018). https://doi.org/10.1002/jgt.22187
17. Bekos, M.A., Bruckdorfer, T., Kaufmann, M., Raftopoulou, C.N.: The book thickness of 1-planar graphs is constant. Algorithmica **79**(2), 444–465 (2017). https://doi.org/10.1007/s00453-016-0203-2
18. Bekos, M.A., Kaufmann, M., Raftopoulou, C.N.: On optimal 2- and 3-planar graphs. In: Aronov, B., Katz, M.J. (eds.) 33rd International Symposium on Computational Geometry (SoCG). LIPIcs, vol. 77, pp. 16:1–16:16. Schloss Dagstuhl, Dagstuhl (2017). https://doi.org/10.4230/LIPIcs.SoCG.2017.16. http://drops.dagstuhl.de/opus/volltexte/2017/7230
19. Bodendiek, R., Schumacher, H., Wagner, K.: Bemerkungen zu einem Sechsfarbenproblem von G. Ringel. Abh. Math. Semin. Univ. Hambg. **53**(1), 41–52 (1983). https://doi.org/10.1007/BF02941309
20. Booth, K.S., Lueker, G.S.: Testing for the consecutive ones property, interval graphs, and graph planarity using PQ-tree algorithms. J. Comput. Syst. Sci. **13**(3), 335–379 (1976). https://doi.org/10.1016/S0022-0000(76)80045-1
21. Borodin, O.V.: Solution of Ringel's problems on the vertex-face coloring of plane graphs and on the coloring of 1-planar graphs. Diskret. Analiz **41**, 12–26 (1984). (in Russian)
22. Borodin, O.V.: A new proof of the 6 color theorem. J. Graph Theory **19**(4), 507–521 (1995). https://doi.org/10.1002/jgt.3190190406
23. Boyer, J.M., Myrvold, W.J.: On the cutting edge: simplified $O(n)$ planarity by edge addition. J. Graph Algorithms Appl. **8**(3), 241–273 (2004). https://doi.org/10.7155/jgaa.00091
24. Brandenburg, F., Eppstein, D., Gleißner, A., Goodrich, M.T., Hanauer, K., Reislhuber, J.: On the density of maximal 1-planar graphs. In: Didimo, W., Patrignani, M. (eds.) Proceedings of 20th Symposium on Graph Drawing (GD). LNCS, vol. 7704, pp. 327–338. Springer (2012). https://doi.org/10.1007/978-3-642-36763-2_29
25. Brandenburg, F.J.: Recognizing optimal 1-planar graphs in linear time. Algorithmica **80**(1), 1–28 (2018). https://doi.org/10.1007/s00453-016-0226-8
26. Cabello, S., Mohar, D.: Adding one edge to planar graphs makes crossing number and 1-planarity hard. SIAM J. Comput. **42**(5), 1803–1829 (2013). https://doi.org/10.1137/120872310
27. Cardinal, J., Hoffmann, M., Kusters, V., Tóth, C.D., Wettstein, M.: Arc diagrams, flip distances, and Hamiltonian triangulations. Comput. Geom. **68**, 206–225 (2018). https://doi.org/10.1016/j.comgeo.2017.06.001
28. Chambers, E.W., Eppstein, D., Goodrich, M.T., Löffler, M.: Drawing graphs in the plane with a prescribed outer face and polynomial area. J. Graph Algorithms Appl. **16**(2), 243–259 (2012). https://doi.org/10.7155/jgaa.00257
29. Colin de Verdière, É.: Computational topology of graphs on surfaces. In: Goodman, J.E., O'Rourke, J., Tóth, C.D. (eds.) Handbook of Discrete and Computational Geometry, 3rd edn, Chap. 23, pp. 605–636. CRC Press, Boca Raton (2017)
30. Colin de Verdière, É., Erickson, J.: Tightening nonsimple paths and cycles on surfaces. SIAM J. Comput. **39**(8), 3784–3813 (2010). https://doi.org/10.1137/090761653
31. Dujmović, V., Eppstein, D., Wood, D.R.: Structure of graphs with locally restricted crossings. SIAM J. Discret. Math **31**(2), 805–824 (2017). https://doi.org/10.1137/16M1062879
32. Eades, P., Hong, S.H., Katoh, N., Liotta, G., Schweitzer, P., Suzuki, Y.: A linear time algorithm for testing maximal 1-planarity of graphs with a rotation system. Theor. Comput. Sci. **513**, 65–76 (2013). https://doi.org/10.1016/j.tcs.2013.09.029
33. Eppstein, D., Gupta, S.: Crossing patterns in nonplanar road networks. In: Proceedings of 25th ACM SIGSPATIAL International Conference on Advances in Geographic Information Systems, pp. 40:1–40:9. ACM, New York, NY (2017). https://doi.org/10.1145/3139958.3139999
34. Eppstein, D., van Kreveld, M.J., Mumford, E., Speckmann, B.: Edges and switches, tunnels and bridges. Comput. Geom. **42**(8), 790–802 (2009). https://doi.org/10.1016/j.comgeo.2008.05.005
35. Fox, J., Pach, J.: Applications of a new separator theorem for string graphs. Comb. Probab. Comput. **23**(1), 66–74 (2014). https://doi.org/10.1017/S0963548313000412

36. Frati, F.: A lower bound on the diameter of the flip graph. Electron. J. Comb. **24**(1), P1.43 (2017). http://www.combinatorics.org/ojs/index.php/eljc/article/view/v24i1p43
37. Fulek, R., Kynčl, J., Pálvölgyi, D.: Unified Hanani-Tutte theorem. Electron. J. Comb. **24**(3), P3.18 (2017). http://www.combinatorics.org/ojs/index.php/eljc/article/view/v24i3p18
38. Grigoriev, A., Bodlaender, H.L.: Algorithms for graphs embeddable with few crossings per edge. Algorithmica **49**(1), 1–11 (2007). https://doi.org/10.1007/s00453-007-0010-x
39. Hajnal, P., Igamberdiev, A., Rote, G., Schulz, A.: Saturated simple and 2-simple topological graphs with few edges. J. Graph Algorithms Appl. **22**(1), 117–138 (2018). https://doi.org/10.7155/jgaa.00460
40. Hong, S., Nagamochi, H.: Re-embedding a 1-plane graph into a straight-line drawing in linear time. In: Hu, Y., Nöllenburg, M. (eds.) Proceedings of 24th Symposium on Graph Drawing and Network Visualization (GD). LNCS, vol. 9801, pp. 321–334. Springer (2016). https://doi.org/10.1007/978-3-319-50106-2_25
41. Hong, S.H., Eades, P., Liotta, G., Poon, S.H.: Fáry's theorem for 1-planar graphs. In: Gudmundsson, J., Mestre, J., Viglas, T. (eds.) Proceedings of 18th Computing and Combinatorics Conference (COCOON), pp. 335–346. Springer, Berlin (2012). https://doi.org/10.1007/978-3-642-32241-9_29
42. Hopcroft, J., Tarjan, R.: Efficient planarity testing. J. ACM **21**(4), 549–568 (1974). https://doi.org/10.1145/321850.321852
43. Kaufmann, M., Pach, J., Tóth, G., Ueckerdt, T.: The number of crossings in multigraphs with no empty lens. In: Biedl, T.C., Kerren, A. (eds.) Proceedings of 26th Symposium on Graph Drawing and Network Visualization (GD). LNCS, vol. 11282, pp. 242–254. Springer, Cham (2018). https://doi.org/10.1007/978-3-030-04414-5_17
44. Kobourov, S.G., Liotta, G., Montecchiani, F.: An annotated bibliography on 1-planarity. Comput. Sci. Rev. **25**, 49–67 (2017). https://doi.org/10.1016/j.cosrev.2017.06.002
45. Korzhik, V.P., Mohar, B.: Minimal obstructions for 1-immersions and hardness of 1-planarity testing. J. Graph Theory **72**(1), 30–71 (2013). https://doi.org/10.1002/jgt.21630
46. Kynčl, J., Pach, J., Radoičić, R., Tóth, G.: Saturated simple and k-simple topological graphs. Comput. Geom. **48**(4), 295–310 (2015). https://doi.org/10.1016/j.comgeo.2014.10.008
47. Kynčl, J.: Enumeration of simple complete topological graphs. Eur. J. Comb. **30**(7), 1676–1685 (2009). https://doi.org/10.1016/j.ejc.2009.03.005
48. Kynčl, J.: Simple realizability of complete abstract topological graphs simplified. In: Giacomo, E.D., Lubiw, A. (eds.) Proceedings of 23rd Symposium on Graph Drawing and Network Visualization (GD). LNCS, vol. 9411, pp. 309–320. Springer (2015). https://doi.org/10.1007/978-3-319-27261-0_26
49. Kynčl, J., Valtr, P.: On edges crossing few other edges in simple topological complete graphs. Discret. Math. **309**(7), 1917–1923 (2009). https://doi.org/10.1016/j.disc.2008.03.005
50. Lasoń, M., Micek, P., Pawlik, A., Walczak, B.: Coloring intersection graphs of arcwise connected sets in the plane. In: Nešetřil, J., Pellegrini, M. (eds.) Proceedings of 7th European Conference on Combinatorics, Graph Theory and Applications, pp. 299–304. Scuola Normale Superiore, Pisa (2013)
51. Liu, C.H., Reddy, M.M., Tóth, C.D.: Simple topological drawings of k-planar graphs. In: Book of Abstracts of the 36th European Workshop on Computational Geometry (EuroCG), pp. 80:1–80:6, Würzburg (2020)
52. McGuinness, S.: Colouring arcwise connected sets in the plane I. Graphs Comb. **16**(4), 429–439 (2000). https://doi.org/10.1007/PL00007228
53. de Mendez, P.O., Oum, S., Wood, D.R.: Defective colouring of graphs excluding a subgraph or minor. Combinatorica **39**(2), 377–410 (2019). https://doi.org/10.1007/s00493-018-3733-1
54. Pach, J., Pinchasi, R., Sharir, M., Tóth, G.: Topological graphs with no large grids. Graphs Comb. **21**(3), 355–364 (2005). https://doi.org/10.1007/s00373-005-0616-1
55. Pach, J., Pinchasi, R., Tardos, G., Tóth, G.: Geometric graphs with no self-intersecting path of length three. Eur. J. Comb. **25**(6), 793–811 (2004). https://doi.org/10.1016/j.ejc.2003.09.019
56. Pach, J., Radoičić, R., Tardos, G., Tóth, G.: A generalization of quasi-planarity. In: Pach, J. (ed.) Towards a Theory of Geometric Graphs, Contemporary Mathematics, vol. 342, pp. 177–183. AMS, Providence (2004)

57. Pach, J., Radoičić, R., Tardos, G., Tóth, G.: Improving the crossing lemma by finding more crossings in sparse graphs. Discret. Comput. Geom. **36**(4), 527–552 (2006). https://doi.org/10.1007/s00454-006-1264-9
58. Pach, J., Radoičić, R., Tóth, G.: Relaxing planarity for topological graphs. In: Győri, E., Katona, G.O.H., Lovász, L., Fleiner, T. (eds.) More Sets, Graphs and Numbers: A Salute to Vera Sós and András Hajnal, pp. 285–300. Springer, Berlin (2006). https://doi.org/10.1007/978-3-540-32439-3_12
59. Pach, J., Shahrokhi, F., Szegedy, M.: Applications of the crossing number. Algorithmica **16**(1), 111–117 (1996). https://doi.org/10.1007/BF02086610
60. Pach, J., Tóth, G.: Graphs drawn with few crossings per edge. Combinatorica **17**(3), 427–439 (1997). https://doi.org/10.1007/BF01215922
61. Pach, J., Tóth, G.: A crossing lemma for multigraphs. Discret. Comput. Geom. **63**, 918–933 (2020) https://doi.org/10.1007/s00454-018-00052-z
62. Radermacher, M., Rutter, I.: Inserting an edge into a geometric embedding. In: Biedl, T., Kerren, A. (eds.) Proceedings of 26th Symposium on Graph Drawing and Network Visualization (GD). Springer, Cham (2018). arxiv:1807.11711
63. Ringel, G.: Ein Sechsfarbenproblem auf der Kugel. Abh. Math. Semin. Univ. Hambg. **29**(1), 107–117 (1965). https://doi.org/10.1007/BF02996313
64. Rok, A., Walczak, B.: Coloring curves that cross a fixed curve. Discret. Comput. Geom. **61**(4), 830–851 (2019). https://doi.org/10.1007/s00454-018-0031-z
65. Schaefer, M.: The graph crossing number and its variants: a survey. Electron. J. Comb. **DS21**, 1–113 (2017). http://www.combinatorics.org/files/Surveys/ds21/ds21v3-2017.pdf. Version 3
66. Schaefer, M.: Crossing Numbers of Graphs. Discrete Mathematics and Its Applications. CRC Press, Boca Raton (2018)
67. Schmidt, J.M.: Mondshein sequences (a.k.a. (2, 1)-orders). SIAM J. Comput. **45**(6), 1985–2003 (2016). https://doi.org/10.1137/15M1030030
68. Suk, A., Walczak, B.: New bounds on the maximum number of edges in k-quasi-planar graphs. Comput. Geom. **50**, 24–33 (2015). https://doi.org/10.1016/j.comgeo.2015.06.001
69. Székely, L.A.: Crossing numbers and hard Erdős problems in discrete geometry. Comb. Probab. Comput. **6**(3), 353–358 (1997). https://doi.org/10.1017/S0963548397002976
70. Tardos, G.: Construction of locally plane graphs with many edges. In: Pach, J. (ed.) Thirty Essays on Geometric Graph Theory, pp. 541–562. Springer, New York (2013). https://doi.org/10.1007/978-1-4614-0110-0_29
71. Tardos, G., Tóth, G.: Crossing stars in topological graphs. SIAM J. Discret. Math. **21**(3), 737–749 (2007). https://doi.org/10.1137/050623693
72. Thomassen, C.: Rectilinear drawings of graphs. J. Graph Theory **12**(3), 335–341 (1988). https://doi.org/10.1002/jgt.3190120306
73. Valtr, P.: On the pair-crossing number. In: Goodman, J.E., Pach, J., Welzl, E. (eds.) Combinatorial and Computational Geometry. MSRI Publications, vol. 52, pp. 545–551. Cambridge University Press, Cambridge (2005)
74. Wei-Kuan, S., Wen-Lian, H.: A new planarity test. Theor. Comput. Sci. **223**(1), 179–191 (1999). https://doi.org/10.1016/S0304-3975(98)00120-0
75. Whitney, H.: Congruent graphs and the connectivity of graphs. Am. J. Math. **54**(1), 150–168 (1932). http://www.jstor.org/stable/2371086

Chapter 3
Quasi-planar Graphs

Eyal Ackerman

Abstract A graph is *k-quasi-planar* if it can be drawn in the plane such that no k of its edges are pairwise crossing. Thus, the class of k-quasi-planar graphs contains all planar graphs and several other classes of beyond-planar graphs. The research of k-quasi-planar graphs began in the early 1990s and has focused mainly on upper-bounding their size which is conjectured to be linear. Recently, with the emergence of interest in beyond-planar graphs within the Graph Drawing community, other properties of k-quasi-planar graphs have also been investigated. In this chapter, we survey the literature on k-quasi-planar graphs. Specifically, we mention the progress made toward determining their maximal size, their relationships to other graph classes and a couple of related algorithmic questions.

3.1 Introduction

We consider graphs without loops and parallel-edges. A *topological graph* is a graph drawn in the plane with its vertices as points and its edges as Jordan arcs that connect the points corresponding to its vertices and do not contain any other vertex as an interior point. It is commonly assumed that every pair of edges in a topological graph has a finite number of intersection points, each of which is either a vertex that is common to both edges, or a crossing point at which one edge passes from one side of the other edge to its other side. Note that an edge may not cross itself, since edges are drawn as Jordan arcs. If every pair of edges intersect at most once, then the topological graph is *simple*. A topological graph is called *x-monotone* if its edges are *x*-monotone curves and *geometric* if they are straight-line segments. In a *convex* geometric graph the vertices are in convex position.

A topological graph is *k-quasi-plane* if it does not contain k pairwise crossing edges (for an integer $k \geq 2$). An abstract graph is *k-quasi-planar* if it can be drawn as a k-quasi-plane graph. A 2-quasi-planar graph is thus a planar graph while

E. Ackerman (✉)
Department of Mathematics, Physics, and Computer Science,
University of Haifa at Oranim, 36006 Tivon, Israel
e-mail: ackerman@sci.haifa.ac.il

© Springer Nature Singapore Pte Ltd. 2020
S.-H. Hong and T. Tokuyama (eds.), *Beyond Planar Graphs*,
https://doi.org/10.1007/978-981-15-6533-5_3

3-quasi-planar graphs are also referred to as *quasi-planar*. Most of the research concerning k-quasi-planar graphs has dealt with their size which should be linear according to a well-known and rather old conjecture.

Conjecture 3.1 ([39]) *Every n-vertex k-quasi-planar graph has $O_k(n)$ edges.*[1,2]

We review the progress toward settling Conjecture 3.1 in Sect. 3.2. With the emergence of interest in beyond-planar graphs within the Graph Drawing community, other properties of k-quasi-planar graphs have been studied. In Sect. 3.3, we explore relationships of k-quasi-planar graphs and other classes of (beyond-planar) graphs. Finally, in Sect. 3.4, we review the very few works that considered k-quasi-planar graphs from an algorithmic point of view. Apart from Conjecture 3.1, several other open problems concerning k-quasi-planar graphs are mentioned along the way.

3.2   The Size of k-Quasi-planar Graphs

Let $f_k(n)$ denote the maximum number of edges in an n-vertex k-quasi-plane graph for $k \geq 2$ and $n > 2$. Thus, Conjecture 3.1 states that $f_k(n) = O_k(n)$.

In the following subsections, we give a short history of the progress made so far toward settling Conjecture 3.1 in the general case as well as for small values of k and for restricted classes of topological graphs. Most of the results are mentioned without a proof. The best known bounds on the size of the various classes of k-quasi-plane graphs are summarized in Table 3.1.

3.2.1   3- and 4-Quasi-planar Graphs

Since 2-quasi-plane graphs are plane graphs, it follows from Euler's polyhedral formula that $f_2(n) = 3n - 6$. Hence, $k = 3$ is the smallest value for which Conjecture 3.1 is nontrivial. Pach [39] observed that a simple application of the crossing lemma [6, 35] implies that $f_3(n) \leq O(n^{3/2})$. Indeed, recall that by the crossing lemma the number of crossings in any drawing of an n-vertex graph with $m \geq 4n$ edges is at least $\frac{m^3}{64n^2}$. Therefore, given a 3-quasi-plane graph with $m \geq 4n$ edges, it follows that it has an edge that crosses at least $\frac{m^2}{32n^2}$ other edges. Since none of these edges are crossing (otherwise there would be three pairwise crossing edges), it follows that $\frac{m^2}{32n^2} \leq 3n$ and thus $m \leq 10n^{3/2}$.

The result of Pach, Shahrokhi and Szegedy [41] for general k implied $O_t(n(\log n)^2)$ and $O_t(n(\log n)^4)$ upper bounds on the size of 3- and 4-quasi-plane graphs, respectively, in which every pair of edges intersect at most t times. Agarwal et al. [4] proved

---

[1]This conjecture was actually phrased for geometric graphs in [39] and was attributed to B. Gärtner.
[2]The notation $O_k(\cdot)$ indicates that the constant hiding in the big-Oh notation depends only on k.

Table 3.1 Best known bounds on the size of k-quasi-plane graphs

Class	Upper bound	Lower bound	References
3-quasi-plane graphs	$8n - 20$	$7n - O(1)$	[3]
Simple topological 3-quasi-plane graphs	$6.5n - 20$	$6.5n - O(1)$	[3]
4-quasi-plane graphs	$72n$		[1]
Convex geometric graphs	$\begin{cases} \binom{n}{2}, & n \le 2k - 1 \\ 2(k-1)n - \binom{2k-1}{2}, & n \ge 2k - 1 \end{cases}$	= upper bound	[14]
Geometric graphs	$O_k(n \log n)$		[48]
x-monotone topological graphs	$O_k(n \log n)$		[27]
Simple topological graphs	$O_k(n \log n)$		[47]
Topological graphs where any two edges may intersect at most t times	$O_{k,t}(n \log n)$		[45]
General topological graphs	$n(\log n)^{O(\log k)}$		[25, 26]

Conjecture 3.1 for simple topological 3-quasi-plane graphs. Pach, Radoičić and Tóth [40] extended their proof to general topological graphs and proved that $f_3(n) \le 65n$. Ackerman and Tardos [3] obtained the bounds $7n - O(1) \le f_3(n) \le 8n - 20$. For simple topological graphs they provided the upper bound $6.5n - 20$ and showed that it is tight up to an additive constant. Their upper bounds were proved using the *discharging method* (see below). Ackerman [1] used this technique to settle Conjecture 3.1 also for 4-quasi-planar graphs by showing that $f_4(n) \le 72n$.

Next, we demonstrate the discharging method that was used in [1, 3] to show the linear size of 3- and 4-quasi-planar graphs. In order to simplify the presentation, we consider geometric 3-quasi-plane graphs and prove that such graphs have a vertex of small degree. Since every subgraph of such a graph is also geometric and 3-quasi-plane, a linear size follows by induction.

Denote by $\delta(G)$ the smallest degree in a graph G.

Theorem 3.1 ([3]) *If G is a geometric 3-quasi-plane graph, then $\delta(G) < 20$.*

Proof We may assume that $G = (V, E)$ is connected as otherwise we can conclude by induction. Suppose for contradiction that $\delta(G) \ge 20$. Let G' be the plane graph we obtain by adding the crossing points of G as vertices and subdividing the edges accordingly. We assign *charges* to the vertices and faces of G' as follows: for every $u \in V(G')$ set $\mathrm{ch}(u) := \deg(u) - 4$ and for every face $f \in F(G')$ set $\mathrm{ch}(f) := |f| - 4$. It follows from Euler's formula that the total charge is

$$\sum_{u \in V(G')} (\deg(u) - 4) + \sum_{f \in F(G')} (|f| - 4) = 2|E(G')| - 4|V(G')| + 2|E(G')| - 4|F(G')| = -8.$$

Fig. 3.1 The second discharging step in the proof of Theorem 3.1. Here $f_3 = f_i$ contributes $\frac{1}{5}$ units of charge to f through e_3'

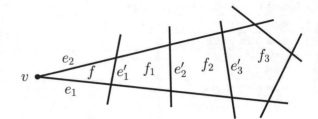

However, we will show next that one can distribute the charges (*discharge*) such that every element has a non-negative charge and hence reach a contradiction.

Note that the only elements with a negative charge are the triangular faces of G' ($|V| \geq 21$ since $\delta(G) \geq 20$ and therefore the size of every face is at least three). Since G has no three pairwise crossing edges, a triangular face must be incident to at least one vertex of G. In the first discharging step, every original vertex $v \in V$ contributes $\frac{4}{5}$ units of charge to each of the $\deg(v)$ faces incident to it (a cut vertex that is incident to a face with multiplicity contributes several times to the same face). Observe that since $\delta(G) \geq 20$, every vertex of G still has a non-negative charge.

The only elements with a negative charge after the first discharging step are the triangular faces that are incident to exactly one vertex of G. Let f be such a face, let $v \in V$ be the only original vertex incident to f and let e_1 and e_2 be the edges of G that are incident to v and f. Denote by e_1' the edge of G' that is incident to f and is not incident to v, and let f_1 be the other face that is incident to e_1'. For $i > 1$ if f_{i-1} and e_{i-1}' are defined, $|f_{i-1}| = 4$ and f_{i-1} is not incident to a vertex in V, then we denote by e_i' the opposite edge to e_{i-1}' in f_{i-1} and denote by f_i the other face but f_{i-1} that is incident to e_i (see Fig. 3.1, for example). Since G' is finite, there is i such that f_i is defined whereas f_{i+1} is undefined. In the second discharging step, f receives $1/5$ units of charge from f_i *through* the edge e_i'. This is repeated for every triangular face f that is incident to exactly one original vertex.

After the second discharging step the charge of every vertex and every triangular face is non-negative. The charge of a face of size four that is not incident to an original vertex remains zero. Each other face contributes at most once through each of its edges whose endpoints are crossing points of G. An easy case-analysis shows that the charge of such a face also remains non-negative after the second discharging step. Therefore, the final charge of every element is non-negative whereas the total charge remains -8, which is a contradiction. Therefore G must have a vertex whose degree is at most 19. □

Corollary 3.1 *Every n-vertex geometric 3-quasi-plane graph has at most $19n$ edges.*

Proof We prove by induction on n. For $n = 1$ the claim clearly holds. Let G be an n-vertex 3-quasi-plane geometric graph for some $n > 1$. By Theorem 3.1 G has a vertex v whose degree is at most 19. Remove v and obtain an $(n-1)$-vertex 3-quasi-plane graph. By the induction hypothesis this graph has at most $19(n-1)$ edges and therefore G has at most $19n$ edges. □

3.2.2 k-Quasi-planar Graphs for $k \geq 5$

Currently, Conjecture 3.1 is still open for $k \geq 5$. Pach [39] observed that his sub-quadratic upper bound for the size of 3-quasi-planar graphs can be generalized to give $f_k(n) = O\left(n^{2-(1/2^{k-1})}\right)$. Pach, Radoičić and Tóth [40] generalized the proof of Pach, Shahrokhi and Szegedy [41] for simple topological graphs and obtained the bound $f_k(n) = O_k(n(\log n)^{4k-12})$. Plugging the result of Ackerman [1] into their proof improved this bound to $f_k(n) = O_k(n(\log n)^{4k-16})$. Fox and Pach [25, 26] improved the exponent of the logarithmic factor from $O(k)$ to $O(\log k)$ and showed that $f_k(n) \leq n(\log n)^{O(\log k)}$.

In fact, Fox and Pach [26] proved a much stronger result, namely, that any set of curves in the plane that does not contain k pairwise intersecting curves can be colored using $(\log n)^{O(\log k)}$ colors such that no curves of the same color intersect. By applying this coloring on the edges (minus their endpoints) of a k-quasi-plane graph G, one can partition G into $(\log n)^{O(\log k)}$ plane graphs and conclude that $|E(G)| \leq (3n - 6)(\log n)^{O(\log k)}$.

It is possible that the above-mentioned $(\log n)^{O(\log k)}$ bound can be further improved to yield a further improvement for the currently best upper bound on $f_k(n)$. However, this approach cannot lead to the linear upper bound of Conjecture 3.1, since Pawlik et al. [44] constructed sets of segments, no three of which pairwise intersect, that cannot be colored with $O(1)$ colors such that no two segments of the same color intersect. This implies that there are 3-quasi-plane graphs that cannot be partitioned into constantly many plane graphs (while maintaining the same embedding).

3.2.3 Restricted Drawings

Better upper bounds than the ones for general k-quasi-plane graphs were obtained for restricted classes of k-quasi-plane graphs, that we discuss in the following. These include convex geometric graphs, x-monotone topological graphs, simple topological graphs and topological graphs in which every pair of edges intersect at most t times for some constant t.

3.2.3.1 Convex Geometric Graphs

Recall that in a convex geometric graph the vertices are in convex position.[3] Gyárfás [29] proved that any set of chords of a circle, no k of which pairwise cross, can be colored with $O_k(1)$ colors such that no two chords of the same color intersect. It follows that an n-vertex convex geometric k-quasi-plane graph has $O_k(n)$ edges. Indeed, consider the edges of such a graph as (open) chords of a circle and color

---

[3]Convex geometric k-quasi-plane graph were also called *outer* k-quasi-plane graphs.

them with $O_k(1)$ colors such that no two crossing edges are of the same color. Then each color class induces an outerplanar graph with at most $2n - 3$ edges and thus the total number of edges is $O_k(n)$.

Capoyleas and Pach [14] gave a tight bound on the maximum number of edges in an n-vertex convex geometric k-quasi-plane graph. They proved that this number is $\binom{n}{2}$ when $n \leq 2k - 1$ and $2(k - 1)n - \binom{2k-1}{2}$ when $n \geq 2k - 1$. In fact, any *maximal*[4] convex geometric k-quasi-plane graph must have this many edges [15, 21, 38].

We remark that the linear size of convex geometric k-quasi-plane graphs also follows from a more general result known as the Marcus–Tardos Theorem [37]. Indeed, Klazar and Marcus [33] pointed out that it is not hard to modify the proof in [37] and obtain a linear bound for the number of edges in an *ordered* graph that does not contain a certain ordered matching (see [37] for details). Since the order of the endpoints of two edges determines whether they are crossing, this result implies that convex geometric k-quasi-planar graphs have at most linearly many edges.

3.2.3.2 Geometric and Simple Topological Graphs

Let $f_{k,t}(n)$ denote the maximum size of an n-vertex k-quasi-plane graph $(n > 2)$ in which each pair of edges intersects at most t times (hence, $f_{k,1}(n)$ denotes the maximum size of a simple topological k-quasi-plane graph).

Already in the first paper that considered k-quasi-planar graphs, Pach [39] obtained a bound of $O\left(n^{2-(1/25(k+1^2))}\right)$ for the size of geometric k-quasi-plane graphs which was better than his $f_k(n) = O\left(n^{2-(1/2^{k-1})}\right)$ bound for general k-quasi-plane graphs. Pach, Shahrokhi and Szegedy [41] proved that $f_{k,t}(n) = O_{k,t}(n(\log n)^{2k-4})$ using a relation between the *bisection width* of a graph and its *crossing number*.

As mentioned before, Agarwal et al. [4] proved that $f_{3,1}(n) = O(n)$. Plugging their result into the proof of Pach et al. [41] implied $f_{k,1}(n) = O_k(n(\log n)^{2k-6})$. Similarly, the bound $f_4(n) = O(n)$ [1] implied $f_{k,1} = O_k(n(\log n)^{2k-8})$.

Valtr [49] proved an $O_k(n \log n)$ bound on the size of geometric k-quasi-plane graphs. Later he extended his result also for x-monotone simple topological graphs [48]. In both cases, Valtr showed that if all the edges intersect a common line, then there are $O_k(n)$ edges. The $O_k(n \log n)$ bound then follows from a standard divide-and-conquer argument.

Theorem 3.2 ([48]) *Let $G = (V, E)$ be a geometric k-quasi-plane graph such that there is a line ℓ that intersects every edge of G. Then $|E| \leq O_k(|V|)$.*

Proof We may assume without loss of generality that ℓ is the y-axis and that every edge of G intersects this line in a distinct point. Denote the edges of G by $e_1, e_2, \ldots, e_m$ according to the order in which they intersect ℓ from bottom to top and let $V = \{v_1, v_2, \ldots, v_n\}$.

---

[4]Call a convex geometric k-quasi-plane graph G *maximal* if it is impossible to insert a new edge to G while maintaining its k-quasi-planarity.

Let S_l (respectively, S_r) be the sequence of length m that we obtain by listing the left (respectively, right) endpoint of each of the edges according to their order. For an integer $t \geq 1$, we say that a sequence is *t-regular* if it does not contain t' consecutive elements such that two of them are identical and $t' \leq t$. A sequence $S = s_1, s_2, \ldots, s_{3t-2}$ is of type *up-down-up(t)* if $s_i \neq s_j$ for every $1 \leq i < j \leq t$ and $s_i = s_{2t-i} = s_{2t-2+i}$. For example, $a, b, c, d, c, b, a, b, c, d$ is of type up-down-up(4).

The proof of the theorem follows from the following lemmas.

Lemma 3.1 ([48]) *For every $t \geq 1$, at least one of the sequences S_l and S_r contains a t-regular subsequence of length at least $\frac{m}{4t}$.* □

Lemma 3.2 ([34]) *For every $t \geq 1$ there is a constant c_t such that if S is a t-regular sequence over an alphabet of size m and S does not contain a subsequence of type up-down-up(t), then $|S| \leq c_t m$.* □

Lemma 3.3 ([48]) *Neither S_l nor S_r contains a subsequence of type up-down-up$((k-1)^3 + 1)$.* □

Proof Assume without loss of generality that S_l contains a subsequence of type up-down-up(t) for $t = (k-1)^3 + 1$. Relabel the vertices of G such that this subsequence becomes $v_1, v_2, \ldots, v_t, v_{t-1}, \ldots, v_1, v_2, \ldots, v_t$ and denote by $e'_1, e'_2, \ldots, e'_{3t-2}$ the edges that correspond to this subsequence. Thus, the appearances of v_i in the subsequence are due to the edges e'_i, e'_{2t-i}, and e'_{2t-2+i} (for $l = 1, t$ two of these edges coincide). Let p_i be the intersection point of e'_i and ℓ (the y-axis) and define three partial orders on $v_1, \ldots, v_t$ as follows:

$\prec_1 = \{(v_i, v_j) \mid i \leq j$ and the ray $\overrightarrow{v_i v_j}$ intersects ℓ below $p_{2t-1}\}$
$\prec_2 = \{(v_i, v_j) \mid i \leq j$ and the ray $\overrightarrow{v_j v_i}$ intersects ℓ above $p_t\}$
$\prec_3 = \{(v_i, v_j) \mid i \leq j$ and $(v_i, v_j) \notin \prec_1 \cup \prec_2\}$

It is not hard to verify that each of these relations is indeed a partial order.[5]

Proposition 3.1 *The following holds*

(a) *If $v_i \prec_1 v_j$ then e'_{2t-2+i} and e'_{2t-2+j} are crossing;*
(b) *if $v_i \prec_2 v_j$ then e'_i and e'_j are crossing; and*
(c) *if $v_i \prec_3 v_j$ then e'_{2t-i} and e'_{2t-j} are crossing.* □

By applying Dilworth's Theorem at most three times it follows that either there is a chain of length k with respect to one of the partial orders, or there are vertices v_i and v_j that are incomparable with respect to each of the partial orders. The latter is impossible by definition of the partial orders, therefore, there must by a chain $v_{i_1} \prec_j v_{i_2} \prec_j \cdots \prec_j v_{i_k}$ for some $j \in \{1, 2, 3\}$. However, by Proposition 3.1 if $j = 1$, then the edges e'_{2t-2+i_s}, $s = 1, 2, \ldots, k$ are pairwise crossing; if $j = 2$, then the

---

[5]For reflexivity define the ray $\overrightarrow{v_i v_i}$ as the ray $\overrightarrow{v_i p_{t+1}}$.

edges e'_{i_s}, $s = 1, 2, \ldots, k$ are pairwise crossing; and if $j = 3$ then the edges e'_{2t-i_s}, $s = 1, 2, \ldots, k$ are pairwise crossing.   □

Let $t = (k + 1)^3 + 1$. By Lemma 3.1 at least one of S_l and S_r contains a t-regular subsequence S of length at least $\frac{m}{4t}$. Since G is k-quasi-plane, if follows from Lemma 3.3 that S is of type up-down-up(t). Therefore, by Lemma 3.2 we have $\frac{m}{4t} \leq |S| \leq c_t n$ and thus $m \leq 4tc_t n = O_k(n)$.   □

Corollary 3.2 ([48]) *Every n-vertex geometric k-quasi-plane graph has $O_k(n \log n)$ edges.*

Valtr's $O_k(n \log n)$ bound remains the best upper bound for geometric graphs. Several researchers made a considerable effort to obtain this bound also for less restricted classes of k-quasi-plane graphs: Fox and Pach [25] proved that $f_{k,t}(n) \leq n \left(C_t \frac{\log n}{\log k} \right)^{O(\log k)}$, where C_t is a constant that depends only on t. Fox, Pach and Suk [27] showed that $f_{k,1}(n) = (n \log n)2^{\alpha(n)^{O_1(k)}}$, where $\alpha(n)$ is the inverse of the Ackermann function and also gave an upper bound of $O_k(n \log n)$ on the size of x-monotone (not necessarily simple) k-quasi-plane graphs. Suk and Walczak [47] improved the former bound to $f_{k,1}(n) = O_k(n \log n)$ and also proved that $f_{k,t}(n) \leq 2^{\alpha(n)^{O_{k,t}(1)}} \log n$. Finally, Rok and Walczak [45] managed to show recently that $f_{k,t}(n) = O_{k,t}(n \log n)$.

3.2.4 Lower Bounds

No lower bounds for $f_k(n)$ are mentioned in the literature, apart from the case of $k = 3$ [3, 4]. As mentioned above, Ackerman and Tardos [3] proved that $f_3(n) \geq 7n - O(1)$ which is not far from their upper bound $8n - 20$. For simple 3-quasi-planar graphs they showed that $6.5n - O(1) \leq f_{3,1}(n) \leq 6.5n - 20$. Brandenburg [13] provided a drawing of K_{10} as a simple 3-quasi-planar graph and this shows that for $n = 10$ the bound $6.5n - 20$ is tight.

Agarwal et al. [4] remark that a 3-quasi-planar graph with roughly $6n$ edges can be obtained by overlaying two edge-disjoint triangulations on the same set of n points. This can be generalized to any fixed k implying that $f_k(n) \geq 3(k - 1)n - O(k)$: Since the *thickness*[6] of the complete graph K_n is $n/6 + O(1)$ [11], there are $k - 1$ edge-disjoint planar subgraphs of K_n each of which has $3n - O(1)$ edges (assuming k is not too large with respect to n). It is also known that any planar graph can be drawn as a plane graph such that its vertices are mapped into any given set of points in the plane according to any given bijection between the vertices and the points [31, 43]. Therefore, these $k - 1$ planar graphs can be embedded simultaneously as plane graphs on the same set of points, and therefore the resulting drawing does not contain k pairwise crossing edges.

---

[6]The *thickness* of a graph G is the minimum number of planar graphs into which G can be decomposed.

Fig. 3.2 A geometric
k-quasi-plane graph with
$3(k-1)n - O(k^2)$ edges:
each of the $k-1$ vertices on
the right is adjacent to every
vertex; the remaining
$n - k + 1$ vertices on the left
are in convex position and
induce $2(k-1)(n - k + 1) - \binom{2k-1}{2}$ edges [14]

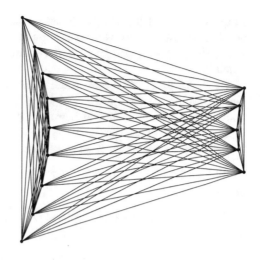

Note that these arguments do not apply when considering geometric or even simple topological graphs. Still, an almost matching lower bound for geometric graphs can be deduced from the tight bound of $2(k-1)n - \binom{2k-1}{2}$ on the size of convex k-quasi-plane graphs [14]: By placing $n - (k-1)$ points in a convex position, such that they all "see" another set of $k-1$ points in a convex position, one can obtain a (geometric) k-quasi-plane graph with $3(k-1)n - O(k^2)$ edges (see Fig. 3.2).

Problem 3.1 Find a nontrivial lower bounds for $f_k(n)$.

For example, it would be interesting if a lower bound of the form $f_k(n) = \Omega(k^{1+\varepsilon}n)$ exists for some constant $\varepsilon > 0$.

3.3   Relationships with Other Classes of Graphs

3.3.1   Beyond-Planar Graphs

The class of $(k$-$)$quasi-planar graphs contains several classes of other beyond-planar graphs. For some classes, this follows immediately from their definition or from an easy counting argument, while for other classes nontrivial redrawing procedures were needed to establish this relationship.

k-**Planar Graphs**
Recall that an abstract graph is k-planar if it can be drawn as a topological graph in which each edge is crossed at most k times. Clearly every k-planar graph can be drawn as a $(k+2)$-quasi-plane graph. Angelini et al. [7] proved that for $k \geq 3$, every

simple topological k-plane graph can be redrawn as a simple topological $(k + 1)$-quasi-plane graph. Their work was complemented by Hoffmann and Tóth [30] who showed that every 2-plane graph can be redrawn as a simple topological 3-quasi-plane graph.

One can ask whether these results can be strengthen by showing that every k-planar graph is k-quasi-planar. For $k = 2$ this is obviously false, since a 2-quasi-planar graph is planar, whereas there are non-planar 2-planar graphs (e.g., *optimal*[7] 2-planar graphs whose size is $5n - 10$). Still, Bekos et al. [12] showed that optimal 2- and 3-planar graphs are 3-quasi-planar.

Problem 3.2 ([7]) Is it true that for every $k \geq 3$ every k-planar graph is k-quasi-planar?

Since the size of k-planar graphs is $O(\sqrt{k}n)$ [42] an even bolder statement might hold.

Problem 3.3 ([7]) Is there a function $q(k) = o(k)$ such that for sufficiently large values of k every k-planar graph is $q(k)$-quasi-planar?

RAC Graphs
A graph that can be drawn with straight-line edges such that every crossing occurs at a right angle is called a *right angle crossing (RAC)* graph. These graphs were introduced by Didimo et al. [20] and were studied in several subsequent works. It is easy to see that any RAC drawing of a graph is 3-quasi-plane. Since a RAC graph has at most $4n - 10$ edges [20], RAC graphs form a proper subset of 3-quasi-planar graphs.

Fan-Planar Graphs
A graph is *fan-planar* if it can be drawn such that for each of its edges e it holds that all the edges that cross e have a common endpoint on the same side of e. It follows immediately that such a graph is 3-quasi-planar. Since a fan-planar graph has at most $5n - 10$ edges [32], fan-planar graphs form a proper subset of 3-quasi-planar graphs.

Planarly Connected Crossing Topological Graphs
Ackerman et al. [2] studied *planarly connected crossing* (PCC) topological graphs. In such a graph for every pair of independent and crossing edges there is a crossing-free edge that connects two of their endpoints. It can be shown that certain drawings of optimal 1-planar graphs and fan-planar graphs posses this property (see [32, Corollary 1]).

Conjecture 3.1, if true, would imply the main result of [2] by which PCC simple topological graphs have a linear size. Indeed, it is easy to see that if G is a PCC simple topological graph, then G is 9-quasi-plane: Suppose for contradiction that G contains a set E' of 9 pairwise crossing edges and let V' be the set of their endpoints. Since G is

---

[7]An n-vertex graph G within a class of graphs $\mathcal{G}$ is *optimal* if there is no other n-vertex graph $G' \in \mathcal{G}$ with more edges than G.

a simple topological graph, no two edges in E' share an endpoint, therefore $|V'| = 18$. Let G' be the subgraph of G induced by V' and let E'' be the crossing-free edges of G'. Clearly (V', E'') is a plane graph. Moreover, all the edges in E' must lie in the same face f of this plane graph, since they are pairwise crossing. It follows that f is incident to every vertex in V' and therefore (V', E'') is an outerplanar graph. Thus, $|E''| \leq 2 \cdot 18 - 3 = 33$. On the other hand, since G' is also PCC topological and no two edges in E' share an endpoint, it follows that $|E''| \geq \binom{9}{2} = 36$, a contradiction.

Gap-Planar Graphs

A graph is k-*gap-planar* if it can be drawn such that every crossing in the drawing is assigned to one of the two edges that define it and no edge is assigned more than k crossings. This graph class was introduced in [10] where it was also shown that for every $k \geq 1$ there is a 3-quasi-planar graph that is not k-gap-planar. Since it is easy to see that every k-gap-planar graph is $(2k + 2)$-quasi-planar, it follows that k-gap-planar graphs form a proper subset of $(2k + 2)$-quasi-planar graphs.

Bar Visibility Graphs

Bar visibility graphs are graphs that can be drawn such that their vertices are represented by disjoint horizontal segments (*bars*) and each of their edges is represented by a vertical segment that intersects the bars that correspond to the endpoints of the edge and no other bars. It is known that this class of graphs is exactly the class of planar graphs [18].

Dean et al. [17] generalized this notion to k-*bar visibility graphs* in which a vertical segment representing an edge (u, v) may intersect at most k other bars except for the bars that represent u and v. Evans et al. [23] proved that every 1-bar visibility graph is 3-quasi-planar. Geneson et al. [28] showed that k-bar visibility graphs in which all the bars have their left endpoint on the y-axis can be represented as convex geometric $(k + 2)$-quasi-plane graphs.

3.3.2 Complete Graphs

Let m_k be the greatest integer such that the complete graph with m_k vertices can be drawn as a k-quasi-plane graph. Let $\overline{m}_k$ and $\tilde{m}_k$ denote similar values for geometric and simple topological k-quasi-plane graphs, respectively.

In [3] appears a drawing of K_9 that is claimed to be a geometric 3-quasi-plane graph. However, there are actually three pairwise crossing edges in that drawing.[8] Still, the drawing can be slightly changed and turned into a 3-quasi-plane graph (see Fig. 3.3). Aichholzer and Krasser [5] showed that K_{10} cannot be drawn as a geometric 3-quasi-plane graph by exploring all the different order-types of ten points in the plane. Thus, we have $\overline{m}_3 = 9$.

---

[8] We thank Roland Schmid for bringing this issue into our attention.

Fig. 3.3 K_9 drawn as a
geometric 3-quasi-plane
graph

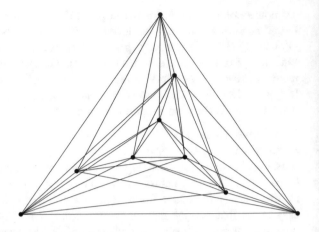

Recall that $f_{3,1}(n) \leq 6.5n - 20$ [3]. Therefore K_{11} cannot be drawn as a simple topological 3-quasi-plane graph. Brandenburg [13] provided a drawing of K_{10} as a simple topological 3-quasi-plane graph, thus showing that $\tilde{m}_3 = 10$.

Since $f_3(n) \leq 8n - 20$ [3], we have $10 \leq m_3 \leq 14$. It would be interesting to determine the exact value of m_3.

Problem 3.4 Determine the greatest integer m_3 such that the complete graph on m_3 vertices can be drawn as a 3-quasi-plane graph.

It is also natural to ask which complete bipartite graphs are, e.g., 3-quasi-planar.

3.4 Computational Aspects of k-Quasi-planar Graphs

There are hardly any results concerning k-quasi-planar graphs from a computational point of view. For example, the following basic questions are open.

Problem 3.5 What is the computational complexity of recognizing k-quasi-planar graphs? Already for $k = 3$ this problem is open.

Problem 3.6 What is the computational complexity of finding a 3-quasi-plane drawing of a given 3-quasi-planar graph?

These questions can also be asked for more restricted classes of k-quasi-planar graphs such as geometric graphs, convex geometric graphs, and graphs that admit drawings that are both k-quasi-planar and l-layered (i.e., each vertex should lie on one of l horizontal lines).

We are only aware of two works that consider algorithmic questions involving k-quasi-planar graphs. In both cases, an original problem asking for a planar embedding is relaxed to allow a k-quasi-plane embedding.

Area Minimization Via Quasi-planarity

It is known that every planar graph can be drawn as a geometric plane graph on a grid of size $O(n^2)$ [16, 46] and that some planar graphs cannot be drawn on a grid of sub-quadratic area [16]. One possible way to obtain drawings of sub-quadratic area is to allow crossings while maintaining some other properties, e.g., $(k-)$quasi-planarity. This line of research was pursued by Di Giacomo et al. [19]. Combining their technique and the recent proof that planar graphs have a bounded *queue number* [22] it follows that every planar graph can be drawn in $O(n)$ area as a k-quasi-plane graph for some absolute constant k.

Di Giacomo et al. [19] also proved that every *partial k-tree* can be drawn as an $O_k(1)$-quasi-plane graph in $O(n)$ area. They also obtained linear size drawings for partial 2-trees, outerplanar graphs and flat series-parallel graphs as 11-, 3- and 5-quasi plane graphs, respectively.

Simultaneous Embedding Via Quasi-planarity

Given two planar graphs G_1 and G_2 with the same vertex set V, a *simultaneous geometric embedding (SGE)* of these graphs is a mapping of V into a planar set of points such that the induced drawing of G_i as a geometric graph is plane, for $i = 1, 2$. It is easy to see that there are pairs of planar graphs that do not have SGE. In examples of pairs of planar graphs that do admit SGE usually one of the graphs is a path (see, e.g., [24]). Di Giacomo et al. [36] initiated the study of *quasi-planar* SGE where the embedded graphs are only required to be 3-quasi-plane. For example, they showed that every path and a tree admit a quasi-planar SGE. This is in contrast to the existence of pairs of a path and a tree that do not admit SGE [9].

In a recent related work Angelini et al. [8] considered another variation of SGE in which the edges are drawn as Jordan curves, the embedded graphs are not necessarily planar and their embedding is required to be 3-quasi-plane. For example, they observed that every triple of a two planar graphs and a tree admits such an embedding. This in turn implies that every pair of a planar graph and a 1-planar graph also admits such an embedding. See [8] for further related results.

References

1. Ackerman, E.: On the maximum number of edges in topological graphs with no four pairwise crossing edges. Discret. Comput. Geom. **41**(3), 365–375 (2009)
2. Ackerman, E., Keszegh, B., Vizer, M.: On the size of planarly connected crossing graphs. J. Graph Algorithms Appl. **22**(1), 311–320 (2018)
3. Ackerman, E., Tardos, G.: On the maximum number of edges in quasi-planar graphs. J. Comb. Theory Ser. A **114**(3):563–571 (2007)
4. Agarwal, P.K., Aronov, B., Pach, J., Pollack, R., Sharir, M.: Quasi-planar graphs have a linear number of edges. Combinatorica **17**(1), 1–9 (1997)

5. Aichholzer, O., Krasser, H.: The point set order type data base: a collection of applications and results. In: Proceedings of the 13th Canadian Conference on Computational Geometry, University of Waterloo, Ontario, Canada, August 13–15, 2001, pp. 17–20 (2001)
6. Ajtai, M., Chvátal, V., Newborn, M., Szemerédi, E.: Crossing-free subgraphs. In: Hammer, P.L., Rosa, A., Sabidussi, G., Turgeon, J. (eds.) Theory and Practice of Combinatorics. North-Holland Mathematics Studies, vol. 60, pp. 9–12. North-Holland, Amsterdam (1982)
7. Angelini, P., Bekos, M.A., Brandenburg, F.J., Da Lozzo, G., Di Battista, G., Didimo, W., Liotta, G., Montecchiani, F., Rutter, I.: On the relationship between k-planar and k-quasi-planar graphs. In: Bodlaender, H.L., Woeginger, G.J. (eds.) Graph-Theoretic Concepts in Computer Science, pp. 59–74. Springer International Publishing, Cham (2017)
8. Angelini, P., Förster, H., Hoffmann, M., Kaufmann, M., Kobourov, S., Liotta, G., Patrignani, M.: The quaSEFE problem. In: Proceedings of the 27th International Symposium on Graph Drawing and Network Visualization, GD'19 (2019)
9. Angelini, P., Geyer, M., Kaufmann, M., Neuwirth, D.: On a tree and a path with no geometric simultaneous embedding. J. Graph Algorithms Appl. **16**(1), 37–83 (2012)
10. Bae, S.W., Baffier, J.-F., Chun, J., Eades, P., Eickmeyer, K., Grilli, L., Hong, S.-H., Korman, M., Montecchiani, F., Rutter, I., Tóth, C.D.: Gap-planar graphs. Theor. Comput. Sci. **745**, 36–52 (2018)
11. Beineke, L.W., Harary, F.: On the thickness of the complete graph. Bull. Am. Math. Soc. **70**(4), 618–620, 07 (1964)
12. Bekos, M.A., Kaufmann, M., Raftopoulou, C.N.: On optimal 2- and 3-planar graphs. In: Aronov, B., Katz, M.J. (eds.) 33rd International Symposium on Computational Geometry, SoCG 2017, July 4–7, 2017, Brisbane, Australia. LIPIcs, vol. 77, pp. 16:1–16:16. Schloss Dagstuhl - Leibniz-Zentrum fuer Informatik (2017)
13. Brandenburg, F.J.: A simple quasi-planar drawing of K_{10}. In: Hu, Y., Nöllenburg, M. (eds.) Graph Drawing. Lecture Notes in Computer Science, vol. 9801, pp. 603–604. Springer, Berlin (2016)
14. Capoyleas, V., Pach, J.: A Turán-type theorem on chords of a convex polygon. J. Comb. Theory Ser. B **56**(1), 9–15 (1992)
15. Chaplick, S., Kryven, M., Liotta, G., Löffler, A., Wolff, A.: Beyond outerplanarity. In: Frati, F., Ma, K.-L. (eds.) Graph Drawing and Network Visualization, pp. 546–559. Springer International Publishing, Cham (2018)
16. De Fraysseix, H., Pach, J., Pollack, R.: How to draw a planar graph on a grid. Combinatorica **10**(1), 41–51 (1990)
17. Dean, A.M., Evans, W.S., Gethner, E., Laison, J.D., Safari, M.A., Trotter, W.T.: Bar k-visibility graphs. J. Graph Algorithms Appl. **11**(1), 45–59 (2007)
18. Battista, G.D., Eades, P., Tamassia, R., Tollis, I.G.: Graph Drawing: Algorithms for the Visualization of Graphs. Prentice-Hall, Upper Saddle River (1999)
19. Di Giacomo, E., Didimo, W., Liotta, G., Montecchiani, F.: h-quasi planar drawings of bounded treewidth graphs in linear area. In: Golumbic, M.C., Stern, M., Levy, A., Morgenstern, G. (eds.) Graph-Theoretic Concepts in Computer Science, pp. 91–102. Springer, Berlin (2012)
20. Didimo, W., Eades, P., Liotta, G.: Drawing graphs with right angle crossings. Theor. Comput. Sci. **412**(39), 5156–5166 (2011)
21. Dress, A., Koolen, J., Moulton, V.: On line arrangements in the hyperbolic plane. Eur. J. Comb. **23**(5), 549–557 (2002)
22. Dujmovic, V., Joret, G., Micek, P., Morin, P., Ueckerdt, T., Wood, D.R.: Planar graphs have bounded queue-number. CoRR (2019). arXiv:abs/1904.04791
23. Evans, W.S., Kaufmann, M., Lenhart, W., Mchedlidze, T., Wismath, S.K.: Bar 1-visibility graphs vs. other nearly planar graphs. J. Graph Algorithms Appl. **18**(5):721–739 (2014)
24. Fowler, J.J., Kobourov, S.G.: Characterization of unlabeled level planar graphs. In: Hong, S.-H., Nishizeki, T., Quan, W. (eds.) Graph Drawing, pp. 37–49. Springer, Berlin (2008)
25. Fox, J., Pach, J.: Coloring K_k-free intersection graphs of geometric objects in the plane. Eur. J. Comb. **33**(5), 853–866 (2012)

26. Fox, J., Pach, J.: Applications of a new separator theorem for string graphs. Comb. Probab. Comput. **23**(1), 66–74 (2014)
27. Fox, J., Pach, J., Suk, A.: The number of edges in k-quasi-planar graphs. SIAM J. Discret. Math. **27**(1), 550–561 (2013)
28. Geneson, J., Khovanova, T., Tidor, J.: Convex geometric $(k + 2)$-quasiplanar representations of semi-bar k-visibility graphs. Discret. Math. **331**, 83–88 (2014)
29. Gyárfás, A.: On the chromatic number of multiple interval graphs and overlap graphs. Discret. Math. **55**(2), 161–166 (1985)
30. Hoffmann, M., Tóth, C.D.: Two-planar graphs are quasiplanar. In: Larsen, K.G., Bodlaender, H.L., Raskin, J. (eds.) 42nd International Symposium on Mathematical Foundations of Computer Science, MFCS 2017, August 21–25, 2017 - Aalborg, Denmark. LIPIcs, vol. 83, pp. 47:1–47:14. Schloss Dagstuhl - Leibniz-Zentrum fuer Informatik (2017)
31. Kainen, P.C.: Thickness and coarseness of graphs. Abh. Math. Semin. Univ. Hambg. **39**(1), 88–95 (1973)
32. Kaufmann, M., Ueckerdt, T.: The density of fan-planar graphs. CoRR (2014). arXiv:abs/1403.6184
33. Klazar, M., Marcus, A.: Extensions of the linear bound in the fűredi-hajnal conjecture. Adv. Appl. Math. **38**(2), 258–266 (2007)
34. Klazar, M., Valtr, P.: Generalized Davenport-Schinzel sequences. Combinatorica **14**(4), 463–476 (1994)
35. Leighton, F.T.: Complexity Issues in VLSI: Optimal Layouts for the Shuffle-Exchange Graph and Other Networks. MIT Press, Cambridge (1983)
36. Liotta, G., Didimo, W., Di Giacomo, E., Meijer, H., Wismath, S.K.: Planar and quasi-planar simultaneous geometric embedding. Comput. J. **58**(11):3126–3140, 07 (2015)
37. Marcus, A., Tardos, G.: Excluded permutation matrices and the Stanley-Wilf conjecture. J. Comb. Theory Ser. A **107**(1), 153–160 (2004)
38. Nakamigawa, T.: A generalization of diagonal flips in a convex polygon. Theor. Comput. Sci. **235**(2), 271–282 (2000)
39. Pach, J.. Notes on geometric graph theory. In: Goodman, J., Pollack, R., Steiger, W. (eds.) Discrete and Computational Geometry: Papers from DIMACS Special Year. DIMACS Series, vol. 6, pp. 273–285. AMS, Providence (1991)
40. Pach, J., Radoičić, R., Tóth, G.: Relaxing planarity for topological graphs. In: Akiyama, J., Kano, M. (eds.) Discrete and Computational Geometry. Lecture Notes in Computer Science, vol. 2866, pp. 221–232. Springer, Berlin (2003)
41. Pach, J., Shahrokhi, F., Szegedy, M.: Applications of the crossing number. Algorithmica **16**(1), 111–117 (1996)
42. Pach, J., Tóth, G.: Graphs drawn with few crossings per edge. Combinatorica **17**(3), 427–439 (1997)
43. Pach, J., Wenger, R.: Embedding planar graphs at fixed vertex locations. Graphs Comb. **17**(4), 717–728 (2001)
44. Pawlik, A., Kozik, J., Krawczyk, T., Lasoń, M., Micek, P., Trotter, W.T., Walczak, B.: Triangle-free intersection graphs of line segments with large chromatic number. J. Comb. Theory Ser. B **105**, 6–10 (2014)
45. Rok, A., Walczak, B.: Coloring curves that cross a fixed curve. Discret. Comput. Geom. **61**(4), 830–851 (2019)
46. Schnyder, W.: Embedding planar graphs on the grid. In: Proceedings of the 1st Annual ACM-SIAM Symposium on Discrete Algorithms, SODA'90, Philadelphia, PA, USA, pp. 138–148. Society for Industrial and Applied Mathematics (1990)
47. Suk, A., Walczak, B.: New bounds on the maximum number of edges in k-quasi-planar graphs. Comput. Geom. **50**, 24–33 (2015)
48. Valtr, P.: Graph drawings with no k pairwise crossing edges. In: Proceedings of the 5th International Symposium on Graph Drawing, GD'97, London, UK, pp. 205–218. Springer (1997)
49. Valtr, P.: On geometric graphs with no k pairwise parallel edges. Discret. Comput. Geom. **19**(3), 461–469 (1998)

Chapter 4
1-Planar Graphs

Yusuke Suzuki

Abstract Topological graph theory discusses, in most cases, graphs embedded in the plane (or other surfaces). For example, such plane graphs are sometimes regarded as the simplest town maps. Now, we consider a town having some pedestrian bridges, which cannot be realized by a plane graph. Its underlying graph can actually be regarded as a 1-*plane* graph. The notion of 1-plane and 1-*planar* graphs was first introduced by Ringel in connection with the problem of simultaneous coloring of the vertices and faces of plane graphs. In particular, in contrast to planarity testing, testing 1-planarity of a given graph is an NP-complete problem. Even though 1-planar graphs have been widely studied recently, we still know relatively little about them. In this chapter, we begin with formally defining 1-plane and 1-planar graphs and mainly focus on "maximal", "maximum," and "optimal" 1-planar graphs, which are relatively easy to treat. This chapter reviews some basic properties of these graphs.

4.1 Definition and Basic Results

A *drawing* of a graph G on the sphere $\mathbb{S}^2$ is a representation of G, where vertices are distinct points in $\mathbb{S}^2$, and edges are Jordan arcs in the sphere joining the points corresponding to their end vertices. (Note that the sphere is the one-point compactification of the Euclidean plane. The above drawing of G on $\mathbb{S}^2$ is equivalent to a drawing of G in the plane, except that none of the faces has a special role in the sphere.) A *crossing point* is a transversal intersection of two arcs on the sphere. In this chapter, we consider only *proper* drawings such that edges are simple arcs without vertices of the graph in their interiors, two arcs having an intersection always cross-transversely, no two adjacent edges cross each other, and no more than two edges cross at a single point.

A graph G is 1-*planar* if it can be drawn on the sphere $\mathbb{S}^2$, so that each edge crosses at most one other edge. The notion of 1-planar graphs was first introduced by

Y. Suzuki (✉)
Department of Mathematics, Niigata University, 8050 Ikarashi 2-no-cho,
Nishi-ku, Niigata 950-2181, Japan
e-mail: y-suzuki@math.sc.niigata-u.ac.jp

© Springer Nature Singapore Pte Ltd. 2020
S.-H. Hong and T. Tokuyama (eds.), *Beyond Planar Graphs*,
https://doi.org/10.1007/978-981-15-6533-5_4

Ringel [25] in connection with the problem of simultaneous coloring of the vertices and faces of plane graphs. For aspects of 1-planar graphs that are not covered in this chapter, refer to a recent survey [16]. Note that all graphs in this chapter are assumed to be simple and connected unless otherwise specified. However, we sometimes consider 1-*planar* (or 1-*plane*) *multigraphs*, i.e., with loops or multiple edges, in our statements and proofs. In some cases, we still refer to "simple graphs" for clarity. By the above definition, notice that every planar graph is 1-planar. We can also regard the drawing as a continuous map $f : G \to \mathbb{S}^2$ which may not be injective, where G is regarded as a one-dimensional topological space. In this chapter, we call the above map f a 1-*embedding* of G into the sphere. In this case, we say that the image $f(G)$ is a 1-*plane graph*; similar to the difference between "planar graph" and "plane graph". (Sometimes, we denote a given 1-plane graph by G, instead of $f(G)$, to simplify notation. Further, we sometimes call the image G (or $f(G)$) a 1-embedding on $\mathbb{S}^2$.) An edge is *crossing* if it crosses another edge in a 1-plane graph G on the sphere, and is *non-crossing* otherwise. In a 1-plane graph, if an edge v_0v_2 crosses another edge v_1v_3 and has a crossing point z, then we say that the arc zv_i is a *half-edge* of G for each $i \in \{0, 1, 2, 3\}$. In the above, v_iz and $v_{i+1}z$ are *consecutive*, where the indices are taken modulo 4. Throughout the chapter, we often use the following fact in our argument.

Proposition 4.1 *Let G be a connected 1-plane multigraph on $\mathbb{S}^2$. Then, each connected component of $\mathbb{S}^2 - G$ is homeomorphic to an open disk (also known as a 2-cell). Further, for any two consecutive half-edges v_0z and v_1z, where $v_0, v_1 \in V(G)$, there exists a connected component of $\mathbb{S}^2 - G$ having v_0 and v_1 on its boundary.*

Proof Suppose that there is a connected component D of $\mathbb{S}^2 - G$ not homeomorphic to a 2-cell. Then, the boundary of D is disconnected and has components $J_1, \ldots, J_k$ with $k \geq 2$, each of which is homeomorphic to a simple closed curve. It is clear that there exists a connected component of G corresponding to J_i for $i \in \{1, \ldots, k\}$, and any two of them are disjoint in G. Therefore, G is disconnected, a contradiction. The second part of the statement holds since the closed set formed by $v_0z \cup v_1z$ is on the boundary of some connected component of $\mathbb{S}^2 - G$ by the 1-planarity. □

A connected component D of $\mathbb{S}^2 - G$ whose boundary contains no crossing point is called a *face* of the 1-plane graph G. In other words, the boundary of a face D of G corresponds to a closed walk consisting of only non-crossing edges of G. A *k-gonal* face of G is a face of G whose boundary walk has a length of exactly k. On the other hand, a connected component D of $\mathbb{S}^2 - G$ whose boundary contains a crossing point is a *fake face*. Note that a fake face is not a face of G vice versa. See Fig. 4.1. It depicts a 1-embedding of a complete graph K_5, or a 1-plane graph isomorphic to K_5; as a result, K_5 is 1-planar. This 1-embedding has one crossing point, four triangular faces, and four triangular fake faces.

Fig. 4.1 1-plane graph K_5

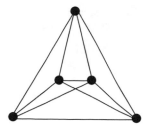

The following is the most important fact giving the upper bound of the number of edges of 1-planar graphs; this had been proved in some papers, e.g., see [1, 24].

Proposition 4.2 *Let G be a simple 1-planar graph with* $|V(G)| \geq 3$. *Then, we have* $|E(G)| \leq 4|V(G)| - 8$.

Proof Let G be a simple 1-plane graph with $|V(G)| \geq 3$. We add edges to G on $\mathbb{S}^2$ to obtain a new 1-plane graph, admitting loops, and multiple edges, which however has neither 1- nor 2-gonal face. The resulting multigraph G' is assumed to be edge maximal with respect to the above property. By Proposition 4.1 and the maximality of G', if G' has a pair of crossing edges v_0v_2 and v_1v_3, then there are four edges v_0v_1, v_1v_2, v_2v_3, and v_3v_0 such that the closed walk $v_0v_1v_2v_3$ bounds a 2-cell that contains no vertex and a unique crossing point. Furthermore, observe that G' is connected and that every face of G' is triangular; if not, we can add a diagonal edge in the face.

Let c denote the number of crossing points of G'. Now we remove a crossing edge from each pair of crossing edges in G' and denote the resulting multigraph by G''; note that we have removed c edges from G'. Clearly, G'' is an embedding without crossing points and each face of G'' is triangular. By Euler's formula, we have $|E(G'')| = 3|V(G'')| - 6$ and $|F(G'')| = 2|V(G'')| - 4$. Furthermore, we have $c \leq |F(G'')|/2$ since each crossing point in G' corresponds to a pair of adjacent triangular faces in G'', and all other triangular faces of G'' are already present in G'. Then we obtain the inequality in the statement as follows:

$$
\begin{aligned}
|E(G)| &\leq |E(G')| \\
&= |E(G'')| + c \\
&\leq |E(G'')| + |F(G'')|/2 \\
&= (3|V(G'')| - 6) + (|V(G'')| - 2) \\
&= 4|V(G'')| - 8 \\
&= 4|V(G)| - 8
\end{aligned}
$$

Therefore, the proposition follows.   □

The following fact is easily obtained from Proposition 4.2.

Proposition 4.3 *A complete graph K_7 with seven vertices is not 1-planar.*

Fig. 4.2 Maximal 1-plane
graph with six vertices

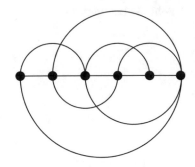

Proof By Proposition 4.2, a 1-planar graph with seven vertices has at most 20 edges. However, K_7 has 21 edges.   □

A 1-planar graph G is *optimal* if it satisfies the equality in Proposition 4.2, i.e., $|E(G)| = 4|V(G)| - 8$ holds. With the terminology defined above, a 1-embedded optimal 1-planar graph is called an *optimal 1-plane graph*.

Let G be a 1-planar graph. For any nonadjacent vertices $u, v \in V(G)$, if $G + uv$ is not 1-planar, then G is *maximal*. On the other hand, a 1-plane graph G is *maximal* if it cannot be augmented to a larger 1-plane graph by adding an edge as an arc to G on the sphere without introducing forbidden crossings. The reader notes the difference between these two notions of maximality, defined for 1-planar graphs and 1-plane graphs. Note that any 1-embedding $f(G)$ of any maximal 1-planar graph G is maximal 1-plane, but the converse does not hold in general. Figure 4.2 depicts a maximal 1-plane graph G. However, the underlying graph of G is not maximal 1-planar since we know that K_6 is 1-planar (see $M(6)$ in Fig. 4.10).

Furthermore, a 1-planar graph G with n vertices is *maximum* if $|E(G)| \geq |E(G')|$ for any other 1-planar graph G' with n vertices. Clearly, every maximum 1-planar graph is maximal. It is easy to see that every optimal 1-planar graph is maximum, but the converse does not hold true. It was proved that there is an optimal 1-planar graph with n vertices if and only if $n = 8$ or $n \geq 10$ (see e.g., [3, 4, 28]). In other words, if n is either 9 or at most 7, then any maximum 1-planar graph with n vertices is not optimal. Especially, if $n \leq 6$, then the maximum 1-planar graph is a complete graph with n vertices (see $M(3)$, $M_1(4)$, $M_2(4)$, $M(5)$, and $M(6)$ shown in Fig. 4.10).

In the remainder of this section, we present some basic properties that hold for 1-planar graphs.

Proposition 4.4 *Let G be a 1-plane graph with n vertices. Then, the number of crossing points is at most $n - 2$.*

Proof Let c denote the number of crossing points of G. For every crossing point z created by two edges $v_0 v_2$ and $v_1 v_3$, we successively add a non-crossing edge $v_i v_{i+1}$ so that $z v_i v_{i+1}$ bounds a fake face of G if such an edge does not already exist for $i \in \{0, 1, 2, 3\}$, where the indices are taken modulo 4. Note that we allow creating multiple edges in the above operation. After that, we remove all crossing edges of

G and denote the resulting plane multigraph by G'. Note that G' has neither a 1- nor a 2-gonal face. Now we have the following equality by Euler's formula where F_k denotes the number of k-gonal face of G'.

$$\sum_{k \geq 3}(k-2)F_k = 2n - 4$$

Thus, we obtain the inequality $F_4 \leq n - 2$. It is clear that $c \leq F_4$ by our construction, and hence we have $c \leq n - 2$. Thus, we got our desired conclusion. □

Proposition 4.5 *Let G be a maximal 1-plane graph and let $\{v_0 v_2, v_1 v_3\}$ be a pair of crossing edges having a crossing point z. Then, the four edges $v_0 v_1$, $v_1 v_2$, $v_2 v_3$ and $v_3 v_0$ are present in G. Furthermore, if G is 4-connected, then $z v_i v_{i+1}$, for $i \in \{0, 1, 2, 3\}$ bounds a fake face with indices taken modulo 4.*

Proof There exists a connected component D of $\mathbb{S}^2 - G$ homeomorphic to an open disk (or a 2-cell region) whose boundary contains two half-edges $v_0 z$ and $v_1 z$ by Proposition 4.1. If $v_0 v_1 \notin E(G)$, then G would not be maximal since we can join v_1 and v_2 by an arc passing through D, a contradiction. Similarly, we can show the existence of the other three edges.

Next, suppose that G is 4-connected. Let D be a 2-cell region bounded by $v_0 v_1$ and the half-edges $v_0 z$ and $v_1 z$. Assume, to the contrary, that D contains a vertex of G. If $v_0 v_1$ is non-crossing, then $\{v_0, v_1\}$ would become a cut set, which separates vertices in D from the others, a contradiction. If $v_0 v_1$ is a crossing edge and crosses $xy \in E(G)$ where y is located in D, then $\{v_0, v_1, x\}$ would become a 3-cut of G, which also separates vertices in D from the others. It contradicts the 4-connectivity condition of G. □

Proposition 4.6 *Let G be a maximal 1-plane graph. Then, every face of G is either triangular or quadrangular. Furthermore, if G has a quadrangular face, then G contains $M_1(4)$, shown in Fig. 4.10, as a subgraph. Moreover, if G is 3-connected, then either every face of G is triangular or G is homeomorphic to $M_1(4)$.*

Proof Let f be a k-gonal face bounded by a closed walk $C = v_0 v_1 \cdots v_{k-1}$ for $k \geq 4$. If C is not a cycle, then $v_i = v_j$ for some $i \neq j$. Under the condition, it is easy to see that v_i is a cut vertex of G. Then, we can join two vertices in different components of $G - v_i$ by an arc passing through f, preserving the simplicity. It contradicts the maximality of G. Thus, C is a cycle.

Since G is maximal, there exist edges $v_i v_j$ for all $\{i, j\}$ with $0 \leq i < j \leq k - 1$ which lie outside of f; otherwise, one could add a new edge inside f. If $k \geq 5$, there would be an edge $v_i v_j$ having at least two crossing points, contrary to the 1-planarity of G; e.g., $v_0 v_2$ must cross $v_1 v_3$ and $v_1 v_4$. Thus, $k = 4$ and the edges $v_0 v_2$ and $v_1 v_3$ cross outside of f. Then, G clearly contains $M_1(4)$ as a subgraph, as required. If G is 3-connected, then G has no vertex other than those in $V(M_1(4))$; otherwise $\{v_i, v_{i+1}\}$ would form a 2-cut for some $i \in \{0, 1, 2, 3\}$. Therefore, we got our desired conclusion. □

4.2   Connectivity

It is well known that every triangulation of the sphere is 3-connected. However, we cannot guarantee the high connectivity of 1-planar graphs even if we assume the maximality to those graphs. We only ensure the following.

Theorem 4.1 ([8]) *Let G be a maximal 1-plane graph with $|V(G)| \geq 3$. Then a subgraph formed by all non-crossing edges is spanning and 2-connected.*

By the above theorem proven by Eades et al., we can immediately obtain the following.

Proposition 4.7 *Every maximal 1-plane graph G with $|V(G)| \geq 3$ is 2-connected.*

The above "2" is the best possible since it is not difficult to construct a maximal 1-plane graph having a vertex of degree 2; insert a vertex of degree 2 in one of the two triangular fake faces sharing a non-crossing edge of a 1-embedded graph shown in Fig. 4.2.

By Proposition 4.2, the average degree of every 1-planar graph is less than 8. This implies that any 1-planar graph has a vertex of degree at most 7. This "7" is also the best possible since Fabrici and Madaras [9] exhibited a 7-regular 1-planar graph as shown in Fig. 4.3.

A *quadrangulation* (resp., *triangulation*) of the sphere is a simple graph embedded on the sphere such that each face is bounded by a 4-cycle (resp., 3-cycle). By the argument in the proof of Proposition 4.2, the graph formed by all non-crossing edges of an optimal 1-plane graph G forms a quadrangulation of the sphere. We call it a *quadrangular subgraph* of G and denote it by $Q(G)$ (see Fig. 4.4). On the other hand, the following holds for crossing edges.

Proposition 4.8 *Let G be an optimal 1-plane graph. Then, a subgraph of G formed by all crossing edges is disconnected.*

Fig. 4.3 7-regular 1-planar graph

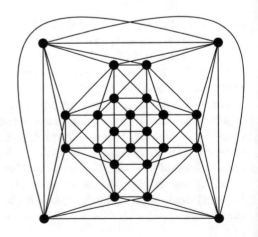

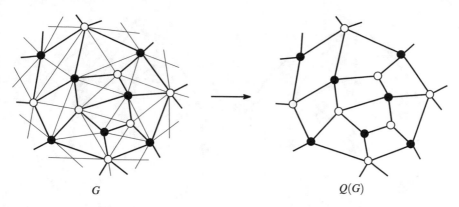

G $Q(G)$

Fig. 4.4 Optimal 1-planar graph and its quadrangular subgraph

Proof Let G be an optimal 1-plane graph. It is well known that every quadrangulation of the sphere is bipartite and hence $Q(G)$ is bipartite. Thus, $V(G)$ can be decomposed into $V_B(G) \cup V_W(G)$ so that every non-crossing edge joins vertices in different sets while every crossing edge joins vertices in the same set. This implies that the subgraph of G formed by all crossing edges has two components having vertex sets $V_B(G)$ and $V_W(G)$, respectively. Therefore, we are done.   □

The following theorem gives us the clear relationship between optimal 1-plane graphs and quadrangulations of the sphere.

Theorem 4.2 ([28]) *Let H be a simple quadrangulation of the sphere. Then there exists a simple optimal 1-plane graph G such that $H = Q(G)$ if and only if H is 3-connected.*

By the above theorem, every optimal 1-planar graph is 3-connected. (In fact, "3" is not the best possible. See the argument below.) Further, we can see that around each vertex of an optimal 1-plane graph, crossing edges and non-crossing edges appear alternately. Hence, each vertex of an optimal 1-planar graph has even degree; i.e., every optimal 1-planar graph is Eulerian. Thus, every optimal 1-planar graph has a vertex of degree 6 and the connectivity cannot be larger than 6. (Recall that the average degree of 1-planar graph is smaller than 8, and that the minimum degree is at least 6 by the simplicity.) In fact, there is an infinite series of 6-connected optimal 1-planar graph obtained as follows: At first, embed a $2k$-cycle $v_1 u_1 v_2 u_2 \cdots v_k u_k$ into the sphere without crossing point and put two vertices a and b in its interior and exterior separated by the cycle, respectively. Next, we add edges av_i and bu_i for $i = 1, \ldots, k$. We call the resulting graph a *pseudo double wheel* and denote it by W_{2k} (see the left-hand side of Fig. 4.5). Since W_2 has multiple edges and W_4 has two vertices of degree 2, the smallest 3-connected pseudo-double wheel is W_6, which is nothing but a cube. We add pairs of crossing edges to all the faces of $W_{2k} (k \geq 3)$, and obtain the optimal 1-plane graph called a X-*pseudo-double wheel* denoted by XW_{2k}. See the right-hand side of Fig. 4.5. We call the vertices a and b *hubs* of XW_{2k}.

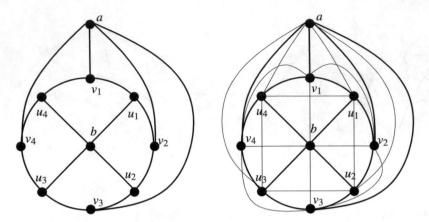

Fig. 4.5 Pseudo-double wheel and X-pseudo-double wheel

Proposition 4.9 *For every $k \geq 3$, XW_{2k} is 6-connected.*

Proof Let G be a X-pseudo-double wheel XW_{2k} with hubs a and b with $k \geq 3$. In fact, $G - \{a, b\}$ is a graph known as the *square* of the cycle of length $2k \geq 6$. In [12], it is proven that $G - \{a, b\}$ is 4-connected. Since both a and b are adjacent to all the vertices in $V(G) - \{a, b\}$ and $|V(G) - \{a, b\}| \geq 6$, G is 6-connected. $\qquad\square$

In fact, throughout the argument in [10, 28], the following theorem had been proven.

Theorem 4.3 ([10, 28]) *The connectivity of an optimal 1-planar graph G is either 4 or 6. If the connectivity is 4 (resp., 6), then there exists a separating 4-cycle (resp., 6-cycle) of $Q(G)$.*

4.3 Planarization

For a given 1-plane graph G, we sometimes consider a plane graph G_P called a *planarization* of G, defined as follows. Let $\{a_1c_1, b_1d_1\}, \{a_2c_2, b_2d_2\}, \ldots, \{a_kc_k, b_kd_k\}$ denote pairs of crossing edges of G. Roughly speaking, we regard a crossing point formed by $\{a_ic_i, b_id_i\}$ as a new vertex z_i. Precisely, our required plane graph G_P has $V(G_P) = V(G) \cup \{z_i | 1 \leq i \leq k\}$ as its vertex set and $E(G_P) = E(G) \cup \{a_iz_i, b_iz_i, c_iz_i, d_iz_i | 1 \leq i \leq k\} \setminus \{a_ic_i, b_id_i | 1 \leq i \leq k\}$ as its edge set. We call z_i a *false vertex* of G_P for $1 \leq i \leq k$, and $v \in V(G) \subset V(G_P)$ a *true vertex*. Clearly, we have $\deg_{G_P}(z_i) = 4$, and edges $a_iz_i, b_iz_i, c_iz_i, d_iz_i$ appear in this order around z_i. The following fact is easily obtained.

Proposition 4.10 *Every face of G_P obtained from a simple 1-plane graph G has at least two true vertices.*

Proof Clearly, G_P is simple if G is simple. Hence, the length of any face of G_P is bounded by a closed walk of length at least three unless $G_P \cong K_2$. If $G_P \cong K_2$, then such two vertices are true and hence the statement holds. Further, two false vertices are not adjacent by our construction of G_P. Thus, we are done.   □

Concerning the connectivity of the planarization G_P of G, the following result is known.

Theorem 4.4 ([9]) *If G is a 3-connected 1-plane graph with the minimum number of crossings taken over all 1-embeddings $f : G \to \mathbb{S}^2$, then G_P is 3-connected.*

Before reading the following proposition, recall that a planar graph is 1-planar by the definition of 1-planarity.

Proposition 4.11 *A planarization G_P of a 1-plane graph G is 5-connected if and only if G is a 5-connected plane graph.*

Proof If a 1-plane graph G has at least one crossing point, then G_P has a vertex of degree 4. In this case, G_P cannot be k-connected for $k \geq 5$. Thus, if G_P is 5-connected, then G has no crossing point. That is, $G = G_P$ and hence G is a 5-connected plane graph. The converse is obvious since $G = G_P$ also holds in this case.   □

By the above fact, the connectivity of the planarization G_P of a 1-plane graph G is at most 4 if G has at least one crossing point. This raises the question of what condition for a 1-plane graph G is sufficient to guarantee the 4-connectivity of G_P? So far, we know the following.

Theorem 4.5 ([13]) *If a 1-plane graph G is 7-connected, then G_P is 4-connected.*

The "7" in the above theorem is the best possible. The 1-plane graph shown in Fig. 4.6 is 6-connected. However, the planarization of the graph clearly has a 3-vertex cut, which consists of three false vertices. Furthermore, the connectivity of a 7-regular 1-planar graph presented in Fig. 4.3 is 7.

As noted above, we can easily construct a maximal 1-plane graph G having a vertex v of degree 2. In this case, it is easy to see that v is degree 2 also in G_P.

Fig. 4.6 6-connected 1-plane graph whose planarization has a 3-cut

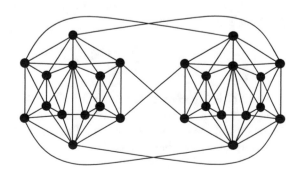

That is, "maximality" does not imply lower bounds on the connectivity. However, for optimal 1-plane graphs, the following theorem holds.

Theorem 4.6 ([13]) *The planarization of an optimal 1-plane graph is 4-connected.*

Using the above result, we can easily obtain the following proposition; a proof was previously published in [13].

Proposition 4.12 *Every optimal 1-planar graph is Hamiltonian.*

Proof Let G be an optimal 1-plane graph and denote the planarization of G by G_P. By Theorem 4.6, G_P is 4-connected and hence G_P has a Hamiltonian cycle C by [29]. Now assume that C passes through a false vertex z corresponding to a crossing point created by a pair of crossing edges $\{v_0v_2, v_1v_3\}$. If a 2-path v_izv_{i+2} is contained in C, then we replace the 2-path by v_iv_{i+2}, which is an edge of G, where the indices are taken modulo 4. On the other hand, if a 2-path v_izv_{i+1} is contained in C, we replace it by an edge v_iv_{i+1}, which is also an edge of G by Proposition 4.5. We do the above replacement for all false vertices contained in C, and obtain a Hamiltonian cycle of G.   □

At the end of this section, we show the following result using the notion of planarization. The proof is based on [6].

Theorem 4.7 ([6]) *A complete bipartite graph $K_{5,4}$ is not 1-planar.*

Proof For the sake of contradiction, suppose that $K_{5,4}$ is 1-planar. Let G be a 1-embedding of $K_{5,4}$, and G_P denotes its planarization. It is known that $cr(K_{5,4}) = 8$ by [15], where $cr(H)$ represents the crossing number of H. Thus, G_P has at least 8 crossing points. This implies that G has at least 16 crossing edges and has at most 4 non-crossing edges.

Now, consider the following equation derived from Euler's formula, where $\deg_H(f)$ denotes the length of the boundary walk of a face f:

$$\sum_{v \in V(G_P)} (\deg_{G_P}(v) - 4) + \sum_{f \in F(G_P)} (\deg_{G_P}(f) - 4) = -8.$$

Clearly, G_P has four vertices of degree 5 and all other true and false vertices have degree 4. Thus, we have,

$$\sum_{f \in F(G_P)} (\deg_{G_P}(f) - 4) = -12.$$

Since G is bipartite, G has no cycle of length 3. Thus, each triangular face has a false vertex on its boundary. Furthermore, by Proposition 4.10, such a triangular face is incident to a non-crossing edge. That is, G_P has at most eight triangular faces. This contradicts the above equation.   □

4.4 Edge Density

As mentioned in Sect. 4.1, every 1-planar graph with n vertices has at most $4n - 8$ edges. In this section, we evaluate the number of edges of those graphs under various additional constraints.

Let $M(\mathcal{G}, n)$ and $m(\mathcal{G}, n)$ denote the maximum and the minimum number of edges taken over all graphs with n vertices in a graph class $\mathcal{G}$, respectively. For example, it is well known that $M(\mathcal{T}, n) = m(\mathcal{T}, n) = 3n - 6$ for the family of maximal planar graphs $\mathcal{T}$ assuming $n \geq 3$; and such graphs are known as triangulations of the sphere. However, we know that $M(\mathcal{P}, n) \neq m(\mathcal{P}, n)$ (resp., $M(\mathcal{P}', n) \neq m(\mathcal{P}', n)$) in general where $\mathcal{P}$ (resp., $\mathcal{P}'$) denotes the family of maximal 1-planar (resp., 1-plane) graphs. In fact, $M(\mathcal{P}, n) = M(\mathcal{P}', n)$ represents the number of edges of a maximum 1-planar graph with n vertices by our definition. That is, in most cases ($n = 8$ and $n \geq 10$), the above value equals to $4n - 8$, which is the number of edges of an optimal 1-planar graph with n vertices. Furthermore, we have $M(\mathcal{P}, n) = \binom{n}{2}$ if $n \leq 6$, whose underlying graph is a complete graph K_n. In the remaining cases (i.e., $n \in \{7, 9\}$), we have $M(\mathcal{P}, n) = 4n - 9$. (See Sect. 4.6. Maximum 1-plane graps with $3 \leq n \leq 7$ vertices which are not optimal are exhibited.)

As we have seen, $m(\mathcal{P}, n)$ and $m(\mathcal{P}', n)$ are more interesting values to discuss. Here, observe that $m(\mathcal{P}, n) \geq m(\mathcal{P}', n)$ for every n by the definitions. At first, we introduce the results concerning $m(\mathcal{P}', n)$. Eades et al. [8] proved that $\frac{9n}{5} - \frac{18}{5} \leq m(\mathcal{P}', n) \leq \frac{7n}{3} - 2$, and Brandenburg et al. [5] improved the above lower bound to $\frac{21n}{10} - \frac{10}{3}$. Further in [5], they construct maximal 1-plane graphs having $\frac{7n}{3} - 3$ edges for any large n. In [5], it was also proved that $m(\mathcal{P}, n) \geq \frac{28n}{13} - \frac{10}{3}$ and that there exist maximal 1-planar graphs having $\frac{45n}{17} - \frac{84}{17}$ edges for any large n. Very recently, both lower bounds were improved to $\frac{20n}{9} - \frac{10}{3}$ by Barát and Tóth [2].

Next, we introduce some results for multipartite graphs. Karpov [14] proved that every bipartite 1-planar graph has at most $3n - 8$ edges for even $n \neq 6$ and at most $3n - 9$ for odd n and for $n = 6$. For tripartite 1-planar graphs, we show the following result here.

Theorem 4.8 *Every tripartite 1-planar graph with n vertices has at most $\frac{7}{2}n - 7$ edges.*

Proof Let G be a tripartite 1-plane graph with n vertices, and let c denote the number of crossing points of G. For any pair of crossing edges $\{v_0 v_2, v_1 v_3\}$ of G, we perform the following operation. Observe that there exists a pair of vertices $\{v_i, v_{i+1}\}$, say $\{v_0, v_1\}$ without loss of generality, such that v_0 and v_1 belong to the same partite set. We remove an edge $v_0 v_2$ from G, and add an edge $v_0 v_1$ so that $v_0 v_1 v_3$ forms a corner of a face or a fake face (see Fig. 4.7). Now denote the resulting multigraph by G'. Note that G' is probably not tripartite. If there exists a pair of multiple edges forming a 2-gonal face of G', then such edges come from left and right pairs of crossing edges of G; note that such edges do not exist in G since each of them joins vertices in the same partite set (see Fig. 4.7 again). Therefore, G' has at most $\frac{c}{2}$ such pairs of multiple edges. We remove an edge from every pair of multiple edges forming a

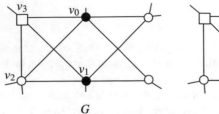 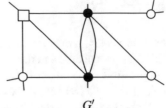

$$G \qquad\qquad\qquad G'$$

Fig. 4.7 Operation in the proof of Theorem 4.8

Fig. 4.8 Tripartite 1-plane
graph with $\frac{7}{2}n - 7$ edges

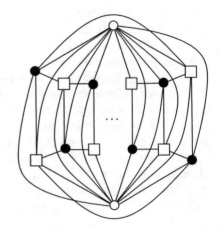

2-gonal face of G', and obtain a plane multigraph G''. Combining with the result in
Proposition 4.4, we obtain the following:

$$
\begin{aligned}
|E(G)| &= |E(G')| \\
&\leq |E(G'')| + \frac{c}{2} \\
&\leq 3n - 6 + \frac{n-2}{2} \\
&= \frac{7}{2}n - 7.
\end{aligned}
$$

Therefore, the theorem follows.   □

The upper bound in the above theorem is sharp. In fact, the graph depicted in
Fig. 4.8 has $4k + 2$ ($k \geq 2$) vertices and $14k$ edges. Furthermore, observe that there
exist infinitely many 4-colorable optimal 1-planar graphs (see [21]). This implies
that the upper bound of the number of edges for 4-partite 1-planar graphs with n
vertices cannot be less than $4n - 8$ if $n \geq 8$ and $n \neq 9$.

4.5   Minors and Subgraphs

For terminology around *minors* of graphs, refer to a general text of graph theory, e.g., [7]. It is well known that a graph G is planar if and only if it contains neither K_5 nor $K_{3,3}$ as a minor. However, 1-planarity cannot be characterized in terms of forbidden minors. In contrast to planar graphs, it is easy to see that every graph is a minor of a 1-planar graph; see [11]. We prove the following stronger result.

Theorem 4.9 ([27]) *For every graph H, there is an optimal 1-planar graph having a topological minor of H.*

Proof We draw a given graph H on the sphere as a proper drawing. Let z be a crossing point of v_0v_1, $v_2v_3 \in E(H)$. We delete v_0v_2 and v_1v_3 from H on the sphere, and add vertices u_i and edges u_iv_i and u_iu_{i+1} for $i \in \{0, 1, 2, 3\}$ where the indices are taken modulo 4. By Proposition 4.1, we may assume that the above-added edges are all non-crossing such that $u_0u_1u_2u_3$ bounds a quadrangular face. We successively apply the above operation for each crossing point of H and denote the resulting plane graph by H' (see the center of Fig. 4.9).

Now, we subdivide edges of H' if necessary, other than those of the 4-cycles around the crossing points above so that the resulting graph becomes bipartite. Furthermore, we add edges so that the resulting graph H'' is a simple quadrangulation of the sphere. (Note that we can add a diagonal edge to any $2l$-gonal face ($l \geq 3$) in the bipartite graph preserving the simplicity by the planarity. See the right-hand side of Fig. 4.9.) If H'' is 3-connected, then there exists an optimal 1-plane graph $\tilde{G}$ with $Q(G) = H''$ by Theorem 4.2 and then G clearly has a topological minor of H. If H'' is not 3-connected, then we apply the following operation to H''. For every face f of H'' bounded by $a_0a_1a_2a_3$, we put a 4-cycle $b_0b_1b_2b_3$ and edges joining a_i and b_i into f for each $i \in \{0, 1, 2, 3\}$; all such edges are assumed to be non-crossing. Then, the resulting quadrangulation becomes 3-connected and the theorem follows, by the same argument as above.   □

For the minors of complete graphs in optimal 1-planar graphs, we can easily obtain the following fact since Mader [19] proved that a graph with n vertices and at least $4n - 9$ edges has a K_6-minor.

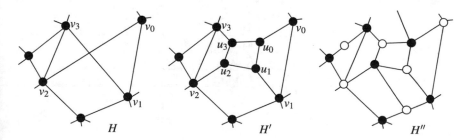

Fig. 4.9 Configurations in the proof of Theorem 4.9

Proposition 4.13 *Every optimal 1-planar graph has a K_6-minor.*

Furthermore, Suzuki proved the following theorem for K_7-minors in optimal 1-planar graphs where XW_8^+ is the unique optimal 1-planar graph that can be obtained from XW_8 by a specific operation.

Theorem 4.10 ([27]) *A 6-connected optimal 1-planar graph G contains a K_7-minor if and only if G is isomorphic to neither XW_{2k} ($k \geq 3$) nor XW_8^+.*

In fact, the characterization for general optimal 1-planar graphs without the connectivity condition to have a K_7-minor is given in the same paper. However, we do not describe it here since it would require several additional conditions.

On the other hand, if G is 1-planar, then any subgraph of G is also 1-planar; in other words, 1-planarity is closed under taking subgraphs. A graph G is a *MN-graph* if G is not 1-planar but for any edge e of G, $G - e$ is 1-planar. For example, Korzhic [17] proved that $K_7 - E(K_3)$ is the unique MN-graph with seven vertices. It easily follows from the above fact that any graph obtained from K_7 by deleting any two nonadjacent edges is 1-planar. Furthermore, it had been proven in [17, 18] that there are infinitely many MN-graphs with a minimum degree of at least 3.

However, if graphs are restricted to complete multipartite graphs, their 1-planarity is completely determined as follows.

Theorem 4.11 ([6]) *Let G be a complete k-partite 1-planar graph with $k \geq 2$. Then, G is isomorphic to a graph in Table 4.1:*

In Table 4.1, $a - b$ (resp., $a-$) represents $\{i \in \mathbb{Z} | a \leq i \leq b\}$ (resp., $\{i \in \mathbb{Z} | a \leq i\}$). For example, $K_{2-3,2,1,1}$ stands for two graphs $K_{2,2,1,1}$ and $K_{3,2,1,1}$. Furthermore, note that $K_{1,1,1,1,1,1}$ is equal to K_6. As we have already seen, any complete 7-partite graph G is not 1-planar since G contains K_7 as its subgraph.

For example, we can see that $K_{5,4}$ is not 1-planar; this fact can also be found as Theorem 4.7 in Sect. 4.3. Furthermore, we also see that $K_{4,3,2}$ is not 1-planar. However, this is clear since $K_{4,3,2}$ contains $K_{5,4}$ as its subgraph. In addition, $K_{4,3,2}$ has 26 edges and it cannot be 1-planar by Theorem 4.8.

Table 4.1 1-Planar complete multipartite graphs

k	1-planar complete k-partite graph
2	$K_{1-,1}$; $K_{2-,2}$; $K_{3-6,3}$; $K_{4,4}$
3	$K_{1-,1,1}$; $K_{2-6,2,1}$; $K_{2-4,2,2}$; $K_{3,3,1}$
4	$K_{1-6,1,1,1}$; $K_{2-3,2,1,1}$; $K_{2,2,2,1-2}$
5	$K_{1-2,1,1,1,1}$; $K_{2,2,1,1,1}$
6	$K_{1,1,1,1,1,1}$

4.6   Re-embeddings of 1-Planar Graphs

Let G be a 1-planar graph. For the precise definition below, assume that every edge $e = uv$ of G has a *middle point* $p \in e - \{u, v\}$ such that p corresponds to the crossing point if e is crossing in a 1-embedding. Two 1-embeddings $f_1, f_2 : G \to \mathbb{S}^2$ are *equivalent* to each other if there exists a homeomorphism $h : \mathbb{S}^2 \to \mathbb{S}^2$ such that $hf_1 = f_2$. If not, they are *inequivalent*. If there is exactly one equivalence class of 1-embeddings of G, we say that G is *uniquely* 1-*embeddable* into the sphere, *up to equivalence*.

For two 1-embeddings f_1 and f_2 of G, if there exists an automorphism $\sigma : G \to G$ and a homeomorphism $h : \mathbb{S}^2 \to \mathbb{S}^2$ such that $hf_1 = f_2\sigma$, they are *weakly equivalent* to each other. Roughly speaking, they have the same picture when we ignore the labeling of vertices.

In this section, we especially discuss "re-embeddability" of maximum 1-planar graphs. The notion of "re-embeddability" of optimal 1-planar graphs was first given by Schumacher [26], who proved that if G is a 5-connected optimal 1-planar graph other than $XW_{2k}(k \geq 3)$, then G is uniquely 1-embeddable into the sphere, up to equivalence. In fact, the 5-connectivity condition is unnecessary in the above result, and Suzuki proved the following theorem.

Theorem 4.12 ([28]) *Let G be an optimal* 1-*planar graph other than $XW_{2k}(k \geq 3)$. Then G is uniquely* 1-*embeddable into the sphere, up to equivalence.*

In fact, $XW_{2k}(k \geq 4)$ has only two 1-embeddings as follows. See the right-hand side of Fig. 4.5 again, and exchange the labels a and b in the figure. Then we obtain another 1-plane graph; e.g., av_1 is non-crossing in the original 1-plane graph while it is crossing in the latter one. Note that the underlying graph of the resulting 1-plane graph is isomorphic to XW_{2k}. That is, the two 1-embeddings of XW_{2k} are inequivalent.

For $k = 3$, XW_6 has exactly eight inequivalent 1-embeddings. In fact, XW_6 is isomorphic to $K_{2,2,2,2}$, and is obtained from a cube H by adding a pair of crossing edges to each face of H; thus, XW_6 has the rich symmetry. Furthermore, it is easy to see that all the inequivalent 1-embeddings of XW_6 are given by the same picture as the above example XW_8. Therefore, it follows that every optimal 1-planar graph is uniquely 1-embeddable into the sphere, up to weak equivalence.

The notion of the above re-embeddings of optimal 1-planar graphs is applied to the construction of maximal 1-planar graphs having small number of edges, which is discussed in Sect. 4.4. Let G be an optimal 1-planar graph with n vertices that is not isomorphic to $XW_{2k}(k \geq 3)$. Let $e = uv$ be a non-crossing edge of G. Then, we add a new vertex of degree 2 to G adjacent to u and v. For each non-crossing edge of G, we do the same operation, and denote the resulting graph by G'. It is easy to check that G' is maximal 1-planar since G is uniquely 1-embeddable into the sphere by Theorem 4.12. Now, G' has $n' = n + (2n - 4) = 3n - 4$ vertices and $(4n - 8) + 2(2n - 4) = 8n - 16$ edges. Consequently, G' has n' vertices and $\frac{8}{3}n' - \frac{16}{3}$ edges. However, the above coefficient is not better than that presented in [5] with a different construction, which was mentioned in Sect. 4.4.

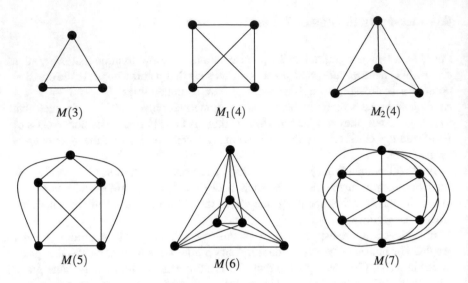

Fig. 4.10 Maximum 1-planar graphs which are not optimal with $n \leq 7$

Next, we consider maximum 1-planar graphs other than optimal 1-planar ones. In fact, the maximum 1-planar graphs that are not optimal in the unlabeled sense are determined as follows.

Theorem 4.13 ([28]) *Let G be a maximum* 1*-planar graph with $n \geq 3$ vertices that is not optimal. Then a* 1*-embedding of G in the sphere is equivalent to one of $M(3)$, $M_1(4)$, $M_2(4)$, $M(5)$, $M(6)$, $M(7)$ and $M_l(9)(l = 1, \ldots, 6)$, up to weak equivalence.*

The 1-plane graphs in the above theorem denoted by $M(3)$, $M_1(4)$, $M_2(4)$, $M(5)$, $M(6)$ and $M(7)$ can be found in Fig. 4.10. (The reader should refer to [28] for $M_l(9)(l = 1, \ldots, 6)$.) Note that the underlying graph of both $M_1(4)$ and $M_2(4)$ is isomorphic to K_4. That is, K_4 has two inequivalent 1-embeddings, up to weak equivalence. Actually, it has been proven in [28] that K_4 is the unique maximum 1-planar graph having such a property; additionally, recall the result of optimal 1-planar graphs discussed above.

Let $f : G \to \mathbb{S}^2$ be a 1-embedding of a graph G into the sphere. An automorphism $\sigma : G \to G$ of G is called a *symmetry* of f if there is a homeomorphism $h : \mathbb{S}^2 \to \mathbb{S}^2$ such that $hf = f\sigma$. The *symmetry group* of f is defined as the set of all symmetries of f and is denoted by $\text{sym}(f)$ or by $\text{sym}(f(G))$. Then $\text{sym}(f)$ is a subgroup of $\text{aut}(G)$, i.e., an *automorphism group* of G, possibly not normal.

Let G be a 1-planar graph and f be its 1-embedding. We denote a set of 1-embeddings that are weakly equivalent to G by $\text{emb}(f)$ or by $\text{emb}(f(G))$; $\text{emb}(f)$ should be a quotient set by the equivalence of 1-embeddings. Then, the following relation is well known: $|\text{emb}(f)| = |\text{aut}(G)|/|\text{sym}(f)|$. Let $\text{Emb}(G)$ denote the quotient set of G's 1-embeddings by the equivalence. If G admits precisely

Table 4.2 The number of 1-embeddings of maximum 1-planar graphs

$f(G)$	$|\text{aut}(G)|$	$|\text{sym}(f)|$	$|\text{emb}(f)|$
$M(3) \cong K_3$	6	6	1
$M_1(4) \cong K_4$	24	24	1
$M_2(4) \cong K_4$	24	8	3
$M(5) \cong K_5$	120	8	15
$M(6) \cong K_6$	720	12	60
$M(7) \cong C_3 + C_4$	48	4	12
$M_1(9)$	4	2	2
$M_2(9)$	4	2	2
$M_3(9)$	4	2	2
$M_4(9)$	432	12	36
$M_5(9)$	2	2	1
$M_6(9)$	1	1	1

k inequivalent 1-embeddings $f_1, \ldots, f_k$, up to weak equivalence, then we have that $\text{Emb}(G) = \text{emb}(f_1) \cup \cdots \cup \text{emb}(f_k)$.

Table 4.2 presents the numbers of 1-embeddings of maximum 1-planar graphs (see the rightmost column). In the table, if G is not isomorphic to K_4, we have $\text{Emb}(G) = \text{emb}(f)$ for some f, as mentioned above. For example, the 1-embedding $M(6)$ of K_6 attains the maximum value $|\text{emb}(M(6))| = 60$, which comes from $|\text{aut}(K_6)| = 6! = 720$ and $|\text{sym}(M(6))| = 12$. If $G \cong K_4$, we have $\text{Emb}(G) = \text{emb}(M_1(4)) \cup \text{emb}(M_2(4))$, and hence $|\text{Emb}(K_4)| = |\text{emb}(M_1(4))| + |\text{emb}(M_2(4))| = 1 + 3 = 4$.

In [18], the notion of a PN-*graph*, defined as a 3-connected planar graph having no 1-embeddings into the sphere with at least one crossing point was introduced. It is well known that every 3-connected planar graph can be uniquely embedded into the sphere (without crossing points). That is, any PN-graph has the unique 1-embedding into the sphere. Note that, in most cases, the unique 1-embedding of a PN-graph is not maximal, and used for constructing 1-planar graphs with our desired property by adding edges; e.g., 3-connected maximal 1-planar graphs having small number of edges (see [13]).

4.7 Difference from Optimal 1-Planar Graphs

For every plane graph G, we can obtain a maximal plane graph by adding edges to G. Recall that such a maximal plane graph is a triangulation of the sphere. However, as we mentioned above, a maximal 1-plane graph is not necessarily optimal. Observe that maximum 1-plane graphs that are not optimal (listed in Theorem 4.13) are such examples. Furthermore, it is easy to see that a 1-plane graph having the subgraph shown in Fig. 4.11 clearly cannot be augmented to an optimal 1-plane graph by adding

Fig. 4.11 1-plane graph
which cannot be augmented
to an optimal one

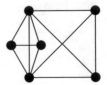

edges to it; note that we deal with only simple graphs in this chapter. Moreover, a
1-plane graph with minimum degree at least 7, e.g., the 7-regular graph shown in
Fig. 4.3, cannot be augmented to an optimal 1-plane graph, either; it is an easy exercise
for the readers.

We define the following family of graphs to relax the condition. A 1-plane graph
G is *near optimal*, if (i) any face of a subgraph H of G formed by all non-crossing
edges is either triangular or quadrangular (i.e., H is known as a *mosaic*), (ii) any
quadrangular face bounded by $v_0v_1v_2v_3$ of H contains the unique crossing point
created by a pair of crossing edges $\{v_0v_2, v_1v_3\}$, and (iii) no two triangular faces of
G share any edge. For example, it is easy to check that $M(6) \cong K_6$ and $M(7)$ in
Fig. 4.10 are near optimal. Furthermore, the 7-regular graph depicted in Fig. 4.3 is
also near optimal. It is clear that every optimal 1-plane graph is near optimal, and
hence this notion can be regarded as a generalization of optimal 1-planar graphs.
Note that any near optimal 1-plane graph has an even number of triangular faces; by
applying the Handshake lemma to the dual of the mosaic H.

Proposition 4.14 *Every near optimal 1-plane graph with n vertices has at least*
$\frac{18}{5}n - \frac{36}{5}$ *edges.*

Proof Let G be a near optimal 1-plane graph with n vertices. Denote the subgraph
of G formed by all non-crossing edges by H. By the above definition (i), H is a
plane graph having only triangular and quadrangular faces. Let F_3 and F_4 denote
the numbers of triangular and quadrangular faces of H, respectively; thus, we have
$|F(H)| = F_3 + F_4$ where $F(H)$ is the set of faces of H. Further, note that $3F_3 +
4F_4 = 2|E(H)|$, and that $3F_3 \leq 4F_4$ by property (iii). Then, we have the following
by substituting these into Euler's formula:

$$2|V(H)| - (3F_3 + 4F_4) + 2(F_3 + F_4) = 4$$
$$2|V(H)| - 4 = F_3 + 2F_4$$
$$6|V(H)| - 12 \leq 4F_4 + 6F_4$$
$$\frac{3}{5}|V(H)| - \frac{6}{5} \leq F_4..$$

Clearly, we have $|E(G)| = 3|V(G)| - 6 + F_4$ and hence the inequality in the
statement follows; observe that $|V(G)| = |V(H)|$.   □

The lower bound in Proposition 4.14 is sharp. See Fig. 4.12. The graph depicted
in the figure is the smallest one attaining the lower bound in the proposition; in the

Fig. 4.12 Near-optimal
1-plane graph with 12
vertices

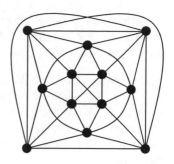

graph, no two fake faces share a non-crossing edge of G. In fact, we can construct an infinite sequence of graphs attaining the lower bound. (The reader should try to construct such an infinite series of graphs.) Observe that if $F_3 = 0$ in the above proof, then G is optimal and has $4n - 8$ edges.

Proposition 4.15 *Every 5-connected maximal 1-plane graph G is near optimal.*

Proof Let G be a 5-connected maximal 1-plane graph. By Proposition 4.5, each crossing point of G lies in a quadrangular face of the subgraph of G formed by all non-crossing edges. Since G is not isomorphic to K_4, any face of G is triangular by Proposition 4.6.

Assume, to the contrary, that G has two triangular faces $v_0 v_1 v_2$ and $v_1 v_2 v_3$ sharing $v_1 v_2$. Since G is a maximal 1-plane graph, there exists an edge joining v_0 and v_3. If $v_0 v_3$ is non-crossing, then G would have a separating 3-cycle $C = v_0 v_1 v_3$ which consists of only non-crossing edges; otherwise, C bounds a face of G and v_1 would have degree 3, contrary to G being 5-connected.

If $v_0 v_3$ is crossing, it crosses another edge $u_1 u_2$. By Proposition 4.5 again, there exists non-crossing edges $v_0 u_1$, $u_1 v_3$, $v_3 u_2$ and $u_2 v_0$. Here, observe that we have $\{v_1, v_2\} \cap \{u_1, u_2\} = \emptyset$; otherwise, G would have a vertex of degree 4, which is either v_1 or v_2. Under the situation, either $v_0 v_1 v_3 u_1$ or $v_0 v_1 v_3 u_2$ is separating, contrary to G being 5-connected. Therefore, the statement holds.   □

Note that Proposition 4.15 implies that every 5-connected 1-plane graph G can be augmented to a near-optimal 1-plane graph by adding edges to G. In the above proposition, the 5-connectivity condition is necessary since the unique 1-embedding $M(5)$ of K_5 is not near-optimal.

To obtain an optimal 1-plane graph, we actually need stronger conditions; e.g., the 5-connectivity condition is not sufficient since $M(6)$ in Fig. 4.10, which is the unique embedding of K_6 up to weak equivalence is maximum; and hence maximal but not optimal. However, we know some graph classes having our desired property. First, it is easy to see that every 3-connected quadrangulation can be augmented to an optimal 1-plane graph by adding pairs of crossing edges by Theorem 4.2. Furthermore, Noguchi and Suzuki proved the following theorem.

Theorem 4.14 ([23]) *Every triangulation T of the sphere contains a spanning quadrangulation as a subgraph. Furthermore, if T is 5-connected, then every spanning quadrangulation subgraph of T is 3-connected.*

The lower bound 5 on the connectivity of T in Theorem 4.14 is the best possible; i.e., there exist infinitely many 4-connected triangulations of the sphere that do not have the property. As a corollary of the above theorem, it follows that every 5-connected triangulation T of the sphere can be augmented to an optimal 1-plane graph by adding edges to T. Moreover, Noguchi and Suzuki proved the following theorem.

Theorem 4.15 ([23]) *Let Q be a quadrangulation of the sphere with $|V(Q)| \geq 6$. Then Q can be augmented to a 4-connected triangulation of the sphere by adding a diagonal edge in every face of Q.*

Using the above result, we can easily prove the following proposition, which was also shown in [23].

Proposition 4.16 *Let G be an optimal* 1*-plane graph. Then G contains a spanning* 4*-connected triangulation as a subgraph.*

Proof By Theorem 4.2, G has a 3-connected quadrangulation Q as its subgraph. Since the cube having 8 vertices is the smallest 3-connected quadrangulation of the sphere, Q satisfies the condition of Theorem 4.15. Thus, we can choose one diagonal edge from each pair of crossing edges in the face of Q, so that the resulting graph becomes a 4-connected triangulation. Thus, we got a conclusion.   □

The "4" in the above proposition is clearly the best possible; recall that there are optimal 1-planar graphs with connectivity 4. By the above proposition, we can prove Proposition 4.12 in Sect. 4.3 more easily by using the result [29] again.

4.8   Open Problems

At the end of this chapter, we show some open problems concerning the topics dealt in the chapter.

1. Is every 6-connected (or 7-connected) 1-planar graph Hamiltonian? In fact, Noguchi [22] constructed a infinite sequence of non-Hamiltonian 5-connected 1-planar graphs.
2. Improve the bounds for the number of edges in maximal 1-planar or 1-plane graphs, mentioned in Sect. 4.4.
3. Characterize optimal 1-planar graphs having no K_n-minor for $n \geq 8$. Furthermore, characterize optimal 1-planar multigraphs having no K_n-minor for $n \geq 5$.

4. Is every 7-connected 1-planar graph uniquely 1-embeddable into the sphere? If this is true, then "7" is the best possible since every X-pseudo double wheel, which is 6-connected by Proposition 4.9, has at least two inequivalent 1-embeddings, up to equivalence.
5. Is the underlying graph of every near-optimal 1-plane graph is maximal 1-planar?
6. Extend the problems in this chapter for 1-embeddings on non-spherical closed surfaces. In [20], it was shown that there is a one-to-one correspondence between simple optimal 1-embeddings of a non-spherical closed surface $\mathbb{F}^2$ and polyhedral quadrangulations of $\mathbb{F}^2$, i.e., 3-connected and 3-representative quadrangulations of $\mathbb{F}^2$. However, little is known about general 1-embeddings on non-spherical surfaces.

Acknowledgements This work was supported by JSPS KAKENHI Grant Number 16K05250.

References

1. Albertson, M.O., Mohar, B.: Coloring vertices and faces of locally planar graphs. Graphs Comb. **22**, 289–295 (2006)
2. Barát, J., Tóth, G.: Improvements on the density of maximal 1-planar graphs. J. Graph Theory **88**, 101–109 (2018)
3. Bodendiek, R., Schumacher, H., Wagner, K.: Bemerkungen zu einem Sechsfarbenproblem von G. Ringel. Abh. Math. Sem. Univ. Hamburg **53**, 41–52 (1983)
4. Bodendiek, R., Schumacher, H., Wagner, K.: Über 1-optimale Graphen. Math. Nachr. **117**, 323–339 (1984)
5. Brandenburg, F.J., Eppstein, D., Gleissner, A., Goodrich, M.T., Hanauer, K., Reislhuber, J.: On the density of maximal 1-planar graphs, Graph Drawing 2012. Lect. Notes Comput. Sci. **7704**, 327–338 (2013)
6. Czap, J., Hudác, D.: 1-planarity of complete multipartite graphs. Discrete Appl. Math. **160**, 505–512 (2012)
7. Diestel, R.: Graph Theory, 5th edn. Springer, Heidelberg (2016)
8. Eades, P., Hong, S., Kato, N., Liotta, G., Schweitzer, P., Suzuki, Y.: A linear time algorithm for testing maximal 1-planarity of graphs with a rotation system. Theoret. Comput. Sci. **513**, 65–76 (2013)
9. Fabrici, I., Madaras, T.: The structure of 1-planar graphs. Discrete Math. **307**, 854–865 (2007)
10. Fujisawa, J., Segawa, K., Suzuki, Y.: The matching extendability of optimal 1-planar graphs. Graphs Comb. **34**, 1089–1099 (2018)
11. Grigoriev, A., Bodlaender, H.L.: Algorithms for graphs embeddable with few crossings per edge. Algorithmica **49**, 1–11 (2007)
12. Hobbs, A.M.: Some Hamiltonian results in power of graphs. Res. Natl. Bur. Stand. B **77B**, 1–10 (1973)
13. Hudác, D., Madaras, T., Suzuki, Y.: On properties of maximal 1-planar graphs. Discuss. Math. Graph Theory **32**, 737–747 (2012)
14. Karpov, D.V.: An upper bound on the number of edges in an almost planar bipartite graphs. J. Math. Sci. **196**, 737–746 (2014)
15. Kleitman, D.J.: The crossing number of $K_{5,n}$. J. Comb. Theory **9**, 315–323 (1970)
16. Kobourov, S.G., Liotta, G., Montecchiani, F.: An annotated bibliography on 1-planarity. Comput. Sci. Rev. **25**, 49–67 (2017)
17. Korzhik, V.P.: Minimal non-1-planar graphs. Discrete Math. **308**, 1319–1327 (2008)

18. Korzhik, V.P., Mohar, B.: Minimal obstructions for 1-immersions and hardness of 1-planarity testing. J. Graph Theory **72**, 30–71 (2013)
19. Mader, W.: Homomorphiesätze für Graphen. Math. Ann. **178**, 154–168 (1968)
20. Nagasawa, T., Noguchi, K., Suzuki, Y.: Optimal 1-embedded graphs on the projective plane which triangulate other surfaces. J. Nonlinear Convex Anal. **19**, 1759–1770 (2018)
21. Nakamoto, A., Noguchi, K., Ozeki, K.: Cyclic 4-colorings of graphs on surfaces. J. Graph Theory **82**, 265–278 (2016)
22. Noguchi, K.: Hamiltonicity and connectivity of 1-planar graphs, preprint
23. Noguchi, K., Suzuki, Y.: Relationship among triangulations, quadrangulations and optimal 1-planar graphs. Graphs Comb. **31**, 1965–1972 (2015)
24. Pach, J., Tóth, G.: Graphs drawn with few crossings per edge. Combinatorica **17**, 427–439 (1997)
25. Ringel, G.: Ein Sechsfarbenproblem auf der Kugel. Abh. Math. Sem. Univ. Hamburg **29**, 107–117 (1965)
26. Schumacher, H.: Zur Struktur 1-planarer Graphen. Math. Nachr. **125**, 291–300 (1986)
27. Suzuki, Y.: K_7-Minors in optimal 1-planar graphs. Discrete Math. **340**, 1227–1234 (2017)
28. Suzuki, Y.: Re-embeddings of maximum 1-planar graphs. SIAM J. Discrete Math. **24**, 1527–1540 (2010)
29. Tutte, W.T.: A theorem on planar graphs. Trans. Amer. Math. Soc. **82**, 99–116 (1956)

Chapter 5
Algorithms for 1-Planar Graphs

Seok-Hee Hong

Abstract A *1-planar* graph is a graph that can be embedded in the plane with at most one crossing per edge. It is known that testing 1-planarity of a graph is NP-complete. This chapter reviews the algorithmic results on 1-planar graphs. We first review a linear time algorithm for testing maximal 1-planarity of a graph if a *rotation system* (i.e., the circular ordering of edges for each vertex) is given. A graph is *maximal 1-planar* if the addition of an edge destroys 1-planarity. Next, we sketch a linear time algorithm for testing outer-1-planarity. A graph is *outer-1-planar* if it has an embedding in which every vertex is on the outer face and each edge has at most one crossing. The 1-plane graphs have two forbidden subgraphs to admit a straight-line drawing. We review a linear time algorithm for constructing a straight-line drawing of 1-plane graphs. Finally, we conclude with reviews on recent related results.

5.1 Introduction

Recent research topics in topological graph theory and graph drawing generalize the notion of planarity to sparse non-planar graphs, called *beyond planar graphs*, either with forbidden edge crossing patterns or with specific types of edge crossings.

This chapter reviews algorithmic results on 1-planar graphs, i.e., graphs that can be embedded with at most one crossing per edge [25]. The 1-planar graphs are introduced by Ringel [26] in the context of simultaneously coloring vertices and faces of planar graphs. Subsequently, the combinatorial aspects of 1-planar graphs have been investigated.

For example, Borodin [6] investigated coloring for 1-planar graphs, and Borodin et al. [7] studied the acyclic colorability of 1-planar graphs; Zhang and Wu [30] studied the edge colorability of 1-planar graphs. In particular, Pach and Toth [25] proved that a 1-planar graph with n vertices has at most $4n - 8$ edges, which is a tight upper bound.

S.-H. Hong (✉)
University of Sydney, Sydney, Australia
e-mail: seokhee.hong@sydney.edu.au

© Springer Nature Singapore Pte Ltd. 2020
S.-H. Hong and T. Tokuyama (eds.), *Beyond Planar Graphs*,
https://doi.org/10.1007/978-981-15-6533-5_5

There are a number of structural results on 1-planar graphs by Fabrici and Madaras [12], and *maximal 1-planar graphs* by Hudak et al. [21] and Suzuki [28]. Suzuki [27] also investigated structural properties of *optimal 1-planar graphs* (i.e., 1-planar graphs with the maximum number of $4n - 8$ edges).

The class of 1-planar graphs is not closed under edge contraction; accordingly, computational problems seem difficult. Grigoriev and Bodlaender [15], and Korzhik and Mohar [22] independently proved that testing 1-planarity of a graph is NP-complete. It remains NP-hard, even if the rotation system is given as part of the input, shown by Auer et al. [2]. Furthermore, Cabello and Mohar [8] showed that NP-hardness holds even if the input graph is an *almost planar graph* (i.e., deletion of an edge makes the resulting graph planar). More recently, Bannister et al. [3] studied the fixed parameter complexity of 1-planarity.

On the positive side, efficient polynomial time algorithms are known for special subclasses of 1-planar graphs. For example, a linear time algorithm is available for testing maximal 1-planarity, if the rotation system is given, by Eades et al. [10]. Hong et al. [16] and Auer et al. [1] independently presented linear time algorithms for testing outer-1-planarity.

The classical *Fáry's Theorem* [13] showed that every *plane graph* (i.e., a planar graph with a given planar embedding) admits a planar straight-line drawing. However, 1-plane graphs (i.e., 1-planar graphs with a given 1-planar embedding) have two forbidden subgraphs to admit a straight-line drawing, shown by Thomassen [29]. Hong et al. [20] presented linear time testing and drawing algorithms to construct such a straight-line drawing of 1-plane graphs if it exists.

This chapter reviews algorithmic results on 1-planar graphs. More specifically, we describe three linear time algorithms in the following sections:

1. Section 5.2: linear time algorithm by Eades et al. [10] for testing maximal 1-planarity of a graph G with a given rotation system.
2. Section 5.3: linear time algorithm by Hong et al. [16] for testing outer-1-planarity of a graph.
3. Section 5.4: linear time algorithm by Hong et al. [20] for constructing a straight-line drawing of a 1-plane graph.

Section 5.5 concludes with reviews on recent progress.

5.2 Testing Maximal 1-Planarity

Eades et al. [10] proved the following main theorem.

Theorem 5.1 *There exists a linear time algorithm that tests whether graph G with a given rotation system Φ has a maximal 1-planar embedding consistent with Φ. If such an embedding exists, it is unique and the algorithm computes the embedding.*

First, it was shown that in any maximal 1-planar embedding, the subgraph induced by planar edges (called the *red graph*) is spanning and biconnected, and that if

Fig. 5.1 Maximal 1-planar
graph

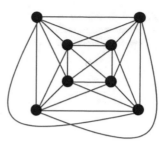

the rotation system does admit a maximal 1-planar embedding, then it is unique.
Figure 5.1 shows an example of a maximal 1-planar graph.

Note that a rotation system Φ does not define crossings between edges. However, for a planar graph, a rotation system uniquely determines a planar embedding.
Therefore, to determine a 1-planar embedding $\xi(G_\Phi)$, it is sufficient to determine a
rotation system of the planar embedding $\xi(G_P)$ of *planarization* G_P of the 1-planar
embedding.

An embedding $\xi(G)$ of a graph G defines the crossing-free (called *red*) edges as
well as the crossing (called *blue*) edges. Denote the subgraph of a graph G induced by
the red edges as *red graph* G_R. Now, we sketch a linear time algorithm to test maximal
1-planarity of a graph with a given rotation system, consisting of the following five
steps:

Algorithm: Testing Maximal 1-Planarity
Input: G_Φ, a graph G with a rotation system Φ.
Output: 1-planar embedding $\xi(G_\Phi)$ or "no".

1. If $|E(G)| > 4n - 8$ or G is not biconnected, then return("no").
2. Compute the red planar subgraph G_R of G_Φ.
3. If G_R is not planar or not biconnected, then return("no").
4. Test 1-planarity of G_Φ, and compute $\xi(G_\Phi)$ and $\xi(G_P)$.
5. Test maximality of $\xi(G_\Phi)$.

Steps 1 and 3 use the Pach–Toth bound [25], a standard biconnectivity algorithm,
and a planarity testing algorithm. In the following, we sketch Steps 2, 4, and 5.

Step 2: Computing the Red Subgraph G_R
Consider graph G as a directed graph, with two directed edges (u, v) and (v, u) for
each pair u, v of adjacent vertices. We say that a directed edge (v_2, v_3) is the *rightmost
continuation* of a directed edge (v_1, v_2), if vertex v_3 is the vertex that precedes v_1 in the
circular ordering of v_2. We say that a walk $v_1, \ldots, v_t$ is a *completed rightmost walk*, if:
(i) for every $i \in \{1, \ldots, t\}$, the directed edge (v_i, v_{i+1}) is the rightmost continuation of
the directed edge (v_{i-1}, v_i), and (ii) for all $i, j \in \{1, \ldots, t\}$, if $(v_i, v_{i+1}) = (v_j, v_{j+1})$,
then $i = j$. See Fig. 5.2, for examples.

Completed rightmost walks characterize the colors of the edges of G_Φ, as in the
following lemma.

S.-H. Hong

Fig. 5.2 Examples of
a rightmost continuation;
b completed rightmost walk

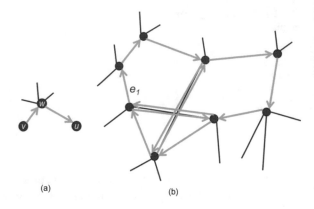

(a)   (b)

Lemma 5.1 *Let G be a 1-plane graph with a given rotation system Φ, whose red graph G_R is spanning and biconnected. An edge e of G is red if and only if there is a completed rightmost walk on G_Φ that traverses e only in one direction.*

We can design an algorithm that takes G_Φ as input, and computes the color of the edges as follows:

(i) simply traverse the graph with rightmost walks;
(ii) by marking edges after the traversal, we can color the edges red or blue.

Step 4: Computing a 1-Planar Embedding of G_Φ

We now test whether there exists a 1-planar embedding of G_Φ consistent with the colors. If such an embedding exists, we compute a planar embedding of the planarization G_P of G_Φ.

After testing biconnectivity and planarity of G_R in Step 3, we have a planar embedding $\xi(G_R)$ of G_R which preserves the given rotation system Φ of G.

Since the 1-planar embedding of G_Φ is unique, if it exists, this implies that we can use the rotation system Φ to identify the red facial cycles and the blue edges inside each red face. Then crossings can be detected by traversing each red face; the traversal detects any edge with more than one crossing.

Lemma 5.2 *There exists a linear time algorithm that tests whether there is a 1-planar embedding of G_Φ that is consistent with Φ such that G_R is the red subgraph. If such an embedding $\xi(G_\Phi)$ exists, it is unique and the algorithm computes the planar embedding $\xi(G_P)$ of the planarization of $\xi(G_\Phi)$.*

Step 5: Testing Maximality of 1-Planar Embedding

Now, we show that maximality of a 1-planar embedding of G with a given rotation system Φ can be tested in linear time. Let $\xi(G)$ be a maximal 1-planar embedding of a graph G, G_P be the planarization of G, and $\xi(G_P)$ be the planar embedding of G_P induced by $\xi(G)$. Let f be a facial cycle in $\xi(G_P)$.

Note that maximal 1-planar graphs have the following properties:

- Each crossing in $\xi(G)$ induces a 4-clique.

Fig. 5.3 Testing maximality
of an 1-planar embedding:
a testing possible addition of
a red edge; **b** testing possible
addition of a blue edge

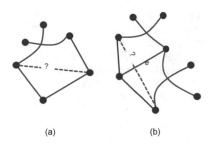

(a)   (b)

- The face f has at most four real vertices, and at most eight vertices (real plus virtual).

We can simply test whether there are two nonadjacent real vertices v_1 and v_2 such that adding the edge (v_1, v_2) does not destroy 1-planarity as follows:

(i) First test each face whether it contains two such v_1 and v_2 (see Fig. 5.3a);
(ii) Then test red edge e whether we can add (v_1, v_2) by crossing e, where v_1 and v_2 are on the faces separated by e (see Fig. 5.3b).

Using the properties of maximal 1-planar graphs above, we can perform these testings in linear time.

5.3 Testing Outer-1-Planarity

We now review a linear time algorithm by Hong et al. [16] to test the outer-1-planarity of a graph G. The following theorem summarizes the main results.

Theorem 5.2 *There is a linear time algorithm to test whether a given graph is outer-1-planar. The algorithm computes an outer-1-planar embedding if it exists.*

To prove Theorem 5.2, a subclass of outer-1-planar graphs, called *one-sided-outer-1-planar (OSO1P)* graph, was introduced as follows. Let G be a graph with vertices s and t, and $G_{+(s,t)}$ be the graph obtained by adding the edge (s, t), if it is not already in G. If $G_{+(s,t)}$ has an outer-1-planar embedding with the edge (s, t) on the outer face, then G is called *one-sided-outer-1-planar (OSO1P)* with respect to (s, t).

A graph is outer-1-planar if and only if its biconnected components are outer-1-planar. Therefore, the algorithm focuses on the biconnected case, using the *SPQR tree* [5] to represent the decomposition of a biconnected graph into triconnected components. We now review the basic terminology of the SPQR tree.

Each node v in the SPQR tree is associated with a graph called the *skeleton* of v, denoted by $\sigma(v)$. There are four types of nodes v:

Fig. 5.4 AOSO1P graph consists of a parallel composition of an OSO1P graph and an OSO1P S-node with a tail: **a** general shape of an AOSO1P graph with respect to (s, t); **b** AOSO1P graph with respect to (s, t); **c** AOSO1P graph with respect to both (s, t) and (t, s)

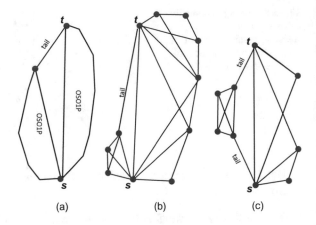

(a)   (b)   (c)

- S-node: $\sigma(v)$ is a simple cycle with at least three vertices;
- P-node: $\sigma(v)$ consists of two vertices connected by at least three edges;
- Q-node: $\sigma(v)$ consists of two vertices connected by a *real* edge and a *virtual* edge; and
- R-node: $\sigma(v)$ is a simple triconnected graph.

A rooted SPQR tree can be obtained by choosing an arbitrary node as its root. Let ρ be the parent node of an internal node v. The graph $\sigma(\rho)$ has exactly one *virtual edge* e in common with $\sigma(v)$, called the *parent virtual edge* of $\sigma(v)$, and a *child virtual edge* in $\sigma(\rho)$. Denote the graph formed by the union of $\sigma(v)$ over all descendants v of ρ by G_ρ.

Let μ be an S-node with parent separation pair (u, v). A *tail at u* for μ is a Q-node child (that is, a real edge) with parent virtual edge (u, x) for some vertex x. A P-node v is called *almost one-sided outer-1-planar (AOSO1P) with respect to the directed edge (s, t)*, if G_v consists of a parallel composition of an OSO1P graph with respect to (s, t) and an S-node μ such that μ has a tail at t and μ is OSO1P with respect to (s, t). See Fig. 5.4 for examples.

If G is an outer-1-planar graph, then $\sigma(v)$ and G_v are outer-1-planar graphs. If G_v is a one-sided outer-1-planar (OSO1P) graph with respect to the parent virtual edge (s, t) of v, then denote v as a one-sided outer-1-planar (OSO1P) node with respect to (s, t).

Step 1: Testing OSO1P and AOSO1P
The algorithm traverses the SPQR tree of G from the leaves toward the root, computing two boolean labels $OSO1P(v, s, t)$ and $AOSO1P(v, s, t)$ which indicate whether v is OSO1P or AOSO1P with respect to (s, t). The label $AOSO1P(v, s, t)$ is computed for each P-node v only.

Note that the only triconnected outer-1-planar graph is K_4 with unique outer-1-planar embedding. Therefore, the R-node case is easy; however, P-node and S-node cases are more involved.

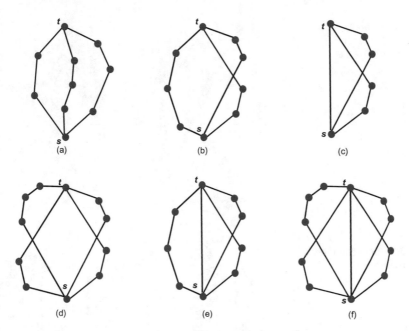

Fig. 5.5 Embeddings of a set of paths sharing endpoints: **a** planar embedding; **b** outer-1-planar embedding of three non-trivial paths; **c** outer-1-planar embedding of three paths, where one path is trivial; **d** outer-1-planar embedding of four non-trivial paths; **e** outer-1-planar embedding of four paths, where one path is trivial; **f** outer-1-planar embedding of five paths, where one path is trivial

Figure 5.5 illustrates structural properties on the possible outer-1-planar embeddings of a set of paths that share endpoints. Let P be a set of paths between two vertices s and t. A path from s to t is called *non-trivial*, if it contains more than two vertices.

We now describe each case (i.e., R-node, P-node, S-node) in detail.

(i) R-node: Let v be an R-node with parent virtual edge (u, v). Then G_v is OSO1P with respect to (u, v) if and only if:

1. $\sigma(v)$ is isomorphic to K_4; and
2. an edge (u, a) of $\sigma(v)$ with $a \neq v$ incident with u represents a child Q-node of v; an edge (v, b) of $\sigma(v)$ with $b \neq u$ represents a child Q-node of v; and (u, a) crosses (v, b); and
3. for each child v' of v, v' is OSO1P with respect to (c, d), where (c, d) is the parent virtual edge of v'.

Figure 5.6a shows an example of an OSO1P R-node: $\sigma(v)$ is K_4, where the inner crossing edges are real edges (i.e., Q-node children), and outer edges are OSO1P child nodes. Figure 5.6b shows a non-OSO1P R-node, where crossing edges are not real edges.

Fig. 5.6 Examples of **a** OSO1P R-node; **b** non-OSO1P R-node

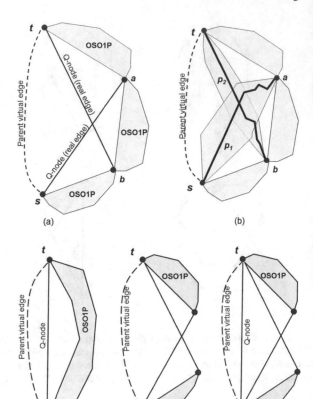

Fig. 5.7 Illustration for an OSO1P P-node

(ii) P-node: Based on the structural properties shown in Fig. 5.5, an OSO1P P-node can have at most three children. Let v be a P-node with parent virtual edge (s, t). Then G_v is OSO1P with respect to (s, t) if and only if

- v has two children, where one is a Q-node (s, t), and the other is OSO1P with respect to (s, t) (see Fig. 5.7a); or
- v has two children, where one is an S-node with tail at s which is OSO1P with respect to (s, t), and the other is an S-node with tail at t which is OSO1P with respect to (s, t) (see Fig. 5.7b); or
- v has three children, where one is a Q-node (s, t), another is an S-node with tail at s which is OSO1P with respect to (s, t), and the other is an OSO1P S-node with tail at t which is OSO1P with respect to (s, t) (see Fig. 5.7c).

It is straightforward to extend the above conditions to test whether a P-node v is AOSO1P.

(iii) S-node: Let v be an S-node with children $v_1, v_2, \ldots, v_k$, where the parent virtual edge of v_i is (s_{i-1}, s_i); see Fig. 5.8a. If each child v_i is OSO1P with respect to

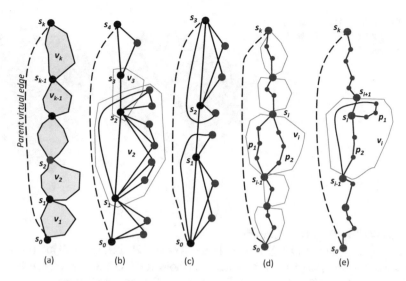

Fig. 5.8 Examples of **a** S-node; **b** OSO1P S-node with a child v_2 that is not OSO1P; **c** S-node that satisfies the necessary conditions, but is not OSO1P; **d** Two paths p_1 and p_2 in G_{v_i}; **e** The path p_1 crosses the edge (s_i, s_{i+1})

(s_{i-1}, s_i), then clearly v is OSO1P with respect to (s_0, s_k); however, the converse is false. Figure 5.8b shows the necessary conditions, where v is OSO1P with respect to (s_0, s_k), however, the child v_2 is not OSO1P with respect to (s_1, s_2). Note that v_3 is a Q-node, and an edge from the skeleton of v_2 crosses this edge.

Let v be an S-node with children $v_1, v_2, \ldots, v_k$, where the parent virtual edge of v_i is (s_{i-1}, s_i), and G_v is OSO1P with respect to (s_0, s_k). Then for $1 \leq i \leq k$:

- v_i is OSO1P with respect to (s_{i-1}, s_i); or
- $i < k$, v_i is AOSO1P with respect to (s_i, s_{i-1}), and v_{i+1} is a Q-node; or
- $i > 1$, v_i is AOSO1P with respect to (s_{i-1}, s_i), and v_{i-1} is a Q-node.

The above conditions are necessary for an S-node to be OSO1P, however not sufficient; e.g., see Fig. 5.8c. The problem is that the Q-node represented by the edge (s_1, s_2) has two crossings. We can express sufficient conditions for an OSO1P S-node recursively, as follows.

Let v be an S-node with children $v_1, v_2, \ldots, v_k$, where the parent virtual edge of v_i is (s_{i-1}, s_i), and let $G(v_1, v_2, \ldots, v_k)$ denote the series composition of graphs $G_{v_1}, G_{v_2}, \ldots, G_{v_k}$. Then G_v is OSO1P with respect to (s_0, s_k) if and only if:

- G_{v_1} is OSO1P with respect to (s_0, s_1) and $G(v_2, v_3, \ldots, v_k)$ is OSO1P with respect to (s_1, s_k); or
- v_1 is a Q-node, G_{v_2} is AOSO1P with respect to (s_1, s_2), and $G(v_3, v_4, \ldots, v_k)$ is OSO1P with respect to (s_2, s_k); or
- G_{v_1} is AOSO1P with respect to (s_1, s_0), v_2 is a Q-node, and $G(v_3, v_4, \ldots, v_k)$ is OSO1P with respect to (s_2, s_k).

Fig. 5.9 Testing O1P at the
root node: **a** Root R-node; **b**
Root P-node

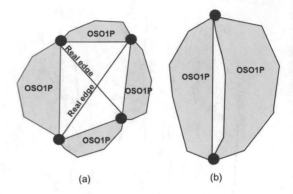

(a)   (b)

Step 2: Testing Outer-1-Planarity

After computing labels $OSO1P(v, s, t)$ and $AOSO1P(v, s, t)$ for all internal nodes
v of the SPQR tree, we can test whether the whole graph G (i.e., the root ρ) is outer-
1-planar. We can require the root node ρ to be an R-node or a P-node.

For the root R-node, the algorithm tests the following conditions (see Fig. 5.9a):
G is outer-1-planar (O1P) if and only if

1. $\sigma(\rho)$ is isomorphic to K_4, and
2. at least two children of ρ are Q-nodes, and
3. for each child node v of $\sigma(\rho)$ with parent virtual edge (a, b), G_v is OSO1P with
   respect to (a, b).

For the root P-node, testing O1P is simpler: G is outer-1-planar if and only if
$\sigma(\rho)$ is a parallel composition of two OSO1P graphs (see Fig. 5.9b).

5.4 Straight-Line Drawing Algorithm for 1-Planar Graphs

The classical *Fáry's Theorem* [13] proved that every *plane graph* (i.e., planar topological embedding of a planar graph) has a planar straight-line drawing. Indeed, planar
straight-line drawing is one of the most popular drawing conventions in Graph Drawing [4, 24]. For example, de Fraysseix et al. [14] showed that planar straight-line
grid drawing can be efficiently constructed in a *quadratic area*.

On the other hand, Thomassen [29] showed that there are two 1-plane graphs that
cannot be drawn with straight-line edges. More specifically, he proved that a 1-plane
graph G admits a straight-line 1-planar drawing if and only if G contains neither the
B graph (see Fig. 5.10a) nor the W graph (see Fig. 5.10b).

Based on Thomassen's characterization, Hong et al. [20] presented linear time
testing and drawing algorithms, proving the following main theorem.

Fig. 5.10 **a** The B graph; **b** the W graph; **c** 1-plane embedding of K_4 containing B subgraph

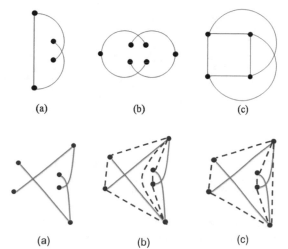

(a)   (b)   (c)

Fig. 5.11 Example of an augmentation: **a** 1-plane graph without the B graph or the W graph; **b** bad augmentation introducing B graph; **c** good augmentation not introducing B graph

(a)   (b)   (c)

Theorem 5.3 *There is a linear time algorithm to test whether a 1-plane embedding contains the B graph or the W graph, and a linear time drawing algorithm to construct a straight-line 1-planar drawing if it exists.*

Here, we mainly explain the drawing algorithm consisting of two steps: an augmentation step and a drawing step.

Step 1: Red-Maximal Augmentation

The first step, called *red-maximal augmentation*, is to augment a 1-plane graph G by adding edges without introducing new crossings while preserving the straight-line drawability of G. Denote the crossing-free edges of a 1-plane graph G as *red* edges.

A *red augmentation* $G' = (V, E')$ of $G = (V, E)$ is a 1-plane graph with $E \subseteq E'$ such that no edge in $E' - E$ has a crossing. A 1-plane graph is *red-maximal* if the addition of any edge makes a crossing. The red-maximal 1-plane graphs have nice properties, which are helpful for the drawing algorithm.

Computing a red-maximal augmentation G^+ of a 1-plane graph G, preserving the absence of B and W subgraphs, consists of two steps: (i) the first step adds edges for each crossing γ with a 4-cycle; (ii) the second step triangulates any remaining faces.

The first step adds edges to a 1-plane graph G without the B subgraph or the W subgraph until each crossing γ is surrounded by a 4-cycle. Note that there are different ways to add the edge (a, b), as shown in Fig. 5.11: Fig. 5.11b introduces the B subgraph, while Fig. 5.11c does not.

Furthermore, there may be many crossing vertices γ that share the same neighbors, (a, b), as shown in Fig. 5.12c. Nevertheless, it is always possible to route the edge (a, b) without introducing the B subgraph, using the orientation of crossings with respect to (a, b): *clockwise* (see Fig. 5.12a) or *anticlockwise* (see Fig. 5.12b). For example, in Fig. 5.12c, the edge (a, b) can be added between γ_j and γ_{j+1} without introducing the B graph.

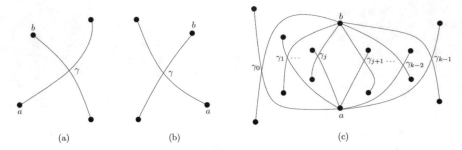

Fig. 5.12 a The crossing γ is *clockwise* with respect to (a, b); **b** γ is *anticlockwise* with respect to (a, b); **c** (a, b) is a separation pair with many crossings: the edge (a, b) can be added between γ_j and γ_{j+1} without introducing the B graph

Fig. 5.13 Example of a red-maximal augmentation: **a** 1-plane embedding; **b** red-maximal augmentation of **a**

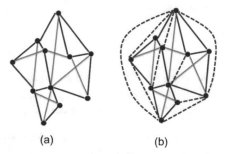

Based on the orientation of crossings, we can add edges to obtain an augmentation such that each crossing is surrounded by a 4-cycle, without introducing the W subgraph and the B subgraph.

The second step is triangulating the remaining faces. Let a and b be two nonadjacent vertices in the 1-plane graph G' after the first step, sharing a face f. We can add the edge (a, b) inside f, without crossing any edge, and without introducing the W subgraph and the B subgraph. Figure 5.13 shows an example of a red-maximal augmentation.

The following lemma summarizes the results of the augmentation step.

Lemma 5.3 *Let G be a 1-plane graph without the B subgraph or the W subgraph. Then there is a red-maximal augmentation G^+ of G without the B subgraph or the W subgraph, which can be computed in linear time.*

Properties of Red-Maximal 1-Plane Graphs

The structure of a red-maximal 1-plane graph is relatively simple; this simplifies the drawing algorithm. Let G^+ be a red-maximal 1-plane graph that does not contain the B subgraph or the W subgraph, and G^* be the planarization of G^+. Then G^+ and G^* have the following properties:

- If f is an internal face of G^* with no crossing vertex, then f is a 3-cycle.

- If f is an internal face of G^* with a crossing vertex, then f is either a 3-cycle, 4-cycle, or 5-cycle.
- If f is the outer face of G^*, then f has no crossing vertices.
- If f is the outer face of G^*, then f is either a 3-cycle or 4-cycle. If f is a 4-cycle, then it induces a 4-clique with a crossing.
- If γ is a crossing between edges (a, c) and (b, d), then there is a path P of red edges from a to b such that the cycle C in G^* formed by the edges (a, γ) and (γ, b), and P contains no vertices strictly inside C.

Step 2: Drawing Algorithm

The input of the drawing algorithm is a red-maximal augmentation G^+ without the B subgraph or the W subgraph. Let G_r be the subgraph of red edges of G^+. Based on the structural properties of G^+, both G^+ and G_r are biconnected. Therefore, the drawing algorithm uses the SPR tree, a simplified version of the SPQR tree [5] without Q-nodes.

Let $\sigma(v)$ denote the *skeleton* of node v in the SPR tree, which has one of the three types: (i) S-node: $\sigma(v)$ is a simple cycle with at least three vertices; (ii) P-node: $\sigma(v)$ consists of two vertices connected by at least three edges; (iii) R-node: $\sigma(v)$ is a simple triconnected graph.

The algorithm uses the SPR tree of the red subgraph G_r of G^+, rooted at a node whose skeleton contains the vertices on the outer face. Let $\sigma(v)^-$ denote a graph after deleting the parent virtual edge from $\sigma(v)$. The algorithm uses a similar approach for star-shaped drawings of planar graphs [17], however, in a simplified way due to the nice properties of the red-maximal augmentation.

More specifically, the algorithm recursively processes each node v in the SPR tree in a top-down manner, from the root node to the leaf nodes, as follows:

1. Construct a *convex drawing* D_v of $\sigma(v)$ in a given convex polygon P_v.
2. Re-insert crossing edges in the corresponding face of D_v with straight-line edges.
3. For each child μ of v, define a convex polygon P_μ and replace the corresponding virtual edge in D_v with a drawing of $\sigma(\mu)$.

The algorithm uses a convex drawing algorithm of Chiba et al. [9] as a subroutine for drawing R-nodes, as follows: It takes a convex polygon P_v and the plane graph $\sigma(v)$ as input, and computes a *convex* drawing D_v of $\sigma(v)$. Since each face of D_v is a convex polygon, we can re-insert the crossing edges using straight lines, without introducing any new crossings.

In fact, the algorithm processes each node v differently, based on its type (i.e., R-node, S-node, and P-node). Here, we give a brief sketch for each case.

(i) R-node: First, construct a convex drawing D_v of $\sigma(v)$ for the root R-node (respectively, $\sigma(v)^-$ for non-root R-node) inside a given convex polygon P_v. Next, re-insert the crossing edges in the corresponding face in D_v as straight-line segments.

After inserting crossing edges, define a drawing region and a convex polygon P_μ for drawing $\sigma(\mu)$ of each child node μ recursively. Figure 5.14 shows an example of a root R-node.

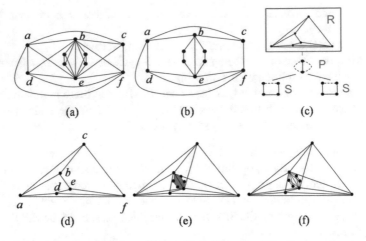

Fig. 5.14 Example of a root R-node v: **a** G^+; **b** G_r; **c** SPR tree of G_r; **d** convex drawing of $\sigma(v)$; **e** re-insert crossing edges in a convex face and define a drawing area and convex polygon P_μ for drawing $\sigma(\mu)$ of a child node μ; **f** re-insert crossing edges inside P_μ

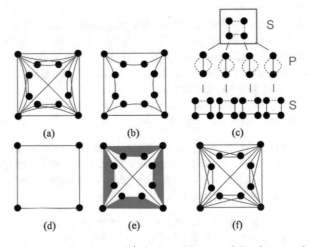

Fig. 5.15 Example of a root S-node v: **a** G^+; **b** G_r; **c** SPR tree of G_r; **d** convex drawing of $\sigma(v)$; **e** re-insert crossing edges in a convex face and define a drawing area and convex polygon P_μ for drawing $\sigma(\mu)$ of child node μ; **f** re-insert crossing edges inside P_μ

(ii) S-node: If v is a root S-node, draw $\sigma(v)$ as a triangle or a rectangle; re-insert the crossing edges, if $\sigma(v)$ is a 4-cycle. Then define a drawing region and a convex polygon P_μ for drawing $\sigma(\mu)$ of each child node μ recursively.

If v is a non-root S-node, then we draw $\sigma(v)^-$ as a path. Then, the main task is to define a drawing area and a convex polygon P_μ for drawing $\sigma(\mu)$ of each child node μ recursively. Figure 5.15 shows an example of the root S-node.

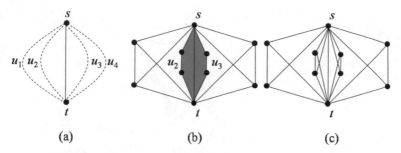

Fig. 5.16 Defining drawing area for the children of a P-node v: **a** $\sigma(v)$; **b** define left trapezoids for μ_1 and μ_2, and right trapezoids for μ_3 and μ_4; re-insert crossing edges in a convex drawing of $\sigma(\mu_1)$ (respectively, $\sigma(\mu_4)$) and define a drawing area and convex polygon P_{μ_2} (respectively, P_{μ_3}) for drawing $\sigma(\mu_2)$ (respectively, $\sigma(\mu_3)$); **c** re-insert crossing edges in the drawing D_{μ_2} (respectively, D_{μ_3})

(iii) P-node: The main task for P-node is to define a drawing area and a convex polygon P_μ for drawing $\sigma(\mu)$ of each child node μ recursively. For R-node child μ, define P_μ as either a triangle or a rhombus; for S-node child μ, define P_μ as either a triangle or a trapezoid, based on the properties of the red-maximal augmentation.

Let vertices s and t be the separation pair of v, and denote the virtual edges between s and t as $u_1, u_2, \ldots, u_m$, in left-to-right order, as in Fig. 5.16a. Denote the corresponding children of v as $\mu_1, \mu_2, \ldots, \mu_m$. Suppose that the edge $e = (s, t)$ occurs between u_k and u_{k+1}. The polygons P_{μ_i} must be drawn based on the ordering: define a *left* triangle (or trapezoid) for $\mu_1, \mu_2, \ldots, \mu_k$, and a *right* triangle (or trapezoid) for $\mu_{k+1}, \mu_{k+2}, \ldots, \mu_m$ to avoid edge crossings. See Fig. 5.16b.

First, draw $\sigma(\mu_1)$ inside the polygon P_{μ_1}, and re-insert crossing edges in the drawing D_{μ_1}. Then, define a drawing area for $\sigma(\mu_2)$ with a convex polygon P_{μ_2}, such that it does not cross any edges already drawn in D_{μ_1}. For an example, see Fig. 5.16b.

Next, draw $\sigma(\mu_2)$ inside the polygon P_{μ_2}, and re-insert crossing edges in the drawing D_{μ_2}. Repeat this process until we process $\sigma(\mu_k)$. Similarly, process $\mu_{k+1}, \mu_{k+2}, \ldots, \mu_m$ symmetrically, starting from μ_m and working toward μ_{k+1}. See Fig. 5.16c for an example.

When replacing each virtual edge in the convex drawing D_v of $\sigma(v)$ with a drawing of $\sigma(\mu)$, where μ is a child node of v, we can define a convex polygon P_μ for the boundary of $\sigma(\mu)$ thin enough not to create any new crossings.

The following lemma summarizes the results of the drawing step.

Lemma 5.4 *Let G^+ be a red-maximal 1-plane graph without the B subgraph or the W subgraph. Then there is a linear time algorithm to construct a straight-line 1-planar drawing of G^+.*

Exponential Area

It was also shown that some 1-plane graphs require exponential area for any straight-line grid 1-planar drawing. More specifically, for all $k > 1$, there is a 1-plane graph

Fig. 5.17 A 1-plane graph
G_k for which every
straight-line grid 1-planar
drawing has exponential
area. Here, the case $k = 6$ is
shown

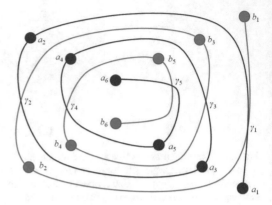

G_k with $2k$ vertices and $2k - 2$ edges such that any straight-line grid (i.e., each vertex has integer coordinates) 1-planar drawing of G_k requires at least 2^{k-1} area. See Fig. 5.17 for an example, where $k = 6$.

5.5 Recent Progress

This chapter reviews the algorithmic results on 1-planar graphs. More specifically, we review three linear time algorithms for testing *maximal* 1-planar graphs with a given rotation system, testing *outer* 1-planar graphs, and drawing 1-plane graphs with straight-line edges.

We now briefly review recent results on related topics, mainly focusing on the algorithmic aspects.

- *Testing full-outer-2-planarity:* A graph is *fully-outer-2-planar* if it admits an outer-2-planar embedding (i.e., each vertex is placed on the outer boundary and no edge has more than two crossings) such that no crossing appears along the outer boundary.

  Hong and Nagamochi [19] showed that triconnected full-outer-2-planar graphs have a constant number of full-outer-2-planar embeddings. Based on these properties, linear time algorithms for testing full-outer-2-planarity of a connected, biconnected, and triconnected graph were presented. The algorithms also produce a full-outer-2-planar embedding of a graph, if it exists.
- *Re-embedding 1-plane graph:* Re-embedding of a 1-plane graph is to change a given 1-planar embedding with B or W subgraph to a new 1-planar embedding without the B subgraph or the W subgraph, by changing the rotation system or the outer face of the given 1-planar embedding, while preserving the same set of pairs of crossing edges.

  Hong and Nagamochi [18] presented a characterization of forbidden configuration (i.e., a given 1-plane graph can be re-embedded into a straight-line drawable 1-

planar embedding if and only if it does not contain the configuration). Based on the characterization, a linear time algorithm for finding a straight-line drawable 1-planar embedding or the forbidden configuration was presented.

- *Straight-line drawings of almost planar graphs:* The almost planar graph consists of a planar graph plus one edge, also called graphs with *1-skewness* (i.e., removal of an edge makes the graph planar).

  Eades et al. [11] presented a characterization of almost planar topological graphs that admit a straight-line drawing. Based on the characterization, linear time algorithms for testing whether an almost planar graph admits a straight-line drawing, and for constructing such a drawing if it exists, were presented. It was also shown that some almost planar graphs require an exponential area for any straight-line grid drawing.

- *Straight-line drawings of general embedded non-planar graphs:* Nagamochi [23] investigated the stretchability problem of a general embedded topological graph. He showed that there is a 3-planar embedding and quasi-planar embedding that admits no straight-line drawing, which cannot be characterized by forbidden configuration.

  He also considered a problem of whether a given embedded graph G admits a straight-line drawing under the same *frame*, which is defined by a fixed biconnected planar spanning subgraph of G. He presented forbidden configurations (i.e., a given embedding admits a straight-line drawing under the same frame if and only if it contains no forbidden configuration) for the problem.

  It was shown that if a given embedding is *quasi-planar* (i.e., no pairwise crossing edges) and its crossing-free edges induce a biconnected spanning subgraph, then the stretchability can be tested using forbidden configurations in polynomial time.

For the last decades, 1-planar graphs have been extensively studied and consequently many combinatorial and algorithmic questions are already solved. Many combinatorial results are also available for k-planar graphs, including structural properties, geometric representations, as well as the relationships between various beyond planar graphs. For details, we refer to corresponding chapters in this book.

However, many fundamental algorithmic questions on the other classes of beyond planar graphs are remained to be solved and deserve further investigation. For details, we refer to open problems listed in corresponding chapters in this book.

Acknowledgements This work is supported by ARC (Australian Research Council) Discovery Project grant.

References

1. Auer, C., Bachmaier, C., Brandenburg, F.J., Gleißner, A., Hanauer, K., Neuwirth, D., Reislhuber, J.: Outer 1-planar graphs. Algorithmica **74**(4), 1293–1320 (2016). https://doi.org/10.1007/s00453-015-0002-1

2. Auer, C., Brandenburg, F.J., Gleißner, A., Reislhuber, J.: 1-planarity of graphs with a rotation system. J. Graph Algorithms Appl. **19**(1), 67–86 (2015). https://doi.org/10.7155/jgaa.00347
3. Bannister, M.J., Cabello, S., Eppstein, D.: Parameterized complexity of 1-planarity. J. Graph Algorithms Appl. **22**(1), 23–49 (2018). https://doi.org/10.7155/jgaa.00457
4. Battista, G.D., Eades, P., Tamassia, R., Tollis, I.G.: Graph Drawing: Algorithms for the Visualization of Graphs. Prentice-Hall (1999)
5. Battista, G.D., Tamassia, R.: On-line maintenance of triconnected components with spqr-trees. Algorithmica **15**(4), 302–318 (1996)
6. Borodin, O.V.: Solution of the Ringel problem on vertex-face coloring of planar graphs and coloring of 1-planar graphs. Metody Diskret. Analiz. **41**, 12–26, 108 (1984)
7. Borodin, O.V., Kostochka, A.V., Raspaud, A., Sopena, E.: Acyclic colouring of 1-planar graphs. Discret. Appl. Math. **114**(1–3), 29–41 (2001)
8. Cabello, S., Mohar, B.: Adding one edge to planar graphs makes crossing number and 1-planarity hard. SIAM J. Comput. **42**(5), 1803–1829 (2013). https://doi.org/10.1137/120872310
9. Chiba, N., Yamanouchi, T., Nishizeki, T.: Linear Time Algorithms for Convex Drawings of Planar Graphs. Progress in Graph Theory, pp. 153–173 (1984)
10. Eades, P., Hong, S., Katoh, N., Liotta, G., Schweitzer, P., Suzuki, Y.: A linear time algorithm for testing maximal 1-planarity of graphs with a rotation system. Theor. Comput. Sci. **513**, 65–76 (2013). https://doi.org/10.1016/j.tcs.2013.09.029
11. Eades, P., Hong, S., Liotta, G., Katoh, N., Poon, S.: Straight-line drawability of a planar graph plus an edge. In: Dehne, F., Sack, J., Stege, U. (eds.) Algorithms and Data Structures - 14th International Symposium, WADS 2015, Victoria, BC, Canada, August 5–7, 2015. Proceedings, Lecture Notes in Computer Science, vol. 9214, pp. 301–313. Springer (2015). https://doi.org/10.1007/978-3-319-21840-3_25
12. Fabrici, I., Madaras, T.: The structure of 1-planar graphs. Discret. Math. **307**(7–8), 854–865 (2007)
13. Fáry, I.: On straight line representations of planar graphs. Acta Sci. Math. Szeged **11**, 229–233 (1948)
14. de Fraysseix, H., Pach, J., Pollack, R.: How to draw a planar graph on a grid. Combinatorica **10**(1), 41–51 (1990). https://doi.org/10.1007/BF02122694
15. Grigoriev, A., Bodlaender, H.L.: Algorithms for graphs embeddable with few crossings per edge. Algorithmica **49**(1), 1–11 (2007). https://doi.org/10.1007/s00453-007-0010-x
16. Hong, S., Eades, P., Katoh, N., Liotta, G., Schweitzer, P., Suzuki, Y.: A linear-time algorithm for testing outer-1-planarity. Algorithmica **72**(4), 1033–1054 (2015). https://doi.org/10.1007/s00453-014-9890-8
17. Hong, S., Nagamochi, H.: An algorithm for constructing star-shaped drawings of plane graphs. Comput. Geom. **43**(2), 191–206 (2010). https://doi.org/10.1016/j.comgeo.2009.06.008
18. Hong, S., Nagamochi, H.: Re-embedding a 1-plane graph into a straight-line drawing in linear time. In: Hu, Y., Nöllenburg, M., (eds.) Graph Drawing and Network Visualization - 24th International Symposium, GD 2016, Athens, Greece, September 19–21, 2016, Revised Selected Papers, Lecture Notes in Computer Science, vol. 9801, pp. 321–334. Springer (2016). https://doi.org/10.1007/978-3-319-50106-2_25
19. Hong, S., Nagamochi, H.: A linear-time algorithm for testing full outer-2-planarity. Discret. Appl. Math. **255**, 234–257 (2019). https://doi.org/10.1016/j.dam.2018.08.018
20. Hong, S.H., Eades, P., Liotta, G., Poon, S.H.: Fáry's theorem for 1-planar graphs. In: Gudmundsson, J., Mestre, J., Viglas, T., (eds.) Proceedings of COCOON 2012, Lecture Notes in Computer Science, vol. 7434, pp. 335–346. Springer (2012)
21. Hudák, D., Madaras, T., Suzuki, Y.: On properties of maximal 1-planar graphs. Discuss. Math. Graph Theory **32**(4), 737–747 (2012). https://doi.org/10.7151/dmgt.1639
22. Korzhik, V.P., Mohar, B.: Minimal obstructions for 1-immersions and hardness of 1-planarity testing. J. Graph Theory **72**(1), 30–71 (2013)
23. Nagamochi, H.: Straight-line drawability of embedded graphs, Technical Report 2013–005. Kyoto University, Japan, Department of Applied Mathematics and Physics (2013)

24. Nishizeki, T., Rahman, M.S.: Planar Graph Drawing. Lecture Notes Series on Computing, vol. 12. World Scientific (2004). https://doi.org/10.1142/5648
25. Pach, J., Tóth, G.: Graphs drawn with few crossings per edge. Combinatorica **17**(3), 427–439 (1997)
26. Ringel, G.: Ein Sechsfarbenproblem auf der Kugel. Abh. Math. Sem. Univ. Hamburg **29**, 107–117 (1965)
27. Suzuki, Y.: Optimal 1-planar graphs which triangulate other surfaces. Discret. Math. **310**(1), 6–11 (2010). https://doi.org/10.1016/j.disc.2009.07.016
28. Suzuki, Y.: Re-embeddings of maximum 1-planar graphs. SIAM J. Discret. Math. **24**(4), 1527–1540 (2010). https://doi.org/10.1137/090746835
29. Thomassen, C.: Rectilinear drawings of graphs. J. Graph Theory **12**(3), 335–341 (1988)
30. Zhang, X., Wu, J.L.: On edge colorings of 1-planar graphs. Inform. Process. Lett. **111**(3), 124–128 (2011). http://dx.doi.org/10.1016/j.ipl.2010.11.001

Chapter 6
Edge Partitions and Visibility Representations of 1-planar Graphs

Giuseppe Liotta and Fabrizio Montecchiani

Abstract This chapter discusses the relationship between edge partitions and visibility representations of 1-planar graphs. Partitioning the edge set of a graph such that each partition set induces a simpler subgraph is a fundamental problem in graph theory, with applications in graph algorithms and graph drawing. For example, it is known that the edge set of every planar graph can be partitioned into two outerplanar graphs. A visibility representation of a graph is a classic drawing paradigm; it maps the vertices of the graph to geometric objects and the edges of the graph to lines of sight between pairs of objects. A classic result shows that every planar graph can be represented as a visibility representation such that the vertices are horizontal bars and the edges are vertical lines of sight between pairs of bars. While both edge partitions and visibility representations have been extensively studied for planar graphs, they recently attracted attention also for 1-planar graphs, i.e., those graphs that can be drawn in the plane such that each edge is crossed at most once. After giving an overview of 1-planarity, we survey the main results concerning edge partitions and visibility representations of 1-planar graphs, and we highlight an interesting interplay between them. In particular, we show how an edge partition of a 1-planar graph G into two planar subgraphs such that one of them has small vertex degree can be used to construct a visibility representation of G in which vertices are orthogonal polygons with few reflex corners each. Finally, we conclude this chapter with a selection of open problems related to the covered topics.

6.1 Introduction

A graph is 1-*planar* if it admits a drawing in the plane such that each edge is crossed at most once. The family of 1-planar graphs represents a natural extension of planar graphs, and it is arguably the most investigated family of beyond-planar graphs, as

G. Liotta · F. Montecchiani (✉)
Dipartimento di Ingegneria, Università degli Studi di Perugia, Via G. Duranti 93,
06125 Perugia, Italy
e-mail: fabrizio.montecchiani@unipg.it

G. Liotta
e-mail: giuseppe.liotta@unipg.it

© Springer Nature Singapore Pte Ltd. 2020
S.-H. Hong and T. Tokuyama (eds.), *Beyond Planar Graphs*,
https://doi.org/10.1007/978-981-15-6533-5_6

Fig. 6.1 **a** An edge partition
of a maximal planar graph G
into three trees (solid,
dashed, and dashed-dotted
edges). **b** A bar visibility
representation of G

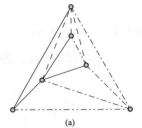

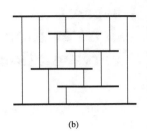

(a)   (b)

witnessed by the annotated bibliography of Kobourov et al. [44]; see also [27]. In particular, 1-planar graphs have been studied in terms of structural properties and in terms of their representations. This chapter, besides giving an overview of 1-planarity, focuses on both the aforementioned aspects, as we survey recent results concerning edge partitions of 1-planar graphs and their interplay with visibility representations. We start by giving a short overview of edge partitions and visibility representations.

Edge partitions. Edge partitions received considerable attention in the literature, especially in the context of planar graphs. In a seminal paper, Colbourn and Elmallah [32] proved that the edge set of every planar graph can be partitioned into two partial 3-trees. Kedlaya [43] and Ding et al. [28] improved this result, as they independently showed how to partition the edges of a planar graph into two partial 2-trees. A further improvement was presented by Gonçalves [37], who proved that the edge set of every planar graph can be partitioned into two outerplanar graphs. This result solves, in a special case, a conjecture by Chartrand, Geller, and Hedetniemi [19]. Furthermore, Schnyder [55] proved that the edges of a maximal planar graph can be partitioned into three trees; see, for example, Fig. 6.1a that shows an edge partition into three trees of a planar triangulation.

Concerning general graphs, various graph parameters are based on the concept of edge partitions. Two examples are the *arboricity* and the *thickness* of a graph, which are defined as the minimum number of forests and of planar graphs, respectively, needed to cover all edges of the graph; see Fig. 6.2a for an example of a graph having thickness 2.

Fig. 6.2 **a** An edge partition
of a graph G with thickness
two into two planar graphs
(solid, dashed). **b** A rectangle
visibility representation of G

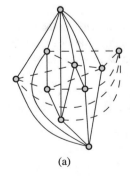

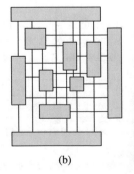

(a)   (b)

Fig. 6.3 The complete graph K_5 represented as **a** an ortho-polygon visibility representation and **b** a bar 1-visibility representation (the bar traversed by a visibility is locally drawn dotted around the intersection)

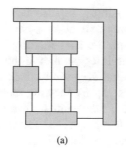 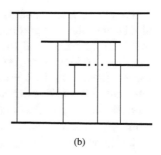

(a)   (b)

Visibility representations. At a high level of generality, a visibility representation of a graph G assigns to each vertex of G a distinct geometric object (such as a bar or a polygon), and to each edge of G an unobstructed line of sight between the pair of objects that correspond to its endpoints. A classical visibility model studied in the literature is the *bar visibility representation*. According to this model, each vertex v of a graph G is mapped to a distinct horizontal segment $b(v)$ (called *bar*) and each edge (u, v) of G corresponds to a vertical unobstructed segment (called *visibility*) having one endpoint on $b(u)$ and the other one on $b(v)$; see, for example, Fig. 6.1b. It is well known that every planar graph can be realized as a bar visibility representation [29, 52, 54, 59–61] in the so-called *weak model*, in which the existence of a visibility between a pair of bars does not necessarily imply the existence of an edge in the graph between the two corresponding vertices.

In order to realize nonplanar graphs, more complex visibility models have been proposed. Two notable examples are rectangle visibility representations and bar k-visibility representations. In a *rectangle visibility representation* of a graph, every vertex is represented as an axis-aligned rectangle and two vertices are connected by an edge using either a horizontal or a vertical visibility [56]; see, for example, Fig. 6.2b. Note that if a graph admits a rectangle visibility representation, then it has thickness at most two, because horizontal (vertical) visibilities do not cross each other. Moreover, if a horizontal and a vertical visibility cross each other, then the crossing occurs at right angles, which is beneficial in terms of readability (see, e.g., [40]). In general, testing whether a graph admits a rectangle visibility representation is an NP-hard problem [56]. Furthermore, as proved by Hutchinson et al. [41], an n-vertex graph G that admits a rectangle visibility representation has at most $6n - 20$ edges, which is tight for each $n \geq 8$.

Ortho-polygon visibility representations generalize rectangle visibility representations by mapping each vertex of the graph to a disjoint orthogonal polygon and each edge to a vertical or horizontal visibility between its end-vertices [24]; see, for example, Fig. 6.3a. A *bar k-visibility representation* is a bar visibility representation in which the lines of sight can intersect the bars; more precisely, a visibility can intersect at most k horizontal bars representing the vertices [23]; see, for example, Fig. 6.3b. An n-vertex graph that admits a bar k-visibility representation has thickness $O(k^2)$ [23] and $O(kn)$ edges [23, 39].

Finally, different models of visibility representations in three dimensions have also been studied. Of particular interest to us are *z-parallel visibility representations*, in which the vertices of the graph are isothetic disjoint rectangles parallel to the xy-plane, and the edges are visibilities parallel to the z-axis. With respect to this model, Bose et al. [14] proved that K_{22} admits such a representation, while K_{56} does not. Štola [57] reduced this gap by showing that K_{51} does not admit any z-parallel visibility representation. If the rectangles can be just unit squares, then K_7 is the largest representable complete graph [35]. Other 3D visibility models are box visibility representations [36], and 2.5D box visibility representations [4].

Structure of this chapter. The remainder of this chapter is organized as follows.

- Section 6.2 contains preliminary definitions and notation that will be used throughout this chapter.
- Section 6.3 gives a brief overview of the complexity of recognizing 1-planar graphs and about structural properties for this family of graphs.
- Section 6.4 surveys recent results on edge partitions of 1-planar graphs. In particular, it offers a glimpse of a technique used to prove one of the presented results that is related to visibility representations.
- Section 6.5 first presents results concerning visibility representations of 1-planar graphs obtained through techniques that make use of edge partitions. Then, it showcases further results on visibility representations of 1-planar graphs whose proofs do not exploit edge partitions.
- Section 6.6 concludes this chapter with a list of open problems related to the research topics addressed in the previous sections.

6.2   Basic Definitions and Notation

A *drawing* Γ of a graph G maps each vertex of G to a distinct point of the plane and each edge of G to a simple open Jordan curve between its endpoints. The curves representing the edges are allowed to cross each other, but they may not intersect vertices except for their endpoints. The curves of any two edges share at most one point (either an endpoint or a crossing point). A drawing is *planar* if no two edges cross and a graph is *planar* if it admits a planar drawing. A drawing divides the plane into topologically connected regions, called *faces*. The infinite region is called the *outer face*. For a planar drawing, the boundary of a face consists of vertices and edges, while for a nonplanar drawing the boundary of a face may contain vertices, crossings, and edges (or parts of edges). An *embedding* of a graph G is an equivalence class of drawings of G that define the same set of faces and the same outer face. A *planar embedding* is an embedding that represents an equivalence class of planar drawings. A *plane graph* is a graph with a fixed planar embedding. A *k-planar drawing* is one in which each edge is crossed at most k times, for some integer $k \geq 1$. A graph is *k-planar* if it admits a k-planar drawing. A *k-planar embedding* is an embedding

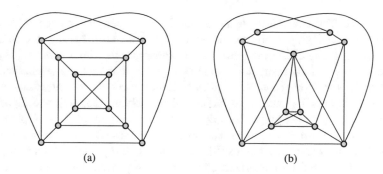

(a)   (b)

Fig. 6.4 a An IC-planar graph and **b** a NIC-planar graph

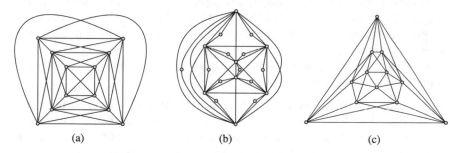

(a)   (b)   (c)

Fig. 6.5 a An optimal 1-planar graph with 12 vertices and 40 edges. **b** A maximal 1-planar graph
with 20 vertices and 44 edges. **c** A maximal NIC-planar graph with 12 vertices and 36 edges

that represents an equivalence class of k-planar drawings, and a *k-plane graph* is a
graph with a fixed k-planar embedding.

A 1-planar graph G with n vertices has at most $4n - 8$ edges [12, 53]. If G has
exactly $4n - 8$ edges, then it is called *optimal*; see Fig. 6.5a for an example. A 1-
planar graph is *IC-planar*, where IC stands for *independent crossings*, if it admits
a 1-planar drawing such that no two crossed edges share an end-vertex; see, for
example, Fig. 6.4a. A 1-planar graph is *NIC-planar*, where NIC stands for *near-
independent crossings*, if it admits a 1-planar drawing such that any two distinct
pairs of crossing edges have at most one end-vertex in common; see, for example,
Fig. 6.4b.

6.3   Recognition and Structural Properties of 1-planar
Graphs

In this section, we survey some basic properties and results about 1-planar graphs
that can be useful for the reader. We point to the annotated bibliography of Kobourov
et al. [44] for a more comprehensive discussion.

Recognition. The class of 1-planar graphs is not closed under edge-contraction, and thus 1-planar graphs are not minor closed [20]. Also, for $n \geq 63$, there are exponentially many distinct minimal non-1-planar graphs with n vertices [45] (i.e., there are exponentially many non-isomorphic graphs that are not 1-planar and such that all their proper subgraphs are 1-planar). Hence, it is not surprising that deciding whether a graph is 1-planar is an NP-complete problem [38, 45], and algorithms that work with 1-planar graphs usually assume the input graph to be 1-plane (i.e., with a given 1-planar embedding). The problem remains NP-complete even for IC-planar [17] and NIC-planar graphs [5]. Bannister et al. [6] studied 1-planarity from the point of view of parameterized complexity and proved that testing whether a graph is 1-planar and finding a corresponding 1-planar drawing is fixed-parameter tractable with respect to vertex cover, tree-depth, and cyclomatic number, while it remains NP-complete for graphs of bounded bandwidth, pathwidth, or treewidth. A 1-planarity testing and embedding algorithm based on a backtracking strategy is described in [11]. On the positive side, recognizing optimal 1-planar graphs can be done in linear time as proved by Brandenburg [16], who exploited a characterization for this family of graphs by Suzuki [58].

Edge density. As already mentioned, a 1-planar graph with n vertices has at most $4n - 8$ edges, which is tight [12, 53]. For example, Fig. 6.5a shows an optimal 1-planar graph with 12 vertices. Every optimal 1-planar graph can be obtained starting from a suitable 3-connected planar quadrangulation and inserting a pair of crossing edges inside each face of this planar graph. If we restrict to those 1-planar graphs that admit a 1-planar straight-line drawing, then we obtain a tight bound of $4n - 9$ edges, as shown by Didimo [26]. In contrast with planar graphs, there exist maximal 1-planar graphs, i.e., 1-planar graphs such that no edge can be added without loosing 1-planarity, with n vertices and $\frac{45}{17}n - \frac{84}{17} < 2.65n$ edges [18]; see, e.g., Fig. 6.5b. A bipartite 1-planar graph with n vertices has at most $3n - 8$ edges if n is even and $n \neq 6$, and at most $3n - 9$ edges otherwise [42] (both bounds are tight). An IC-planar graph with n vertices has at most $3.25n - 6$ edges and this bound is tight [62]. A NIC-planar graph with n vertices has at most $3.6n - 7.2$ edges and this bound is also tight [5, 22]; see, e.g., Fig. 6.5c. Table 6.1 summarizes some of these bounds.

Further structural properties. The chromatic number of a 1-planar graph is at most six [13], which is a tight bound. For example, the complete graph K_6 is 1-planar and

Table 6.1 Density of 1-planar graphs

GRAPH FAMILY	MAX. NUM. EDGES	REFERENCES
1-planar	$4n - 8$	[12, 53]
Straight-line 1-planar	$4n - 9$	[1, 26]
Bipartite 1-planar	$3n - 8$ for even $n \neq 6$, $3n - 9$ otherwise	[42]
IC-planar	$3.25n - 6$	[62]
NIC-planar	$3.6n - 7.2$	[22]

requires six colors in any vertex coloring. For IC-planar graphs, a tight bound of five is known for their chromatic number [46]. An n-vertex 1-planar graph has $O(1)$ book thickness [2, 7], $O(1)$ queue number [9, 31], $O(\sqrt{n})$ pathwidth and treewidth [30], and $O(1)$ expansion [51].

6.4   Edge Partitions of 1-plane Graphs

An *edge partition* of a 1-plane graph G is an edge coloring of G with two colors, say *red* and *blue*, such that both the graph formed by the red edges, called the *red graph*, and the graph formed by the blue edges, called the *blue graph*, are plane. In the figures that follow, the red edges are conveniently represented with dashed curves, while the blue edges are solid.

We first give an overview of results concerning edge partitions of 1-planar graphs, and then we provide a sketch of proof for one of these results, which is then used in the construction of visibility representations of 1-planar graphs.

6.4.1   *Summary of Results*

An edge partition of a 1-plane graph can be constructed by coloring as red an edge for each pair of crossing edges, and by coloring as blue the remaining edges. Depending on the strategy employed to color the edges, one can derive different properties for the red graph.

Edge partitions with acyclic red graph. Czap and Hudák [21] proved that every optimal 1-plane graph admits an edge partition such that the red graph is a forest. This result has been later extended to all 1-plane graphs by Ackerman [1], and given the result by Schnyder [55] mentioned earlier, it follows that the edge set of a 1-plane graph can be partitioned into a set of at most four forests.

Theorem 1 ([1]) *Every 1-plane graph admits an edge partition such that the red graph is a forest.*

Figure 6.6a shows an edge partition of the optimal 1-plane graph of Fig. 6.5a such that the red graph is a forest.

More recently, Lenhart et al. [47] investigated edge partitions of optimal 1-plane graphs such that the maximum vertex degree of the red graph is bounded by a constant. They observed that if G is an n-vertex optimal 1-plane graph with an edge partition whose red graph G_R is a forest, then G_R has n vertices (i.e., it is a spanning subgraph of G) and it is composed of exactly two trees. Based on this finding, they proved that for any given integer $c > 0$, there exists an optimal 1-plane graph such that in any edge partition with the red graph being a forest, the maximum vertex degree of the red graph is at least c, (i.e., the red graph has unbounded vertex degree).

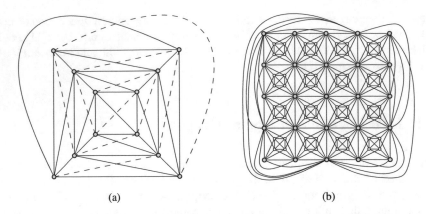

Fig. 6.6 **a** An edge partition of an optimal 1-plane graph such that the red graph (the dashed edges) is a forest of paths. **b** An optimal 1-plane such that the red graph of any of its edge partitions has a maximum vertex degree at least four

Edge partitions with small vertex degree. If we drop the acyclicity requirement on the red graph, then every optimal 1-plane graph admits an edge partition such that the red graph has maximum vertex degree four, and degree four is sometimes necessary [47]. Note that the graph in Fig. 6.6a is such that the vertex degree of the red graph is at most two, while Fig. 6.6b shows an optimal 1-plane graph such that in any edge partition the red graph has maximum vertex degree at least four [47]. This is because for each 4-cycle of gray (bigger) vertices in the graph of Fig. 6.6b, there are at least four red edges incident to some of its vertices. Let n be the number of gray vertices, since there are $n - 2$ 4-cycles, we have that on average each gray vertex is adjacent to more than 3 red edges.

Edge partitions of nonoptimal 1-plane graphs with small vertex degree have been studied by Di Giacomo et al. [25]. They observed a connection between the connectivity of the graph and the existence of edge partitions with a small vertex degree. In particular, they proved that every 3-connected 1-plane graph admits an edge partition such that the red graph has maximum vertex degree six, and degree six is sometimes needed. The next theorem summarizes results in [25, 47] (note that optimal 1-planar graphs are always 3-connected).

Theorem 2 ([25, 47]) *Every* 3-*connected* 1-*plane graph G admits an edge partition such that the red graph has maximum vertex degree at most four if G is optimal, and at most six otherwise. Both bounds on the maximum vertex degree of the red graph are worst-case optimal within the respective families of optimal* 1-*plane and* 3-*connected* 1-*plane graphs.*

On the other hand, for every $n > 0$ there exists an $O(n)$-vertex 2-connected 1-plane graph such that in any edge partition the red graph has maximum vertex degree $\Omega(n)$ [25]. More recently, Di Giacomo et al. [24] proved that every NIC-plane graph admits an edge partition such that the red graph has maximum vertex degree three,

Table 6.2 Edge partitions of 1-planar graphs such that the red graph is acyclic or has bounded vertex degree

GRAPH FAMILY	RED GRAPH	REFERENCES
1-planar	Forest	[1]
Optimal 1-planar	Planar with max vertex degree 4	[47]
3-connected 1-planar	Planar with max degree 6	[25]
NIC-planar	Planar with max degree 3	[24]

Table 6.3 Complexity of testing for edge partitions of 1-planar graphs

GRAPH FAMILY	RED GRAPH	COMPLEXITY	REFERENCES
1-planar	Planar with max degree 2	NP-complete	[24]
1-planar	Planar with max degree 1	$O(n^2)$	[24]

and that this bound on the vertex degree is worst-case optimal. Furthermore, deciding whether a 1-plane graph admits an edge partition such that the red graph has maximum vertex degree two is NP-complete. On the positive side, deciding whether an n-vertex 1-plane graph admits an edge partition such that the red graph has maximum vertex degree one can be done in $O(n^2)$ time. It is unknown whether the complexity of this last problem can be reduced to linear time.

Tables 6.2 and 6.3 summarize some of the aforementioned results.

6.4.2 A Glimpse of a Proof Technique

We now present a sketch of the proof given by Di Giacomo et al. [24] to show that every 3-connected 1-plane graph admits an edge partition such that the red graph has maximum vertex degree six (Theorem 2). This result will be used in Section 6.5 to construct visibility representations of 3-connected 1-plane graphs.

Let G be a 3-connected 1-plane graph. Note that if two edges e_1 and e_2 of G cross one another, then one can always add the missing edges (if any) so to ensure that the four end-vertices of e_1 and e_2 induce a complete graph and the graph remains 1-plane. The four edges of this complete graph distinct from e_1 and e_2 are called *cycle edges*; see, e.g., Fig. 6.7. The proof assumes the existence of four cycle edges for each pair of crossing edges of G, and it relies on two main properties of these edges:

a. Although a cycle edge can be crossed in G, no two cycle edges cross one another.
b. Every edge of G is the cycle edge of at most two pairs of crossing edges.

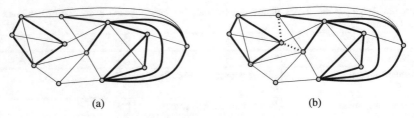

Fig. 6.7 **a** A 3-connected 1-plane graph G; the cycle edges are bold. **b** The graph obtained from G by adding the missing cycle edges (dotted)

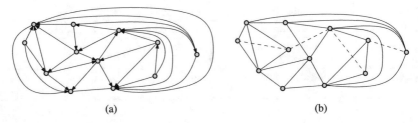

Fig. 6.8 **a** Orienting the edges of G_p^+. **b** An edge partition of the graph in Fig. 6.7a such that the red graph (the dashed edges) has maximum vertex degree no greater than six (two in this example)

Let G_p be the plane graph obtained from G by removing an edge for each pair of crossing edges. The edges to be removed can be chosen arbitrarily with the only rule that they cannot be cycle edges: This choice is always feasible by (a). Let G_p^+ be a plane graph obtained by augmenting G_p so as to become a maximal plane graph. Schnyder [55] proved that the internal edges of G_p^+ can be oriented such that each internal vertex has exactly three outgoing edges and the vertices of the outer face have no outgoing edge; see, e.g., Fig. 6.8a. Consider such an orientation of the edges of G_p^+ and arbitrarily orient the edges of the outer face. By construction, all edges of G_p^+ are now oriented, and every vertex of G_p^+ has at most three outgoing edges. Next, for any two crossing edge (u, v) and (w, z) of G, one can show that both $\{u, v\}$ or both $\{w, z\}$ have an outgoing edge in G_p^+ that is a cycle edge of (u, v) and (w, z). This fact can be used to partition the edge set of G as follows. For each pair of crossing edges (u, v) and (w, z) of G, color red the edge connecting the pair $\{u, v\}$ or $\{w, z\}$ for which this fact holds. By this choice, each end-vertex of a red edge has one outgoing edge among the cycle edges of (u, v) and (w, z). Since every vertex is incident to at most three outgoing edges in G_p^+, and since each edge is the cycle edge of at most two pairs of crossing edges by (b), this procedure assigns the red color to at most six edges for each vertex, as desired; see, e.g., Fig. 6.8b.

6.5   Visibility Representations of 1-plane Graphs

In this section, we discuss results concerning visibility representations of 1-plane graphs. We begin with results about rectangle and ortho-polygon visibility representations. In particular, one of these results is obtained through a technique that makes use of a suitable edge partition, and we report a sketch of its proof. We conclude with a collection of results concerning other types of visibility representations whose proofs do not take advantage of edge partitions.

6.5.1   Rectangle and Ortho-Polygon Visibility Representations

We begin with a definition that will be useful in the following. Given a rectangle visibility representation, we can extract a drawing from it as follows. For each vertex v, place a point inside the corresponding rectangle $r(v)$ and connect it to all the attachment points of the visibilities on the boundary of $r(v)$; this can be done without creating any crossing and preserving the circular order of the edges around the vertices. An embedded graph G has an *embedding-preserving* rectangle visibility representation Γ if the drawing extracted from Γ preserves the embedding of G; see, for example, Fig. 6.9.

One of the first attempts to represent 1-plane graphs as visibility representations is due to Biedl et al. [10], who characterized the 1-plane graphs that admit an embedding-preserving rectangle visibility representation. In particular, they proved the following result.

Theorem 3 ([10]) *Let G be an n-vertex 1-plane graph. Graph G admits an embedding-preserving rectangle visibility representation if and only if it contains no B-configuration, no W-configuration, and no T-configuration as a subgraph (see*

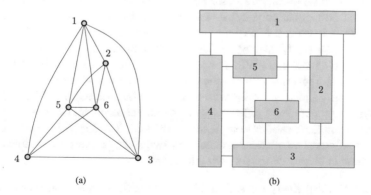

Fig. 6.9 a A 1-plane graph G and **b** an embedding-preserving rectangle visibility representation of G

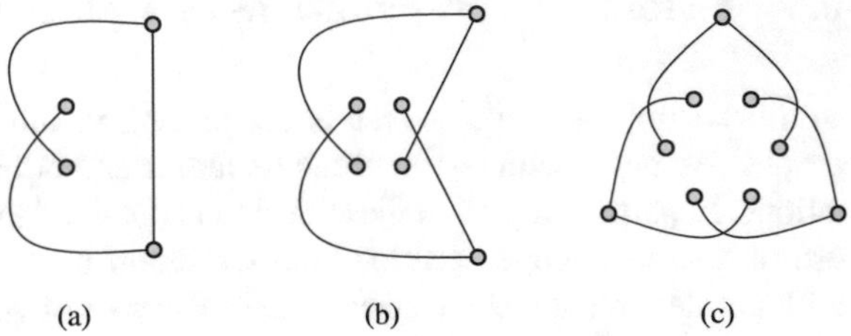

Fig. 6.10 a A
B-configuration. b A
W-configuration. c A
T-configuration

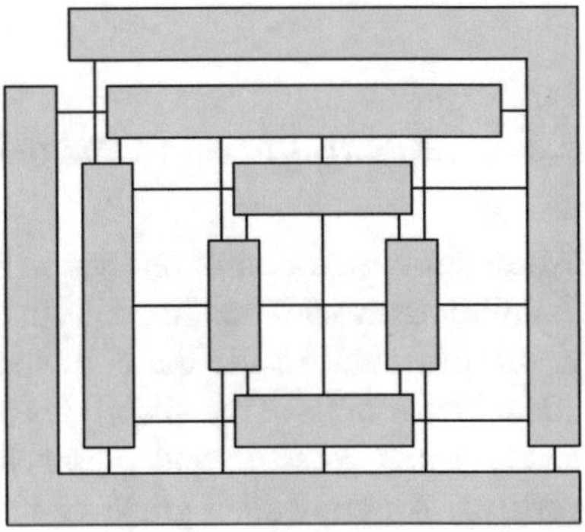

(a)   (b)   (c)

Fig. 6.11 An
embedding-preserving
OPVR with vertex
complexity one

Fig. 6.10). Also, there is an O(n)-time algorithm that tests whether G admits an embedding-preserving rectangle visibility representation, and, in the positive case, it computes one.

Biedl et al. also exhibited infinitely many 1-planar graphs such that none of their 1-planar embeddings can be realized as an embedding-preserving rectangle visibility representation; Fig. 6.6a shows one of these graphs.

Motivated by this negative result, Di Giacomo et al. [25] introduced ortho-polygon visibility representations (*OPVRs*). We recall that an OPVR of an embedded graph G is an embedding-preserving visibility representation of G that maps each vertex to a disjoint orthogonal polygon and each edge to a vertical or horizontal visibility between its end-vertices. (The definition of *embedding-preserving* visibility representation can be easily adapted to OPVRs.) In particular, they proved that every 1-plane graph admits an OPVR. For example, Fig. 6.11 is an OPVR of the graph in Fig. 6.6a. Moreover, in order to measure the visual complexity of an OPVR, Di Giacomo et al. defined an OPVR with *vertex complexity k*, as an OPVR such that k is the maximum number of reflex corners over all vertex polygons in the representation. For example, the vertex complexity of the OPVR in Fig. 6.11 is one, because all vertex polygons are rectangles except for two L-shaped vertices. In this respect, Di Giacomo et al. proved the following theorem.

Theorem 4 ([25]) *Every 3-connected 1-plane graph G with n vertices admits an OPVR with vertex complexity at most 12, which can be computed in O(n) time.*

Di Giacomo et al. also proved the existence of 3-connected 1-plane graphs requiring vertex complexity at least two in any (embedding-preserving) OPVR. On the negative side, there exists 2-connected 1-plane graphs that require $\Omega(n)$ vertex complexity [25]. Very recently, Liotta et al. [49] improved both the upper bound on the vertex complexity of Theorem 4 and the above mentioned lower bound, reducing the gap to only one unit. On the one hand, they exhibited 3-connected 1-plane graphs such that any embedding-preserving OPVR of these graphs has vertex complexity at least four. On the other hand, they showed that vertex complexity five is always sufficient for OPVRs of 3-connected 1-plane graphs.

6.5.2   A Glimpse of a Proof Technique

Theorem 4 can be proved by taking advantage of edge partitions with a small vertex degree. The high-level idea of the proof is as follows.

Let G be a 3-connected 1-plane graph with n vertices. We already observed that G admits an edge partition such that both the blue graph G_B and the red graph G_R are plane graphs, and G_R has maximum vertex degree at most six; see, for example, Fig. 6.12a. We first augment G_B to a maximal plane graph (if needed), and then construct a bar visibility representation γ_B; see, for example, Fig. 6.12b. Assume that two vertices u and v are connected by a red edge and let $\gamma_B(u)$ and $\gamma_B(v)$ be the horizontal bars representing vertices u and v in γ_B, respectively. We attach a vertical bar to $\gamma_B(u)$ and a vertical bar to $\gamma_B(v)$ such that each vertical bar shares an endpoint with the horizontal bar and the two vertical bars can see each other horizontally. This makes it possible to draw the horizontal red edge (u, v); see, for example, Fig. 6.12c. Once all red edges have been added to γ_B, every vertex v that has some incident red edge is represented as a "rake"-shaped object consisting of one horizontal bar and at most six vertical bars (we have a vertical bar for each red edge incident to v and there are at most six such edges). This "rake"-shaped object can then be used as the skeleton of an orthogonal polygon that has two reflex corners per vertical bar and hence no more than 12 reflex corners in total; see, for example, Fig. 6.12d.

6.5.3   Further Results on Visibility Representations of 1-plane Graphs

In 2014, Brandenburg [15] and Evans et al. [33] studied bar 1-visibility representations and proved that every 1-planar graph admits this kind of representation (see, e.g., Fig. 6.3b). Both papers are based on constructive linear-time algorithms that take as input a 1-plane graph G and that may change the embedding of G in order to construct the final representation. Hence, it remains unknown whether embedding-preserving bar 1-visibility representations always exist for 1-plane graphs. In what

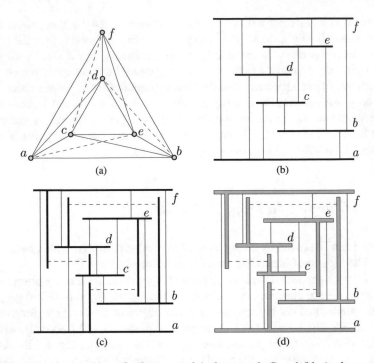

Fig. 6.12 a An edge partition of a 3-connected 1-plane graph G, red (blue) edges are dashed (solid); **b** A bar visibility representation γ_B of G_B; **c** Insertion of the red edges into γ_B; **d** An embedding-preserving OPVR of G

follows, we briefly sketch the algorithm by Brandenburg [15]. Let G be a 1-plane graph. A 1-plane multigraph G' is computed from G by adding edges in such a way that: the four end-vertices of each crossing induce a complete graph; no edge can be added to G' without introducing additional crossings; if two vertices are connected by a set of parallel edges, then all of them are uncrossed and non-homotopic. Consider now the plane multigraph P obtained from G' by removing all pairs of crossing edges, and denote by $\mathcal{O}$ an orientation of P such that $P_{\mathcal{O}}$ is a planar st-multigraph. The algorithm by Tamassia and Tollis [59] is applied to compute a bar visibility representation of $P_{\mathcal{O}}$. Finally, all pairs of crossing edges are reinserted through a post-processing step that extends the length of some bars so to introduce the missing visibilities by traversing at most one bar each (and by also ensuring that each bar is traversed by at most one visibility).

Liotta and Montecchiani [48] studied L-visibility representations, in which every vertex is represented by a horizontal and a vertical segment sharing an endpoint (i.e., by an L-shape in the set $\{\llcorner, \lrcorner, \urcorner, \ulcorner\}$), and each edge is drawn as either a horizontal or a vertical visibility segment joining the two L-shapes corresponding to its two end-vertices; see, e.g., Figure 6.13. They proved that every IC-plane graph G admits an L-visibility representation, which can be computed in linear time. Analogously

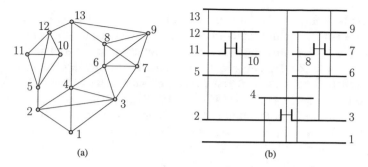

Fig. 6.13 **a** An IC-plane graph G and **b** an L-visibility representation of G

as for the results in [15, 33], the algorithm may change the embedding of G, but the final representation is such that each visibility is crossed at most once and no two crossed visibilities are incident to the same L-shape. See also [34] for other results about L-visibility representations.

More recently, Angelini et al. [3] studied 3D visibility representations of 1-planar graphs. In particular, they studied z-parallel visibility representations (ZPRs). Recall that, in a ZPR of a graph, the vertices are isothetic disjoint rectangles parallel to the xy-plane and the edges are visibilities parallel to the z-axis. Angelini et al. proved that every 1-planar graph has a ZPR γ. In addition, γ is 1-*visible*, i.e., there is a

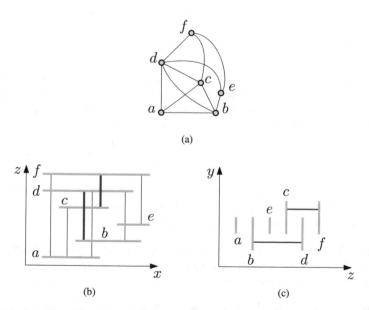

Fig. 6.14 **a** A 1-planar graph G. **b** The intersection of a 1-visible ZPR γ of G with the plane $Y = 0$; the red (bold) visibilities traverse a bar. **c** The projection to the yz-plane of γ (only the red visibilities are shown)

plane that is orthogonal to the rectangles of γ and such that its intersection with γ defines a bar 1-visibility representation of G. An example of a 1-visible ZPR is shown in Fig. 6.14. From a high-level perspective, the proof of Angelini et al. works as follows. Let G be a 1-plane graph. First, a bar 1-visibility representation γ_1 of G is constructed. This representation is used as the intersection of the final ZPR γ with the plane $Y = 0$ (see, e.g., Fig. 6.14a). In particular, each bar b of γ_1 is transformed into a rectangle R_b by computing the y-coordinates of its top and bottom sides, so that each visibility in γ_1 that traverses a bar b can be represented as a visibility in γ that passes above or below R_b (see, e.g., Fig. 6.14b). This is done by using two suitable acyclic orientations of the edges of G.

6.6   Concluding Remarks and Open Problems

We conclude this chapter by mentioning a recent work of Bekos et al. [8], who extended the study of edge partitions to k-planar graphs, with $k > 1$. In particular, Bekos et al. focused on optimal 2-plane and optimal 3-plane graphs, which are 2-plane graphs and 3-plane graphs with maximum density, i.e., with exactly $5n - 10$ and $5.5n - 11$ edges, respectively. They proved that, differently from 1-plane graphs (Theorem 1), it is not possible to partition the edges of a simple optimal 2-plane graph G into a 1-plane graph and a forest, while, on the positive side, it is possible to partition the edge set of G into a 1-plane graph and two plane forests (and it can be done in linear time). Moreover, there exist efficient algorithms to partition the edges of a simple optimal 2-plane graph into a 1-plane graph and a plane graph with maximum vertex degree 12, or with maximum vertex degree 8 if the optimal 2-plane graph is such that its crossing-free edges form a graph with no separating triangles. Besides these two upper bounds, Bekos et al. exhibited infinitely many simple optimal 2-plane graphs such that in any partition of their edges composed of a 1-plane graph and a plane graph, the latter has vertex degree at least 6. Concerning optimal 3-plane graphs, they showed that every such a graph can be decomposed into a 2-plane graph and two plane forests, if its crossing-free edges form a biconnected graph. Finally, they observed that every k-plane graph ($k \geq 2$) can be partitioned into a $(k - 1)$-plane graph and a plane graph.

Several interesting open problems concerning edge partitions and visibility representations of beyond-planar graphs can be studied. We mention here some of those that, in our opinion, are among the most interesting.

Open Problem 1 *Let G be a 2-plane graph. Is it possible to color the edges of G with three colors, such that one color induces a plane graph, and each of the two other colors induces a plane forest?*

Open Problem 2 *Let G be a 2-plane graph. Is it possible to color the edges of G with two colors, such that one color induces a 1-plane graph and the other color induces a plane graph with a small vertex degree?*

Note that a positive answer to the first question would imply that the edge set of any 2-plane graph can be partitioned into a set of five *plane* forests (while the Nash–Williams formula [50] already implies the existence of a decomposition into five forests, because 2-planar graphs have at most $5n - 10$ edges). Thus, a more general question is the following.

Open Problem 3 *Let G be a k-plane graph. Is it possible to color the edges of G with $k + 3$ colors, such that each color induces a plane forest?*

Concerning this last question, we remark that, given a k-plane graph, a decomposition into $3k + 1$ forests can already be derived from the fact that every such a graph can be decomposed into a $(k - 1)$-plane graph and a plane graph [8], and that every 1-plane graph can be decomposed into a plane graph and a plane forest [1].

The study of OPVRs can be extended to k-planar graphs with $k > 1$. Note that an OPVR may not always exist for k-planar graphs with $k > 1$.

Open Problem 4 *Study upper and lower bounds on the vertex complexity of OPVRs of representable k-plane graphs $(k > 1)$.*

References

1. Ackerman, E.: A note on 1-planar graphs. Discret. Appl. Math. **175**, 104–108 (2014)
2. Alam , M.J., Brandenburg, F.J., Kobourov, S.G.: On the book thickness of 1-planar graphs (2015). CoRR, arXiv:1510.05891
3. Angelini, P., Bekos, M.A., Kaufmann, M., Montecchiani, F.: On 3D visibility representations of graphs with few crossings per edge. Theor. Comput. Sci. **784**, 11–20 (2019)
4. Arleo, A., Binucci, C., Di Giacomo, E., Evans, W.S., Grilli, L., Liotta, G., Meijer, H., Montecchiani, F., Whitesides, S., Wismath, S.K.: Visibility representations of boxes in 2.5 dimensions. Comput. Geom. **72**, 19–33 (2018)
5. Bachmaier, C., Brandenburg, F.J., Hanauer, K., Neuwirth, D., Reislhuber, J.: NIC-planar graphs. Discret. Appl. Math. **232**, 23–40 (2017)
6. Bannister, M.J., Cabello, S., Eppstein, D.: Parameterized complexity of 1-planarity. J. Graph Algorithms Appl. **22**(1), 23–49 (2018)
7. Bekos, M.A., Bruckdorfer, T., Kaufmann, M., Raftopoulou, C.N.: The book thickness of 1-planar graphs is constant. Algorithmica **79**(2), 444–465 (2017)
8. Bekos, M.A., Di Giacomo, E., Didimo, W., Liotta, G., Montecchiani, F., Raftopoulou, C.: Edge partitions of optimal 2-plane and 3-plane graphs. Discret. Math. **342**(4), 1038–1047 (2019)
9. Bekos, M.A., Förster, H., Gronemann, M., Mchedlidze, T., Montecchiani, F., Raftopoulou, C.N., Ueckerdt, T.: Planar graphs of bounded degree have bounded queue number. SIAM J. Comput. **48**(5), 1487–1502 (2019)
10. Biedl, T.C., Liotta, G., Montecchiani, F.: Embedding-preserving rectangle visibility representations of nonplanar graphs. Discret. Comput. Geom. **60**(2), 345–380 (2018)
11. Binucci, C., Didimo, W., Montecchiani, F.: An experimental study of a 1-planarity testing and embedding algorithm. In: WALCOM 2020, LNCS, vol. 12049, pp. 329–335. Springer (2020)
12. R. Bodendiek, Schumacher, H., Wagner, K.: Bemerkungen zu einem Sechsfarbenproblem von G. Ringel. Abhandlungen aus dem Mathematischen Seminar der Universitaet Hamburg **53**(1), 41–52 (1983)

13. Borodin, O.V.: Solution of the ringel problem on vertex-face coloring of planar graphs and coloring of 1-planar graphs. Metody Diskret. Analiz **108**, 12–26 (1984)
14. Bose, P., Everett, H., Fekete, S.P., Houle, M.E., Lubiw, A., Meijer, H., Romanik, K., Rote, G., Shermer, T.C., Whitesides, S., Zelle, C.: A visibility representation for graphs in three dimensions. J. Graph Algorithms Appl. **2**(3), 1–16 (1998)
15. Brandenburg, F.J.: 1-visibility representations of 1-planar graphs. J. Graph Algorithms Appl. **18**(3), 421–438 (2014)
16. Brandenburg, F.J.: Recognizing optimal 1-planar graphs in linear time. Algorithmica **80**(1), 1–28 (2018)
17. Brandenburg, F.J., Didimo, W., Evans, W.S., Kindermann, P., Liotta, G., Montecchiani, F.: Recognizing and drawing ic-planar graphs. Theor. Comput. Sci. **636**, 1–16 (2016)
18. Brandenburg, F.-J., Eppstein, D., Gleißner, A., Goodrich, M.T., Hanauer, K., Reislhuber, J.: On the density of maximal 1-planar graphs. In: GD 2012, LNCS, vol. 7704, pp. 327–338. Springer (2013)
19. Chartrand, G., Geller, D., Hedetniemi, S.: Graphs with forbidden subgraphs. J. Comb. Theory Ser. B **10**(1), 12–41 (1971)
20. Chen, Z., Kouno, M.: A linear-time algorithm for 7-coloring 1-plane graphs. Algorithmica **43**(3), 147–177 (2005)
21. Czap, J., Hudák, D.: On drawings and decompositions of 1-planar graphs. Electr. J. Comb. **20**(2), P54 (2013)
22. Czap, J., Šugerek, P.: Drawing graph joins in the plane with restrictions on crossings. Filomat **31**(2), 363–370 (2017)
23. Dean, A.M., Evans, W.S., Gethner, E., Laison, J.D., Safari, M.A., Trotter, W.T.: Bar k-visibility graphs. J. Graph Algorithms Appl. **11**(1), 45–59 (2007)
24. Di Giacomo, E., Didimo, W., Evans, W.S., Liotta, G., Meijer, H., Montecchiani, F., Wismath, S.K.: New results on edge partitions of 1-plane graphs. Theor. Comput. Sci. **713**, 78–84 (2018)
25. Di Giacomo, E., Didimo, W., Evans, W.S., Liotta, G., Meijer, H., Montecchiani, F., Wismath, S.K.: Ortho-polygon visibility representations of embedded graphs. Algorithmica **80**(8), 2345–2383 (2018)
26. Didimo, W.: Density of straight-line 1-planar graph drawings. Inf. Process. Lett. **113**(7), 236–240 (2013)
27. Didimo, W., Liotta, G., Montecchiani, F.: A survey on graph drawing beyond planarity. ACM Comput. Surv. **52**(1), 4:1–4:37 (2019)
28. Ding, G., Oporowski, B., Sanders, D.P., Vertigan, D.: Surfaces, tree-width, clique-minors, and partitions. J. Comb. Theory Ser. B **79**(2), 221–246 (2000)
29. Duchet, P., Hamidoune, Y.O., Vergnas, M.L., Meyniel, H.: Representing a planar graph by vertical lines joining different levels. Discret. Math. **46**(3), 319–321 (1983)
30. Dujmović, V., Eppstein, D., Wood, D.R.: Structure of graphs with locally restricted crossings. SIAM J. Discret. Math. **31**(2), 805–824 (2017)
31. Dujmovic, V., Joret, G., Micek, P., Morin, P., Ueckerdt, T., Wood, D.R.: Planar graphs have bounded queue-number. In: FOCS 2019. IEEE Computer Society, pp. 862–875 (2019)
32. Elmallah, E.S., Colbourn, C.J.: Partitioning the edges of a planar graph into two partial k-trees. Congr. Num. 69–80 (1988)
33. Evans, W.S., Kaufmann, M., Lenhart, W., Mchedlidze, T., Wismath, S.K.: Bar 1-visibility graphs vs. other nearly planar graphs. J. Graph Algorithms Appl. **18**(5), 721–739 (2014)
34. Evans, W.S., Liotta, G., Montecchiani, F.: Simultaneous visibility representations of plane st-graphs using l-shapes. Theor. Comput. Sci. **645**, 100–111 (2016)
35. Fekete, S.P., Houle, M.E., Whitesides, S.: New results on a visibility representation of graphs in 3D. In: Brandenburg, F. (ed.) GD 1995, LNCS, vol. 1027, pp. 234–241. Springer (1995)
36. Fekete, S.P., Meijer, H.: Rectangle and box visibility graphs in 3D. Int. J. Comput. Geometry Appl. **9**(1), 1–28 (1999)
37. Gonçalves, D.: Edge partition of planar graphs into two outerplanar graphs. In: STOC 2005, pp. 504–512. ACM (2005)

38. Grigoriev, A., Bodlaender, H.L.: Algorithms for graphs embeddable with few crossings per edge. Algorithmica **49**(1), 1–11 (2007)
39. Hartke, S.G., Vandenbussche, J., Wenger, P.S.: Further results on bar k-visibility graphs. SIAM J. Discret. Math. **21**(2), 523–531 (2007)
40. Huang, W., Eades, P., Hong, S.: Larger crossing angles make graphs easier to read. J. Vis. Lang. Comput. **25**(4), 452–465 (2014)
41. Hutchinson, J.P., Shermer, T.C., Vince, A.: On representations of some thickness-two graphs. Comput. Geom. **13**(3), 161–171 (1999)
42. Karpov, D.V.: An upper bound on the number of edges in an almost planar bipartite graph. J. Math. Sci. **196**(6), 737–746 (2014)
43. Kedlaya, K.S.: Outerplanar partitions of planar graphs. J. Comb. Theory Ser. B **67**(2), 238–248 (1996)
44. Kobourov, S.G., Liotta, G., Montecchiani, F.: An annotated bibliography on 1-planarity. Comput. Sci. Rev. **25**, 49–67 (2017)
45. Korzhik, V.P., Mohar, B.: Minimal obstructions for 1-immersions and hardness of 1-planarity testing. J. Graph Theory **72**(1), 30–71 (2013)
46. Král', D., Stacho, L.: Coloring plane graphs with independent crossings. J. Graph Theory **64**(3), 184–205 (2010)
47. Lenhart, W.J., Liotta, G., Montecchiani, F.: On partitioning the edges of 1-plane graphs. Theor. Comput. Sci. **662**, 59–65 (2017)
48. Liotta, G., Montecchiani, F.: L-visibility drawings of IC-planar graphs. Inf. Process. Lett. **116**(3), 217–222 (2016)
49. Liotta, G., Montecchiani, F., Tappini, A.: Ortho-polygon visibility representations of 3-connected 1-plane graphs. In: GD 2018, LNCS, vol. 11282, pp. 524–537. Springer (2018)
50. Nash-Williams, C.S.A.: Edge-disjoint spanning trees of finite graphs. J. Lond. Math. Soc. **s1-36**(1), 445–450 (1961)
51. Nešetřil, J., de Mendez, P.O., Wood, D.R.: Characterisations and examples of graph classes with bounded expansion. Eur. J. Comb. **33**(3), 350–373 (2012)
52. Otten, R.H.J.M., Wijk, J.G.V.: Graph representations in interactive layout design. In: IEEE ISCSS, pp. 914–918. IEEE (1978)
53. Pach, J., Tóth, G.: Graphs drawn with few crossings per edge. Combinatorica **17**(3), 427–439 (1997)
54. Rosenstiehl, P., Tarjan, R.E.: Rectilinear planar layouts and bipolar orientations of planar graphs. Discret. Comput. Geom. **1**, 343–353 (1986)
55. Schnyder, W.: Embedding planar graphs on the grid. In: Johnson, D.S. (ed.) SODA 1990, pp. 138–148. SIAM (1990)
56. Shermer, T.C.: On rectangle visibility graphs. III. External visibility and complexity. In: CCCG 1996, pp. 234–239. Carleton University Press (1996)
57. Štola, J.: Unimaximal sequences of pairs in rectangle visibility drawing. In: Tollis, I.G., Patrignani, M. (eds.) GD 2008, LNCS, vol. 5417, pp. 61–66. Springer (2009)
58. Suzuki, Y.: Optimal 1-planar graphs which triangulate other surfaces. Discret. Math. **310**(1), 6–11 (2010)
59. Tamassia, R., Tollis, I.G.: A unified approach to visibility representations of planar graphs. Discret. Comput. Geom. **1**(1), 321–341 (1986)
60. Thomassen, C.: Plane representations of graphs. In: Progress in Graph Theory, pp. 43–69. AP (1984)
61. Wismath, S.K.: Characterizing bar line-of-sight graphs. In: SoCG 1985, pp. 147–152. ACM (1985)
62. Zhang, X., Liu, G.: The structure of plane graphs with independent crossings and its applications to coloring problems. Open Math. **11**(2), 308–321 (2013)

Chapter 7
k-Planar Graphs

Michael A. Bekos

Abstract A topological graph is called *k-planar*, for $k \geq 0$, if each edge has at most k crossings; hence, by definition, 0-planar topological graphs are plane. An abstract graph is called *k-planar* if it is isomorphic to a k-planar topological graph, i.e., if it can be drawn on the plane with at most k crossings per edge. While planar and 1-planar graphs have been extensively studied in the literature and their structure has been well understood, this is not the case for k-planar graphs, with $k \geq 2$. These graphs have a more complex structure, which is significantly more difficult to comprehend. As an example, we mention that tight (possibly up to additive constants) bounds on the edge-density of k-planar graphs are only known for small values of k (that is, for $k \in \{0, 1, 2, 3, 4\}$), even though their existence yields corresponding improvements on the leading constant of the lower bound on the number of crossings of a graph, provided by the well-known Crossing Lemma. In this chapter, we focus on k-planar graphs, with $k \geq 2$, and review the known combinatorial and algorithmic results from the literature. We also identify several interesting open problems in the field.

7.1 Introduction

A topological graph is *k-planar*, for $k \geq 0$, if each edge has at most k crossings; for an example, refer to Fig. 7.1a which depicts a 3-planar topological graph that is not 2-planar (to see the latter observe that, e.g., the bold edge is crossed three times). Accordingly, a graph is *k-planar*, if it is isomorphic to the underlying abstract graph of a k-planar topological graph, i.e., if it can be drawn on the plane with at most k crossings per edge. Equivalently, one can define k-planar graphs in terms of the following forbidden configuration: "an edge is crossed by $k + 1$ or more edges"; for example, Fig. 7.1b shows a crossing configuration that is forbidden in a 3-planar topological graph (since the bold-drawn edge is crossed four times).

Observe that, by definition, every 0-planar topological graph is in fact a plane graph. In addition, every k-planar graph is also $(k + 1)$-planar, which naturally

M. A. Bekos (✉)
Wilhelm-Schickard-Institut für Informatik, Universität Tübingen, Tübingen, Germany
e-mail: bekos@informatik.uni-tuebingen.de

© Springer Nature Singapore Pte Ltd. 2020
S.-H. Hong and T. Tokuyama (eds.), *Beyond Planar Graphs*,
https://doi.org/10.1007/978-981-15-6533-5_7

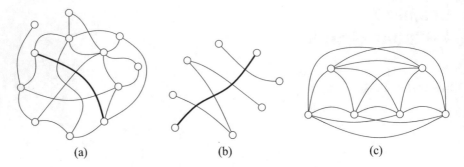

Fig. 7.1 Illustration of: (**a**) a 3-planar topological graph, (**b**) a crossing configuration that is forbidden in a 3-planar topological graph, and of (**c**) the complete graph on six vertices K_6 as 1-planar topological graph

defines a hierarchy in k-planarity. The results in the literature are, in a sense, inversely proportional to this hierarchy. In particular, there is a tremendous amount of results for planar graphs; to mention only some of the most important landmarks, we refer to the characterization of planar graphs in terms of forbidden minors, due to Kuratowski [1], to the existence of linear-time algorithms to test graph planarity [2–4], to the Four-Color Theorem [5, 6], and to the Euler's polyhedron formula (see, e.g., [7]), which can be used to show that n-vertex planar graphs have at most $3n - 6$ edges.

The class of 1-planar graphs, which is the next in the hierarchy, has also been extensively studied in the literature. Early works date back to 1960s [8] and continued over the years; see, e.g., [9–17]. More precisely, the class of 1-planar graphs was initially introduced by Ringel [8], who proved, as a generalization of the Four-Color Theorem, that every 1-planar graph has chromatic number at most 7 and conjectured that this bound could be lowered to 6. Ringel's conjecture was settled by Borodin [10], who showed that indeed the chromatic number of 1-planar graphs is at most 6 and that this bound is tight, as for example, the complete graph on six vertices (which is 1-planar; see, e.g., Fig. 7.1c) requires 6 colors. It is also worth mentioning that, in contrast to the existence of linear-time algorithms to test whether a graph is planar, testing whether a graph is 1-planar is an NP-complete problem [18, 19], and remains NP-complete even if the input graph has bounded bandwidth, pathwidth, or treewidth [20], or it can be obtained from a planar graph by adding a single edge [21]. Efficient recognition algorithms are known only for subclasses of 1-planar graphs; see, e.g., [22, 23]. From a graph drawing perspective, notable is also a result of Thomassen [24], who characterized the 1-planar topological graphs that admit the corresponding 1-planar embedding-preserving straight-line drawings in terms of two forbidden configurations (see also [25]). For a survey on 1-planarity, the reader is referred to [26].

An immediate observation emerging from this short overview is that planar and 1-planar graphs have been extensively studied in the literature and their structure has been well understood. However, this is not the case for k-planar graphs, with

$k \geq 2$, as the results for these graphs are significantly fewer. In the remainder of this chapter, we review the most important combinatorial and algorithmic results from the literature and we also identify several interesting open problems in the field.

7.2   Examples of k-Planar Graphs with High Density

In this section, we present examples of k-planar graphs with high edge-density, for different values of $k \geq 1$. Intuitively, the number of edges of a k-planar graph cannot be quadratic with respect to the number of its vertices, when k is fixed, since the number of crossings along each edge is fixed. For more details on the edge-density of k-planar graphs, refer to Sect. 7.3.

We start our discussion with the case $k = 1$, as the corresponding constructions for $k = 2$ and $k = 3$, that we will present, are similar. In the literature, the 1-planar graphs with a fixed number of vertices and maximum edge-density are referred to as *optimal* 1-planar graphs and they have been characterized [16, 27] as the graphs obtained by drawing a pair of crossing edges in the interior of each face of a 3-connected *quadrangulation*, i.e., of a planar graph whose faces are all of length four; see Fig. 7.2a for example. Since by Euler's polyhedron formula, a quadrangulation with n vertices has exactly $2(n - 2)$ edges and $n - 2$ faces, it follows that the graphs obtained by the aforementioned procedure have exactly $2(n - 2) + 2 \cdot (n - 2) = 4n - 8$ edges. In addition, they contain neither parallel edges nor self-loops, since the underlying quadrangulations are 3-connected. The reader is also referred to the work by Brinkmann et al. [28], who describe how one can generate all 3-connected quadrangulations with n vertices by means of two different operations, and to the

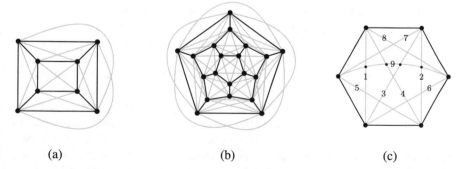

(a)   (b)   (c)

Fig. 7.2 Illustration of: (**a**) a 1-planar topological graph with $n = 8$ vertices and $4n - 8 = 24$ edges obtained by adding a pair of crossing in the interior of each face of the cube graph, (**b**) a 2-planar topological graph with $n = 20$ vertices and $5n - 10 = 90$ edges obtained by adding a pentagram in the interior of each face of the dodecahedral graph, and of (**c**) the fact that if one draws nine edges in the interior of a face of length six, then at least one edge will have inevitably four crossings

work by Brandenburg [23], who gives a linear-time recognition algorithm for optimal 1-planar graphs.

The characterization of the optimal 2-planar graphs, i.e., those 2-planar graphs with maximum density, is similar; see [29]. These graphs are obtained by drawing a *pentagram* (that is, five mutually crossing edges) in the interior of each face of a 3-connected *pentagulation*, i.e., of a planar graph whose faces are all of length five; see Fig. 7.2b for an example. Similarly to the case of optimal 1-planar graphs, one can show that a pentagulation with n vertices has exactly $5(n-2)/3$ edges and $2(n-2)/3$ faces. This directly implies that the graphs obtained by drawing a pentagram in the interior of each face of a 3-connected pentagulation with n vertices have exactly $5(n-2)/3 + 5 \cdot 2(n-2)/3 = 5n - 10$ edges. Note that Hasheminezhad et al. [30], describe eight different operations to generate all pentagulations with n vertices (and therefore, all different optimal 2-planar graphs). However, to the best of our knowledge, there is no polynomial time algorithm to recognize optimal 2-planar graphs (note that in general it is NP-complete to decide whether a graph is k-planar [31]). This brings us to the first open problem of this section.

Open Problem 1 *Can optimal 2-planar graphs be recognized in polynomial time?*

At this point, we can make an observation. The densest n-vertex 0-planar graphs have $3n - 6$ edges (as they are maximal planar). Accordingly, the densest n-vertex 1- and 2-planar graphs (that is, the optimal 1- and 2-planar graphs, respectively) have $4n - 8$ and $5n - 10$ edges, respectively. So, one would naturally expect that the densest n-vertex 3-planar graphs have $6n - 12$ edges. Also, following the corresponding constructive approaches for 1- and 2-planar graphs that we gave above, one would also expect that examples of densest 3-planar graphs can be derived by drawing nine edges in the interior of each face of a 3-connected planar graph whose faces are all of length six. However, neither of the two expected properties hold. Indeed, it is not difficult to see that if one draws nine edges in the interior of a face of length six, then some of the drawn edges will inevitably have four crossings, which is forbidden by 3-planarity (see, e.g., Fig. 7.2c). In addition, there does not exist 3-connected planar graphs whose faces are all of length six, as otherwise the dual[1] of such a graph would be a 6-regular planar graph, which contradicts Euler's polyhedron formula.

In fact, as we will shortly see in Sect. 7.3, an n-vertex 3-planar graph can have at most $5.5n - O(1)$ edges, while the bound of $6n - 12$ edges holds for n-vertex 4-planar graphs. However, by appropriately adjusting the constructions that we gave earlier for optimal 1- and 2-planar graphs, we can still derive 3- and 4-planar graphs that have $5.5n - O(1)$ and $6n - O(1)$ edges, respectively. We describe two different constructions for the case of 3-planar graphs; the corresponding ones for 4-planar graphs are analogous. Since there do not exist 3-connected planar graphs whose faces are all of length six, a first idea is to start from a 3-connected planar graph, whose faces are all of length six except for few that have length five.

---

[1]Recall that the *dual* G^* of a plane graph G is defined as follows: G^* has a vertex for each face of G and for every two vertices of G^* there is an edge connecting them if and only if there corresponding faces of G share an edge.

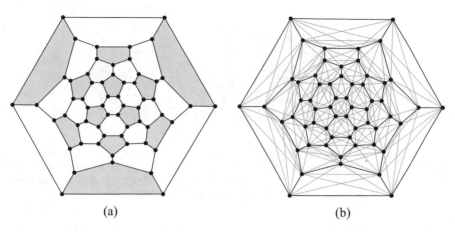

(a)   (b)

Fig. 7.3 Illustration of: **a** the football graph (also known as truncated icosahedron), which has twenty faces of length six, twelve faces of length five (gray colored), 60 vertices, and 90 edges, and **b** the 3-planar topological graph with $n = 60$ vertices and $5.5n - 20 = 310$ edges obtained by adding eight edges in the interior of each face of length six, and five edges in the interior of each face of length five of the football graph

In fact, it is not difficult to construct 3-connected planar graphs, whose faces are all of length six except for exactly twelve that have length five; these graphs are also known as *fullerene*. Figure 7.3a shows such an example with 60 vertices and 90 edges; this graph is known as *football graph* or *truncated icosahedron*. By Euler's polyhedron formula, an n-vertex fullerene graph has exactly $3n/4$ edges and $n/2 + 2$ faces. It follows that if we add eight edges in the interior of each face of length six, and five edges in the interior of each face of length five, then the resulting graph is 3-planar with exactly $3n/4 + 8 \cdot (n/2 + 2 - 12) + 5 \cdot 12 = 5.5n - 20$ edges. The corresponding bound for 4-planar graphs is derived by adding nine (instead of eight) edges in the interior of each face of length six.

A slightly improved lower bound construction is due to Pach et al. [32]. The idea here is to relax the 3-connectivity constraint in the underlying planar graph. On one hand, this implies that we will not be able to add all edges in the faces of the underlying planar graph, since parallel edges will be inevitably introduced. On the other hand, by relaxing the 3-connectivity constraint, the construction of a planar graph whose faces are all of length six is possible. The construction is illustrated in Fig. 7.4. By identifying the topmost vertices with their corresponding bottommost ones (that is, by wrapping the construction around a cylinder), the faces of the underlying planar graph (drawn in black in Fig. 7.4) are all of length six. However, in order to avoid introducing parallel edges, in each of the two faces corresponding to the bases of the cylinder, only six (instead of eight) edges can be drawn. Hence, the derived graph has $5.5n - 11 - 4 = 5.5n - 15$ edges. The corresponding bound for 4-planar graphs is $6n - 18$; see also [33]. In view of these two results, we state the following open problem.

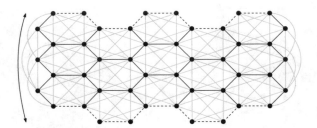

Fig. 7.4 Illustration of a 3-planar topological graph with $n = 36$ vertices and $5.5n - 15 = 183$ edges. The construction is wrapped around a cylinder by identifying the topmost vertices with their corresponding bottommost ones. To avoid introducing parallel edges, in each of the two faces corresponding to the bases of the cylinder only six edges can be drawn

Open Problem 2 *Is there a 3-planar (a 4-planar) graph with n vertices and more than 5.5n − 15 (6n − 18, respectively) edges?*

For values of k greater than 3, Pach and Tóth [34] suggest a construction, which yields k-planar graphs with n vertices and approximately $1.92\sqrt{kn}$ edges; however, their construction is only limited to high values of k. Without entering all the details of the construction, we mention that the vertices of the constructed graphs are arranged on a $\sqrt{n} \times \sqrt{n}$ grid, whose points have been slightly perturbed so to be in general position (in order to avoid edge overlaps). Then, two vertices are connected by an edge if and only if their distance is at most d, where d is selected such that no edge is crossed more than k times. Indeed, the authors prove that if d is set to $\sqrt[4]{3k/2}(1 - o(1))$, then no edge is crossed more than k times and the graph has approximately $\frac{d^2\pi}{2} n = \sqrt{3k/8}\pi\, n \approx 1.92\sqrt{kn}$ edges.

7.3 Density Results and Their Implications to the Crossing Lemma

There exists several results for the edge-density of k-planar graphs for different values of k; for an overview refer to Table 7.1. As we will shortly see, the bounds for 1-, 2-, 3-, and 4-planar graphs have led to successive improvements on the upper bound on the edge-density of general k-planar graphs, from $4.108\sqrt{kn}$ [34], to $3.95\sqrt{kn}$ [32] and to $3.81\sqrt{kn}$ [33], and on the leading constant of the lower bound on the number of crossings of a graph, provided by the well-known Crossing Lemma, from $\frac{1}{100} = 0.01$ [35, 36] to $\frac{1}{64} \approx 0.0156$ [7], to $\frac{1}{33.75} \approx 0.0296$ [34], to $\frac{1}{31.1} \approx 0.0322$ [32], to $\frac{1}{29} \approx 0.0345$ [33]. In the following, we present three different techniques for obtaining the upper bounds on the number of edges of 1-, 2-, 3-, and 4-planar graphs.

In Sect. 7.2, we have already explained how to construct 1- and 2-planar graphs with $4n - 8$ and $5n - 10$ edges, respectively. We now show that these two bounds are also upper bounds on the number of edges of 1- and 2-planar graphs, respectively;

Table 7.1 Bounds on the number of edges of k-planar graphs with n vertices for different values of k; the ones marked with an asterisk (∗) are asymptotically tight; the remaining ones are either tight or tight up to small additive constants (recall the constructions from Sect. 7.2)

Model	General		Bipartite	
	Bound	Ref.	Bound	Ref.
1-planar:	$4n - 8$	[34]	$3n - 8$	[37]
2-planar:	$5n - 10$	[34]	$3.5n - 12$	[38]
3-planar:	$5.5n - 10.5$	[29, 32]	–	–
4-planar:	$6n - 12$	[33]	–	–
k-planar:	$3.81\sqrt{k}n$ ∗	[33]	$3.005\sqrt{k}n$ ∗	[38]

we sketch the proof given by Pach and Tóth [34], which holds for k-planar graphs with $k \in \{1, 2, 3, 4\}$. However, while for $k = 1$ and $k = 2$ the obtained bounds are tight, for the corresponding bounds for $k = 3$ and $k = 4$ there exist improvements, as we will see later in this section.

Consider a k-planar n-vertex graph G with $k \in \{1, 2, 3, 4\}$ and denote by G_p a spanning subgraph of G with the largest number of edges, called *maximal-planar substructure*, such that in the drawing of G_p (that is inherited from the one of G) no two edges cross each other. With a slight abuse of notation, let $G - G_p$ be the graph obtained from G by removing only the edges of G_p and let e be an edge of $G - G_p$. Since G_p is maximal, edge e must cross at least one edge of G_p. The part of edge e between a vertex of e and the nearest crossing with an edge of G_p is referred to as *half-edge*; see Fig. 7.5 for an illustration. Clearly, the number of edges of $G - G_p$ equals half of the number of half-edges, since each edge of $G - G_p$ contains two half-edges. By k-planarity, each half-edge *is contained in* a face f of G_p and crosses at most $k - 1$ other half-edges (and a boundary edge of f).

The crucial part in the proof by Pach and Tóth [34], is the following upper bound on the number of half-edges, denoted by $h(f)$, contained in a face f of G_p, whose boundary is connected and consists of $|f| \geq 3$ edges

$$h(f) \leq (|f| - 2)(k + 1) - 1. \tag{7.1}$$

Using this upper bound, the number of half-edges of a general face f (i.e., whose boundary is not necessarily connected) can be related to the number of triangles in a triangulation of f, denoted by $t(f)$, and to the number of boundary edges of f, denoted by $|f|$, as follows:

$$h(f) \leq t(f)k + |f| - 3. \tag{7.2}$$

In fact, if the boundary of f is connected, then the number of triangles in a triangulation of f is $|f| - 2$, that is, $t(f) = |f| - 2$. Hence, Eq. 7.2, directly follows from Eq. 7.1. On the other hand, if the boundary of f is not connected, then $t(f) \geq |f|$ holds. In this case, the bound follows by the observation that the number of half-edges

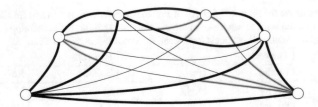

Fig. 7.5 Illustration of a topological 4-planar graph G. Its maximal-planar substructure G_p is drawn bold. The half-edges contained in the three faces of G_p incident to its unbounded face are highlighted in gray

contained in f cannot be more than $k|f|$, because, by k-planarity, every boundary edge of f has at most k crossings. Hence, $h(f) \leq |f|k \leq t(f)k + |f| - 3$.

Since each edge of $G - G_p$ contains two half-edges, it follows by Eq. 7.2, that the number of edges of $G - G_p$ is at most

$$\frac{1}{2} \sum_{f \in F_p} (t(f)k + |f| - 3), \tag{7.3}$$

where F_p denotes the set of faces of G_p. On the other hand, since in order to triangulate a face f of G_p, one needs at least $|f| - 3$ edges, it follows that the number of edges of G_p is at most

$$3n - 6 - \sum_{f \in F_p} (|f| - 3). \tag{7.4}$$

Combining Eqs. 7.3 and 7.4, with the fact in a triangulation of G_p there exist in total $2n - 4$ triangles (that is, $\sum_{f \in F_p} |f| = 2n - 4$), it follows that the total number of edges of G is at most

$$3n - 6 + \frac{1}{2} \sum_{f \in F_p} (t(f)k - (|f| - 3)) \leq 3n - 6 + (n - 2)k = (k + 3)(n - 2).$$

We summarize this bound in the following theorem.

Theorem 1 (Pach and Tóth [34]) *For $k \in \{1, 2, 3, 4\}$, a k-planar graph with n vertices has at most $(k + 3)(n - 2)$ edges.*

To derive the bound of $5.5n - 11$ edges on the edge-density of 3-planar graphs with n vertices, Pach et al. [32], propose a similar proof as the one we described above, which is, however, more technical. Here, we describe an alternative proof suggested by Bekos et al. [39], that is based on structural properties of these graphs and also holds for multigraphs containing neither homotopic parallel edges nor homotopic self-loops. Note that a similar proof has been proposed by Bae et al. [40], to derive the upper bound of $5n - 10$ edges on the edge-density of gap-planar graphs.

Among all possible 3-planar graphs with n vertices and maximum density, Bekos et al. [39], choose one, called *crossing-minimal*, with the following two properties:

(i) its maximal-planar substructure has maximum number of edges among all possible maximal-planar substructures of all 3-planar graphs with n vertices and maximum density, and

(ii) the number of crossings is minimum over all corresponding 3-planar such graphs subject to (i).

With a slightly technical proof, it can be proved that the maximal-planar substructure G_p of a crossing-minimal 3-planar graph G with n vertices and maximum density is a triangulation. Hence, the number of edges of G_p is exactly $3n - 6$. It follows that in order to count the number of edges of $G - G_p$, it suffices to count the total number of half-edges contained in the faces of G_p.

By 3-planarity, at most three half-edges are contained in each (triangular) face of G_p. Now, observe that if each face of G_p contained exactly three half-edges, then $G - G_p$ would have $3/2(2n - 4) = 3n - 6$ edges, since G_p has exactly $2n - 4$ triangular faces. This would imply that the total number of edges of G is $6n - 12$, which, however, is an overestimation. To adjust the bound, Bekos et al. [39], show that each face of G_p containing exactly three half-edges can be uniquely associated to a neighboring face of G_p containing at most two half-edges.

To see this, consider a face $\langle v_1, v_2, v_3 \rangle$ of G_p. We say that this face is of *type* (τ_1, τ_2, τ_3) if and only if for each $i = 1, 2, 3$ vertex v_i is an endvertex of τ_i half-edges contained in it. Without loss of generality, we may further assume that $\tau_1 \geq \tau_2 \geq \tau_3$. Here, we only describe how the association is performed when $\langle v_1, v_2, v_3 \rangle$ is of type $(2, 1, 0)$; the $(3, 0, 0)$ case is slightly more technical, while the $(1, 1, 1)$ case cannot occur due to the fact that G is of maximum density (for more details refer to [39]). Since v_2 is the vertex of one half-edge contained in $\langle v_1, v_2, v_3 \rangle$, edge (v_1, v_3) is crossed at least once. This allows the association of $\langle v_1, v_2, v_3 \rangle$ with the triangular face T of G_p neighboring $\langle v_1, v_2, v_3 \rangle$ along (v_1, v_3). More precisely, since the half-edge contained in $\langle v_1, v_2, v_3 \rangle$ that is incident to v_2 has three crossings in $\langle v_1, v_2, v_3 \rangle$, it is clear that there exist no half-edge contained in T having as endpoint either v_1 or v_3. In particular, T may contain at most one additional half-edge incident to the vertex of T that is different from v_1 and v_3, which implies that T contains at most two half-edges, as desired.

From the above association, it follows that if we denote by t_i the number of triangular faces of G_p containing exactly i half-edges, $0 \leq i \leq 3$, then $t_3 \leq t_0 + t_1 + t_2$. This implies that $t_3 \leq (2n - 4)/2 = n - 2$, since the number of faces of G_p is $2n - 4$. Hence, the number of edges of $G - G_p$ are

$$
\begin{aligned}
(t_1 + 2t_2 + 3t_3)/2 &= (t_1 + t_2 + t_3) + (t_3 - t_1)/2 \\
&\leq (2n - 4 - t_0) + t_3/2 \\
&\leq 2n - 4 + (n - 2)/2 \\
&\leq 5/4(2n - 4).
\end{aligned}
$$

So, the total number of edges of G are

$$3n - 6 + 5(2n - 4)/4 = 11n/2 - 11.$$

Note that in [29], it is shown that the aforementioned upper bound can be achieved only if G is a multi-graph containing parallel edges or self-loops. We summarize the discussion above in the following theorem.

Theorem 2 (Bekos et al. [29, 39]) *A 3-planar graph with n vertices has at most $5.5n - 10.5$ edges.*

We conclude this section by briefly discussing the case $k = 4$. In this case, the best-known upper bound is due to Ackerman [33], who employed a charging technique to show that a 4-planar graph $G = (V, E)$ with n vertices cannot have more than $6n - 12$ edges. According to this technique, graph G is first *planarized*, that is, it is transformed into a plane graph $G' = (V', E')$ by replacing each crossing of 4-planar drawing of G with a dummy vertex. Denote by F' the set of faces of G' and for a face $f \in F'$ let $V(f)$ be the set of non-dummy vertices on the boundary of f. Recall that by Euler's polyhedron formula, $|V'| + |F'| - |E'| = 2$ holds. Initially, each face $f \in F'$ is assigned a charge equal to $|f| + |V(f)| - 4$. Therefore, the sum of the charges over all faces of G' is

$$\sum_{f \in F'} (|f| + |V(f)| - 4) = 2|E'| + \sum_{u \in V} \deg(u) - 4|F'|$$
$$= 2|E'| + \sum_{u \in V'} \deg(u) - \sum_{u \in V'-V} \deg(u) - 4|F'|$$
$$= 2|E'| + 2|E'| - 4(|V'| - n) - 4|F'|$$
$$= 4n + 4(|E'| - |V'| - |F'|)$$
$$= 4n - 8.$$

In subsequent steps, the charge is redistributed such that eventually the charge of each face of G' is nonnegative and the charge of each non-dummy vertex $u \in V$ is $\deg(u)/3$. Then, the upper bound on the number of edges of G is derived as follows:

$$\frac{2}{3}|E| = \sum_{u \in V} \deg(u)/3 \le 4n - 8 \Rightarrow |E| \le 6n - 12.$$

We summarize this bound in the following theorem.

Theorem 3 (Ackerman [33]) *A 4-planar graph with n vertices has at most $6n - 12$ edges.*

In view of the above results, we state the following two open problems.

Open Problem 3 *What is the maximum number of edges of a 5-planar graph with n vertices? In particular, does there exist a 5-planar graph with n vertices and more than $6.33n - O(1)$ edges?*

Open Problem 4 *What is the maximum number of edges of a bipartite 3-planar graph with n vertices?*

Note that an answer to Open Problem 3 may yield to a further improvement on the leading constant of the lower bound on the number of crossings of a graph, provided by the Crossing Lemma, as will see in the next section. On the other hand, an answer to Open Problem 4 may yield to an improvement on the leading constant of the corresponding lower bound for bipartite graphs.

7.3.1 Two Important Implications

In the following, we describe two important implications of the currently best-known upper bound on the edge-density of 4-planar graphs. The first one is on the well-known Crossing Lemma, which provides a lower bound on the number of crossings of a graph G, denoted by $\mathrm{cr}(G)$. The roots of this result date back to 1973, when Erdős and Guy conjectured that there exists a positive constant c such that, if G has at least a certain number of edges, then

$$\mathrm{cr}(G) \geq c \cdot \frac{m^3}{n^2}.$$

The first proofs were by Leighton [36] and by Ajtai et al. [35], who independently answered in affirmative the conjecture, when the leading constant c is 0.01. An improvement on the leading constant from 0.01 to $\frac{1}{64} \approx 0.0156$ was presented in [7]. The main ingredient in the proof is a simple probabilistic argument, which later was reused by Pach and Tóth [34], by Pach et al. [32] and by Ackerman [33], to progressively further improve the leading constant to $\frac{1}{33.75} \approx 0.0296$, to $\frac{1}{31.1} \approx 0.0322$ and to $\frac{1}{29} \approx 0.0345$, respectively. The technique is summarized in the proof of the following theorem, which is a slightly weaker version of the corresponding one of [33], in order to avoid a rather technical part in the proof.

Theorem 4 (Ackerman [33]) *Let G be a graph with $n \geq 3$ vertices and m edges, such that $m \geq \frac{141}{20}n$. Then*

$$\mathrm{cr}(G) \geq \frac{2000}{59643} \cdot \frac{m^3}{n^2} \approx \frac{1}{29} \cdot \frac{m^3}{n^2}.$$

Proof The proof consists of two main steps. In the first step, a weaker lower bound on the number of crossings $\mathrm{cr}(G)$ of G is guaranteed by exploiting the bounds on the edge-density of 0-, 1-, 2-, 3-, and 4-planar graphs. Note that we assume that $m > 3n - 6$, as otherwise there is nothing to prove. Intuitively, the idea is the following. By Euler's polyhedron formula, it follows that if $m > 3n - 6$, then G has an edge crossed by at least one other edge. It follows from [32] that if $m > 4n - 8$ (if $m > 5n - 10$), then G has an edge crossed by at least two (by at least three, respectively) other edges. Also, by [32], we know that if $m > 5.5n - 11$, then G has an edge crossed by at least four other edges. Finally, it follows from [33], that if $m > 6n - 12$ then G has an edge crossed by at least five other edges. We obtain by induction on the number of edges of G that $\mathrm{cr}(G)$ is at least

$$m - (3n - 6) + m - (4n - 8) + m - (5n - 10) + m - (5.5n - 11) + m - (6n - 12).$$

Hence

$$\mathrm{cr}(G) \geq 5m - \frac{47}{2}(n - 2). \tag{7.5}$$

In the second step, the aforementioned lower bound is used in a probabilistic argument. Consider a drawing of G with $\mathrm{cr}(G)$ crossings and let $p = \frac{141n}{20m} \leq 1$. Next, construct a random subgraph H_p of G as follows. Choose independently every vertex of G with probability p, and denote by H_p the subgraph of G induced by the chosen vertices. Let also n_p, m_p, and c_p be the random variables corresponding to the number of vertices, of edges and of crossings of H_p. Then, it is not difficult to see that the expected values of these variables are as follows:

$$E(n_p) = p \cdot n \qquad E(m_p) = p^2 \cdot m \qquad E(c_p) = p^4 \cdot \mathrm{cr}(G).$$

By Eq. 7.5, it follows that

$$\mathrm{cr}(H_p) \geq 5m_p - \frac{47}{2}(n_p - 2). \tag{7.6}$$

By taking expectations in Eq. 7.6, and by the linearity of expectations, we have

$$p^4 \mathrm{cr}(G) \geq 5p^2 m - \frac{47}{2}pn \qquad \Rightarrow \qquad \mathrm{cr}(G) \geq \frac{5m}{p^2} - \frac{47n}{2p^3}.$$

The proof follows by plugging $p = \frac{141n}{20m}$ (which is at most 1 by our assumption) to the last inequality, that is

$$\mathrm{cr}(G) \geq \frac{2000}{59643} \cdot \frac{m^3}{n^2}.$$

This concludes the proof.   □

Note that, by exploiting properties of the crossing-free structure of G, Ackerman [33], presents a slightly improved leading constant that is exactly $\frac{1}{29}$ and holds when $m \geq 6.95n$; for details the interested reader is referred to [32], where the technique above has been used for the first time. We also note that, if one establishes an upper bound on the edge-density of 5-planar graphs (see Open Problem 3), then the

second step of the proof of Theorem 4, will start with a lower bound on cr(G) that
is different from the one of Eq. 7.5. This is expected to lead to a further improvement of the leading constant. Finally, we note that following the two steps presented
in the proof of Theorem 4, and based on the corresponding upper bounds on the
edge-density of bipartite 1-planar [37] and 2-planar graphs [38] (see also Table 7.1),
Bekos et al. [38], have derived a different leading constant for the lower bound on the
number of crossings for bipartite graphs, which is given in the following theorem.

Theorem 5 (Bekos et al. [38]) *Let G be a bipartite graph with $n \geq 3$ vertices and
m edges, such that $m \geq \frac{17}{4}n$. Then*

$$\text{cr}(G) \geq \frac{16}{289} \cdot \frac{m^3}{n^2} \approx \frac{1}{18.1} \cdot \frac{m^3}{n^2}.$$

The second implication that we will present in this section is on the edge-density
of k-planar graphs, for general values of $k \geq 5$. As already stated, the bounds for
1-, 2-, 3-, and 4-planar graphs have led to successive improvements on the upper
bound on the number of edges of general k-planar graphs, from $4.108\sqrt{kn}$ [34], to
$3.95\sqrt{kn}$ [32] and to $3.81\sqrt{kn}$ [33].

Theorem 6 (Ackerman [33]) *Let G be a k-planar graph with $n \geq 3$ vertices and m
edges, for some $k \geq 1$. Then*

$$m \leq \sqrt{\frac{29}{2}k}\, n \leq 3.81\sqrt{kn}.$$

Proof For $k \in \{1, 2, 3, 4\}$, the bounds of this theorem are weaker than the corresponding ones by Pach and Tóth [34], by Pach et al. [32] and by Ackerman [33] (see
also Table 7.1). So, without loss of generality we can assume that $k > 4$. We can
further assume that $m \geq 6.95n$, as otherwise there is nothing to prove. By [33], it
follows that

$$\text{cr}(G) \geq \frac{1}{29} \cdot \frac{m^3}{n^2}. \tag{7.7}$$

On the other hand, the fact that G is k-planar trivially implies that

$$\text{cr}(G) \leq \frac{mk}{2}. \tag{7.8}$$

Combining Eqs. 7.7 and 7.8, we obtain that

$$\frac{1}{29} \cdot \frac{m^3}{n^2} \leq \frac{mk}{2} \quad \Rightarrow \quad m \leq \sqrt{\frac{29}{2}k}\, n \leq 3.81\sqrt{kn}.$$

This concludes the proof.   □

With similar arguments, Bekos et al. [38], proved a slightly improved upper bound on the number of edges of bipartite k-planar graphs.

Theorem 7 (Bekos et al. [38]) *Let G be a k-planar bipartite graph with $n \geq 3$ vertices and m edges, for some $k \geq 1$. Then*

$$m \leq \frac{17}{8}\sqrt{2k}\, n \leq 3.006\sqrt{kn}.$$

7.4 Interesting Subclasses

In this section, we present results for some important subclasses of k-planar graphs, with $k \geq 2$. We start with the class of 2- and 3-planar graphs with n vertices and maximum density. These graphs are called *optimal* in the literature. The characterizations of optimal 2- and 3-planar graphs extend the corresponding one for optimal 1-planar graphs. Recall that optimal 1-planar graphs with n vertices have exactly $4n - 8$ edges and can be obtained by adding a pair of crossing edges in the interior of each face of an n-vertex quandragulation (see also Sect. 7.2). Analogously, optimal 2- and 3-planar graphs with n vertices have exactly $5n - 10$ and $5.5n - 11$ edges, respectively. The corresponding characterization for optimal 2-planar graphs, which also hold for multigraphs containing neither homotopic parallel edges nor homotopic self-loops, is as follows:

Theorem 8 (Bekos et al. [29]) *A graph G is optimal 2-planar if and only if G is isomorphic to the underlying abstract graph of a 2-planar topological graph H containing neither homotopic parallel edges nor homotopic self-loops, such that the graph induced by the uncrossed edges of H spans all vertices of H, and each of its faces has length five containing five mutually crossing edges in its interior in H.*

To obtain the aforementioned characterization, Bekos et al. [29], exploit several structural properties of a 2-planar topological graph H with n vertices that is isomorphic to G chosen as follows:

(i) the uncrossed edges of H are maximized overall 2-planar topological graphs with n vertices that are isomorhic to G, and

(ii) the number of crossings of H is minimized overall 2-planar topological graphs with n vertices that are isomorhic to G subject to (i).

Graph H has several interesting properties. Since H is 2-planar optimal, any edge that is crossed twice in H lies in the interior of a 5-cycle consisting explicitly of uncrossed edges. On the other hand, H cannot contain edges that are crossed exactly once, otherwise, the fact that H is optimal is led to a contradiction. Next, it can be shown that the graph induced by the uncrossed edges of H is connected. To see this, assume to the contrary that this graph has at least two connected components, say c_1 and c_2. Since H is connected, there exist edges connecting c_1 and c_2, which cannot be uncrossed. However, since these edges are crossed, they belong in the interior of a 5-cycle consisting explicitly of uncrossed edges, which implies that c_1 and c_2 cannot be distinct; a contradiction. To complete the characterization, it suffices to show that each face of the graph induced by the uncrossed edges of H has length five. In fact, faces of length one or two imply the presence of homotopic self-loops and homotopic parallel edges, respectively, which is not possible. A face of length four would contradict the fact that H is optimal, since, e.g., the edge that triangulates this face can be safely added to H without deviating its 2-planarity and without introducing homotopic parallel edges. With slightly more complicated arguments a face of length three can also be excluded, yielding the characterization.

The most intriguing open problem raised by the aforementioned characterization, besides the corresponding recognition question posed in Open Problem 1, is the following one, which is motivated by the fact that the proof of Theorem 8, depends on the choice of the initial topological graph H. So, it is natural to ask whether this dependency can be eliminated.

Open Problem 5 *Given an optimal 2-planar graph G, does there exist a 2-planar topological graph H whose underlying abstract graph is isomorphic to G, such that H does not have the structural properties of the characterization of Theorem 8?*

Note that the corresponding characterization of optimal 3-planar graphs is analogous to the one of Theorem 8 and its proof uses similar arguments.

Theorem 9 (Bekos et al. [29]) *A graph G is optimal 3-planar if and only if G is isomorphic to the underlying abstract graph of a 3-planar topological graph H containing neither homotopic parallel edges nor homotopic self-loops, such that the graph induced by the uncrossed edges of H spans all vertices of H, and each of its faces has length six containing eight crossing edges in its interior in H.*

Auer et al. [41] studied how sparse a *maximal* 2-planar graph can be, that is, what is the (least) number of edges that a 2-planar graph can have when the addition of any edge (which is not present in the graph) would deviate 2-planarity. Interestingly enough, they prove that such a graph can be considerably sparser than a planar graph. To this end, their main result is the existence of graphs with n vertices and $\frac{387}{147}n + O(1)$ edges, for infinitely many values of n.

The key observation in their construction is that the average degree of a vertex of a 2-planar graph with n vertices and maximum density (that is, with $5n - 10$ edges) is slightly less than 10. Hence, by lowering the average vertex-degree, the edge-density decreases. To achieve this, Auer et al. [41] introduced *hermits*, which are vertices of degree 2 that are "enclosed" by crossing edges preventing their connections to other vertices. The first member in the suggested family of graphs is obtained from the football graph, illustrated in Fig. 7.3a, by

  (i) completing all faces of length five to K_5s,
 (ii) drawing in each face of length six its three diagonals, and
(iii) by attaching degree-2 hermits connecting the vertices of each edge of the football graph;

note that the maximality of this graph follows from the fact that the graph without the hermits has a unique 2-planar embedding. For $\kappa > 1$, the κ-th member in the family is obtained by taking κ copies of the first member, and by identifying the same two vertices from each copy, which are chosen to be adjacent along a face of length five in the underlying football graph. The result is summarized in the following theorem.

Theorem 10 (Auer et al. [41]) *For infinitely many values of n, there exist maximal 2-planar graphs with n vertices and $\frac{387}{147}n + O(1)$ edges.*

The following open problem follows naturally from the result by Auer et al. [41].

Open Problem 6 *How sparse can a 3-planar graph with n vertices be?*

Chaplick et al. [42], and Hong and Nagamochi [43], study the problem of testing whether a graph G is *outer k-planar*, that is, whether G can be drawn as a k-planar topological graph, whose vertices lie on the outer boundary (*outer constraint*). The work by Hong and Nagamochi [43], focuses on the special case $k = 2$, under the additional constraint that no edge-crossings appear along the outer boundary (*full constraint*); these graphs are refereed to as *fully-outer 2-planar graphs*. Note that, in contrast to outer 1-planar graphs, which are in fact planar [23], (fully-)outer 2-planar graphs are not necessarily planar (e.g., the complete graph on five vertices K_5 and the complete bipartite graph $K_{3,3}$ are both fully-outer 2-planar graphs; see Fig. 7.6a and b).

An algorithm by Hong and Nagamochi [43], mainly focuses on 3-connected graphs, due to the following two properties:

 (i) a graph is fully-outer 2-planar if and only if its biconnected components are fully-outer 2-planar, while
 (ii) a biconnected graph is fully-outer 2-planar if and only if in its SPQR-tree (see, e.g., [44, 45]), every P-node has at most two virtual edges, and the skeleton of each R-node is fully-outer 2-planar.

Hence, the main difficulty in the problem of testing whether a graph is fully-outer 2-planar lies in testing whether a 3-connected component is fully-outer 2-planar.

To cope with 3-connected input graphs, Hong and Nagamochi [43] exploit several structural properties. In particular, they prove that a fully-outer 2-planar graph G, which is 3-connected,

 (P.1) does not contain three mutually crossing edges, unless G is the complete bipartite graph $K_{3,3}$;
 (P.2) the maximum degree is at most four;
 (P.3) cannot contain the compete graph K_4 as a subgraph, unless G has less than seven vertices.

Based on Properties (P.1)–(P.3), they further show that a fully-outer 2-planar 3-connected graph G must contain either a $(3, 3)$-rim (see Fig. 7.6c), or a $(3, 4)$-rim (see Fig. 7.6d) or a 4-rim (see Fig. 7.6e), each of which is defined on a set B containing either three vertices (as in the case of a $(3, 3)$-rim or a $(3, 4)$-rim; see Fig. 7.6c and d) or

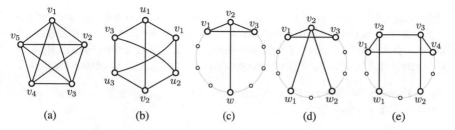

Fig. 7.6 Illustration of: (**a**) the complete graph K_5 on five vertices $\{v_1, \ldots, v_5\}$ as fully-outer 2-planar, (**b**) the complete bipartite graph $K_{3,3} = \{u_1, u_2, u_3\} \times \{v_1, v_2, v_3\}$ as fully-outer 2-planar, (**c**) (3, 3)-rim defined on $B = \{v_1, v_2, v_3\}$, in which vertex v_2 is of degree three, (**d**) a (3, 4)-rim defined on $B = \{v_1, v_2, v_3\}$, in which vertex v_2 is of degree four, and (**e**) a 4-rim defined on $B = \{v_1, v_2, v_3, v_4\}$, in which vertices v_2 and v_3 are of degree three

four vertices (as in the case of a 4-rim; see Fig. 7.6e). If no such rim can be identified in G, then the instance is reported as negative. Otherwise, it is used to transform instance (G, B) into a smaller instance (G', B'), which is solved recursively, where B' is a different rim. The base of the recursion, corresponds to a graph with at most nine vertices, whose fully-outer 2-planarity can be tested (e.g., by a brute-force method) in constant time. Since each transformation requires constant time, the overall time complexity of the algorithm is linear.

Theorem 11 (Hong and Nagamochi [43]) *There is a linear-time algorithm that tests whether a given graph is fully-outer 2-planar, and computes a fully-outer 2-planar embedding of the graph, if it exists.*

Note that for values of k greater than 2, Chaplick et al. [42], exploit several interesting properties of outer k-planar graphs. In particular, they prove that an outer k-planar graph

(i)   has a balanced separator of size at most $2k + 3$,
(ii) is $(\lfloor \sqrt{4k + 1} \rfloor + 1)$-degenerate, and therefore
(iii) is $(\lfloor \sqrt{4k + 1} \rfloor + 2)$-colorable.

For every constant k, the small balanced separators guaranteed by (i) allow for testing outer k-planarity in quasi-polynomial time [42]. We summarize this result in the following theorem.

Theorem 12 (Chaplick et al. [42]) *For every constant k, there is a quasi-polynomial time algorithm that tests whether a given graph is outer k-planar.*

The following open problem follows naturally from Theorems 11 and 12.

Open Problem 7 *Is it possible to decide in polynomial time whether a graph is (fully-)outer 3-planar?*

Since Chaplick et al. [42] give some partial results on the edge-density of outer k-planar graphs (in particular, on the size of the largest outer k-planar clique), it is interesting to ask for a closed formula on the maximum edge-density of a (fully-)outer k-planar graph (that is an improvement of the one for general k-planar graphs).

Open Problem 8 *What is the maximum number of edges of a (fully-)outer k-planar graph with n vertices?*

As a first answer to Open Problem 8, note that for $k = 1$, Auer et al. [46], have shown that an outer 1-planar graph with n vertices has at most $2.5n - 4$ edges. Combining this result with the fact that an n-vertex outerplanar graph has at most $2n - 3$ edges, yields the following lower bound on the number of crossings $cr(G)$ of an outer k-planar graph G

$$cr(G) \geq 2m - 4.5n - 7.$$

Plugging this trivial lower bound to the second step in the probabilistic proof of Theorem 6 (with $p = \frac{27n}{8m}$), yields that the crossing number of an outer k-planar graph G with n vertices and $m \geq \frac{27}{8}n$ edges satisfies

$$cr(G) \geq \frac{128}{2187} \cdot \frac{m^3}{n^2}. \tag{7.9}$$

Thus, by combining Eq. 7.9 with Eq. 7.8, we obtain that an outer k-planar graph with n vertices has at most $\frac{27\sqrt{3k}}{16} n \leq 2.93\sqrt{k} \, n$ edges, which is only a slight improvement on the upper bound for general k-planar graphs by Ackerman [33].

7.5   Relationship with k-quasi-planarity

We conclude this chapter by mentioning an interesting relationship with the class of *k-quasi-planar graphs*, which are topological graphs in which no k edges pairwise cross (note that 3-quasi-planar graphs are also called *quasi-planar* in the literature). It can be easily observed that, for $k \geq 1$, every k-planar graph is $(k + 2)$-quasi-planar. Indeed, if a k-planar graph G were not $(k + 2)$-quasi-planar, then any topological graph isomorphic to G would contain $k + 2$ pairwise crossing edges. However, this would imply that any of these edges is crossed at least $k + 1$ times, thus contradicting the fact that G is k-planar. This simple relationship was further strengthened by Angelini et al. [47], and Hoffmann and Tóth [48], who showed that for $k \geq 2$, every k-planar graph is $(k + 1)$-quasi-planar. Note that this result cannot be extended to the case $k = 1$, as a 2-quasi-planar graph is by definition planar but not every quasi-planar is planar.

Theorem 13 (Angelini et al. [47], Hoffmann and Tóth [48]) *For $k \geq 2$, every k-planar graph is $(k + 1)$-quasi-planar.*

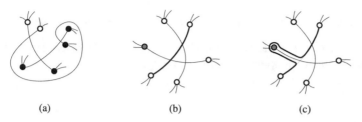

Fig. 7.7   Illustration of: (**a**) an tangled 3-crossing, (**b**) an untangled 3-crossing, and (**c**) the rerouting of the bold-drawn edge $g(X)$ of Fig. 7.7b around the green-colored vertex $f(X)$ of the 3-crossing X of Fig. 7.7b

The core of both approaches by Angelini et al. [47] and by Hoffmann and Tóth [48], is the elimination of all sets of $k + 1$ pairwise crossing edges (called $(k + 1)$-*crossings* for short, in the following) by appropriately redrawing an edge of each of them; note that the $(k + 1)$-crossings are pairwise disjoint edge sets. To achieve this, each $(k + 1)$-crossing is first *untangled*, that is, all its $2k + 2$ endvertices become incident to a common face; see Fig. 7.7a and b for an illustration of a tangled and of an untangled 3-crossing, respectively. Once all $(k + 1)$-crossings have been untangled, the idea is to appropriately define two injective functions f and g, which associate every $(k + 1)$-crossing X of the graph with a vertex $f(X)$ and with an edge $g(X)$, respectively, such that an endvertex of edge $g(X)$ and vertex $f(X)$ are consecutive along the face of X containing all the $2k + 2$ vertices of X. Then, for each $(k + 1)$-crossing X, edge $g(X)$ is *rerouted around* vertex $f(X)$, that is, edge $g(X)$ is redrawn so to pass close to vertex $f(X)$, in such a way that $g(X)$ crosses the all edges incident to $f(X)$ except for the ones in $E[X]$; see Fig. 7.7c, for an illustration. This operation is called *global rerouting*.

In a high-level description, the existence of function f is guaranteed by Hall's theorem applied to an auxiliary bipartite graph whose bipartite sets are the $(k + 1)$-crossings, and the vertices of the graph. Angelini et al. [47] then prove that if $k > 2$, then the topological graph obtained after the global rerouting is $(k + 1)$-quasi-planar. However, in the case $k = 2$, the global rerouting may yield new 3-crossings. Hoffmann and Tóth [48] describe a more complicated technique to eliminate all 3-crossings that is still based on the rerouting idea, but it also takes advantage of a specific crossing pattern in the graph, which allows one to eliminate possible new 3-crossings that may appear using recursion. It is worth noting that with their approach the topological graph obtained this way is also *simple*, in the sense that any two edges intersect in at most one point, which is either a common endpoint or a proper crossing (assuming, of course, that the initial 2-planar topological graph was simple). The approach by Angelini et al. [47], on the other hand needs one additional post-processing step to guarantee this property. We conclude this section with the following open problem.

Open Problem 9 *For $k \geq 3$, is every k-planar graph k-quasi-planar?*

Note that for $k = 2$ the answer is trivially negative, as 2-quasi-planar graphs are planar but there are nonplanar quasi-planar graphs.

References

1. Kuratowski, C.: Sur le problème des courbes gauches en topologie. Fundamenta Mathematicae **15**(1), 271–283 (1930)
2. Boyer, J.M., Myrvold, W.J.: On the cutting edge: simplified O(n) planarity by edge addition. J. Graph Algorithms Appl. **8**(2), 241–273 (2004)
3. de Fraysseix, H., de Mendez, P.O., Rosenstiehl, P.: Trémaux trees and planarity. Int. J. Found. Comput. Sci. **17**(5):1017–1030 (2006)
4. Hopcroft, J.E., Tarjan, R.E.: Efficient planarity testing. J. ACM **21**(4), 549–568 (1974)
5. Appel, K., Haken, W.: Every planar map is four colorable. part I: discharging. Illinois J. Math. **21**(3):429–490 (1977)
6. Appel, K., Haken, W., Koch, J.: Every planar map is four colorable. part II: Reducibility. Illinois J. Math. 21(3):491–567 (1977)
7. Aigner, M., Ziegler, G. M.: Proofs from THE BOOK (3. ed.). Springer, Berlin (2004)
8. Ringel, G.: Ein Sechsfarbenproblem auf der Kugel. Abhandlungen aus dem Mathematischen Seminar der Universität Hamburg **29**, 107–117 (1965)
9. Ackerman, E.: A note on 1-planar graphs. Discrete Appl. Math. **175**, 104–108 (2014)
10. Borodin, O.V.: Solution of the ringel problem on vertex-face coloring of planar graphs and coloring of 1-planar graphs. Metody Diskret. Analiz **41**, 12–26 (1984)
11. Borodin, O.V.: A new proof of the 6 color theorem. J. Graph Theory **19**(4), 507–521 (1995)
12. Chen, Z.-Z., Grigni, M., Papadimitriou, C.H.: Planar map graphs. In: Vitter, J.S. (ed.) STOC, pp. 514–523. ACM (1998)
13. Chen, Z.-Z., Grigni, M., Papadimitriou, C.H.: Recognizing hole-free 4-map graphs in cubic time. Algorithmica **45**(2), 227–262 (2006)
14. Suzuki, Y.: Re-embeddings of maximum 1-planar graphs. SIAM J. Discrete Math. **24**(4), 1527–1540 (2010)
15. Von Bodendiek, R., Schumacher, H., Wagner, K.: Bemerkungen zu einem Sechsfarbenproblem von G. Ringel. Abhandlungen aus dem Mathematischen Seminar der Universität Hamburg **53**(1):41–52 (1983)
16. Von Bodendiek, R., Schumacher, H., Wagner, K.: Über 1-optimale Graphen. Mathematische Nachrichten **117**(1), 323–339 (1984)
17. Zhang, X., Jianliang, W.: On edge colorings of 1-planar graphs. Inf. Process. Lett. **111**(3), 124–128 (2011)
18. Grigoriev, A., Bodlaender, H.L.: Algorithms for graphs embeddable with few crossings per edge. Algorithmica **49**(1), 1–11 (2007)
19. Korzhik, V.P., Mohar, B.: Minimal obstructions for 1-immersions and hardness of 1-planarity testing. J. Graph Theory **72**(1), 30–71 (2013)
20. Bannister, M.J., Cabello, S., Eppstein, D.: Parameterized complexity of 1-planarity. J. Graph Algorithms Appl. **22**(1), 23–49 (2018)
21. Cabello, S., Mohar, B.: Adding one edge to planar graphs makes crossing number and 1-planarity hard. SIAM J. Comput. **42**(5), 1803–1829 (2013)
22. Brandenburg, F.J.: On 4-map graphs and 1-planar graphs and their recognition problem (2015). arXiv:1509.03447
23. Brandenburg, F.J.: Recognizing optimal 1-planar graphs in linear time. Algorithmica **80**(1), 1–28 (2018)
24. Thomassen, C.: Rectilinear drawings of graphs. J. Graph Theory **12**(3), 335–341 (1988)

25. Hong, S.-H., Eades, P., Liotta, G., Poon, S.-H.: Fáry's theorem for 1-planar graphs. In Gud-mundsson, J., Mestre, J., Viglas, T. (eds.) COCOON, LNCS, vol. 7434, pp. 335–346. Springer, Berlin (2012)
26. Kobourov, S.G., Liotta, G., Montecchiani, F.: An annotated bibliography on 1-planarity. Comput. Sci. Rev. **25**, 49–67 (2017)
27. Suzuki, Y.: Optimal 1-planar graphs which triangulate other surfaces. Discrete Math. **310**(1), 6–11 (2010)
28. Brinkmann, G., Greenberg, S., Greenhill, C.S., McKay, B.D., Thomas, R., Wollan, P.: Generation of simple quadrangulations of the sphere. Discrete Math. **305**(1–3), 33–54 (2005)
29. Bekos, M.A., Kaufmann, M., Raftopoulou, C.N.: On optimal 2- and 3-planar graphs. In: Aronov, B., Katz, M.J. (eds.) SoCG, LIPIcs, vol. 77, pp. 16:1–16:16. Schloss Dagstuhl - Leibniz-Zentrum fuer Informatik (2017)
30. Hasheminezhad, M., McKay, B.D., Reeves, T.: Recursive generation of simple planar 5-regular graphs and pentangulations. J. Graph Algorithms Appl. **15**(3), 417–436 (2011)
31. Urschel, J.C., Wellens, J.: Testing k-planarity is NP-complete (2019). arXiv:1907.02104
32. Pach, J., Radoicic, R., Tardos, G., Tóth, G.: Improving the crossing lemma by finding more crossings in sparse graphs. Discrete Comput. Geom. **36**(4), 527–552 (2006)
33. Ackerman, E.: On topological graphs with at most four crossings per edge (2015). arXiv:1509.01932
34. Pach, J., Tóth, G.: Graphs drawn with few crossings per edge. Combinatorica **17**(3), 427–439 (1997)
35. Ajtai, M., Chvátal, V., Newborn, M.M., Szemerédi, E.: Crossing-free sub-graphs. In: Hammer, P.L., Rosa, A., Sabidussi, G., Turgeon, J. (eds.) Theory and Practice of Combinatorics, Number 12 in North-Holland Mathematics Studies, pp. 9–12. North-Holland (1982)
36. Frank Thomson Leighton: Complexity Issues in VLSI: Optimal Layouts for the Shuffle-exchange Graph and Other Networks. MIT Press, Cambridge (1983)
37. Czap, J., Przybylo, J., Skrabul'áková, E.: On an extremal problem in the class of bipartite 1-planar graphs. Discuss. Math. Graph Theory **36**(1), 141–151 (2016)
38. Angelini, P., Bekos, M.A., Kaufmann, M., Pfister, M., Ueckerdt, T.: Beyond-planarity: Turán-type results for non-planar bipartite graphs. In: Hsu, W.-L., Lee, D.-T., Liao, C.-S. (eds.) ISAAC, LIPIcs, vol. 123, pp. 28:1–28:13. Schloss Dagstuhl (2018)
39. Bekos, M.A., Kaufmann, M., Raftopoulou, C.N.: On the density of non-simple 3-planar graphs. In: Hu, Y., Nöllenburg, M. (eds.) Graph Drawing and Network Visualization, LNCS, vol. 9801, pp. 344–356. Springer, Berlin (2016)
40. Bae, S.W., Baffier, J.-F., Chun, J., Eades, P., Eickmeyer, K., Grilli, L., Hong, S.-H., Korman, M., Montecchiani, F., Rutter, I., Tóth, C.D.: Gap-planar graphs. Theor. Comput. Sci. **745**, 36–52 (2018)
41. Auer, C., Brandenburg, F.-J., Gleißner, A., Hanauer, K.: On sparse maximal 2-planar graphs. In Didimo, W., Patrignani, M. (eds.) Graph Drawing, LNCS, vol. 7704, pp. 555–556. Springer, Berlin (2012)
42. Chaplick, S., Kryven, M., Liotta, G., Löffler, A., Wolff, A.: Beyond planarity. In: Frati, F., Ma, K.-L. (eds.) Graph Drawing and Network Visualization, LNCS, vol. 10692, pp. 546–559. Springer, Berlin (2017)
43. Hong, S.-H., Nagamochi, H.: Testing full outer-2-planarity in linear time. In: Mayr, E.W. (ed.) WG, LNCS, vol. 9224, pp. 406–421. Springer, Berlin (2015)
44. Di Battista, G., Tamassia, R.: On-line maintenance of triconnected components with spqr-trees. Algorithmica **15**(4), 302–318 (1996)
45. Gutwenger, C., Mutzel, P.: A linear time implementation of spqr-trees. In: Marks, J. (ed.) Graph Drawing, LNCS, vol. 1984, pp. 77–90. Springer, Berlin (2000)
46. Auer, C., Bachmaier, C., Brandenburg, F.J., Gleißner, A., Hanauer, K., Neuwirth, D., Reislhuber, J.: Outer 1-planar graphs. Algorithmica **74**(4), 1293–1320 (2016)
47. Angelini, P., Bekos, M.A., Brandenburg, F.J., Da Lozzo, G., Di Battista, G., Didimo, W., Liotta, G., Montecchiani, F., Rutter, I.: On the relationship between k-planar and k-quasi-planar graphs. In: Bodlaender, H.L., Woeginger, G.J. (eds.) WG, LNCS, vol. 10520, pp. 59–74. Springer, Berlin (2017)

48. Hoffmann, M., Tóth, C.D.: Two-planar graphs are quasiplanar. In: Larsen, K.G., Bodlaender, H.L., Raskin, J.-F. (eds.) MFCS, LIPIcs, vol. 83, pp. 47:1–47:14. Schloss Dagstuhl - Leibniz-Zentrum fuer Informatik (2017)

Chapter 8
Fan-Planar Graphs

Michael A. Bekos and Luca Grilli

Abstract A *fan-planar* graph is a graph that admits a drawing, in which each edge can cross only edges with a common endvertex, and this endvertex is on the same side of the edge. Hence, by definition, fan-planar graphs extend the class of 1-planar graphs, but still form a proper subclass of 3-quasiplanar graphs, as they cannot contain three mutually crossing edges. Similarly to several other classes of beyond-planar graphs, fan-planar graphs have a linear number of edges, it is NP-hard to recognize them (both in general and in the fixed rotation system setting), while polynomial-time recognition and drawing algorithms are known only for special variants of them. In this chapter, we review known combinatorial and algorithmic results on fan-planar graphs and we identify several open problems in the field.

8.1 Introduction

A *fan-planar graph* is a graph that can be drawn on the plane such that for each edge e, either e is not involved in any crossing or its crossing edges form a *fan*, i.e., a set of edges with a common endvertex that is on the same side of e; see Fig. 8.1a. Such a drawing is called *fan-planar*. Fan-planar graphs were introduced by Kaufmann and Ueckerdt [22] as a generalization of 1-*planar graphs*, which are the graphs that admit a drawing on the plane with at most one crossing per edge. One can equivalently define a fan-planar drawing in terms of the following forbidden crossing configurations:

Configuration I: an edge is crossed by two *independent* edges, i.e., two edges that do not share a common endvertex (see Fig. 8.1b), and

Configuration II: an edge e is crossed by two adjacent edges, which have their common endvertex on different sides of e (see Fig. 8.1c).

M. A. Bekos
Wilhelm-Schickard-Institut für Informatik, Universität Tübingen, Tübingen, Germany
e-mail: bekos@informatik.uni-tuebingen.de

L. Grilli (✉)
Dipartimento di Ingegneria, Università degli Studi di Perugia, Perugia, Italy
e-mail: luca.grilli@unipg.it

© Springer Nature Singapore Pte Ltd. 2020
S.-H. Hong and T. Tokuyama (eds.), *Beyond Planar Graphs*,
https://doi.org/10.1007/978-981-15-6533-5_8

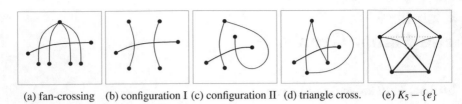

(a) fan-crossing   (b) configuration I (c) configuration II (d) triangle cross.   (e) $K_5 - \{e\}$

Fig. 8.1 Illustration of: **a** a fan-crossing; **b** an edge that is crossed by two independent edges; **c** an edge that is crossed by two edges having their common endvertex on different sides of it; **d** an edge that is crossed by three edges of a triangle; and **e** a 1-fbp drawing of the complete graph K_5 minus one edge (drawn dotted)

According to Brandenburg [13], a graph that admits a drawing, in which Configuration I is forbidden but Configuration II is allowed, is called *adjacency-crossing* graph (as each edge is crossed by edges that are pairwise adjacent). Observe that if each edge is drawn as a straight-line segment, then Configuration II cannot occur. Hence, every graph that admits a straight-line drawing in which Configuration I does not occur, is fan-planar. In the more general setting, however, where every edge is not necessarily drawn as a straight-line segment (but rather as a curve), forbidding Configuration II ensures that an edge cannot be crossed by three edges of a triangle; see Fig. 8.1d. An adjacency-crossing graph that does not allow an edge to be crossed by three edges of a triangle is called *fan-crossing* graph [13].

We emphasize that the aforementioned definitions apply also to multi-graphs. However, in order to avoid graphs with few vertices and infinitely many edges (details are given in Sect. 8.2), an extra restriction is usually imposed, that is, both the interior and the exterior regions defined by any pair of parallel edges contain at least one vertex. Such parallel edges are commonly referred to as *non-homotopic*.

An interesting subclass of fan-planar graphs, referred to as 1-*sided* 1-*fan-bundle-planar* graphs or 1-*fbp* graphs for short, was introduced by Angelini et al. [3]. Inspired from the powerful technique of edge bundling (see, e.g., [21, 24]), in a 1-*fbp drawing* of a graph, the edges of a fan are grouped into a *bundle*, so that the crossings between an edge and all the edges of a fan become a single crossing between this edge and the corresponding bundle; see Fig. 8.1d. Additionally, it is required that (i) each bundle is crossed by at most one other bundle (1-planarity restriction), and (ii) each edge can be bundled with other edges only on one of its endvertices (1-sided restriction). A 1-*fan-bundle-planar* graph (or 1-*fbp* graph for short) is a graph that admits a 1-fbp drawing. Restrictions (i) and (ii) together imply that 1-fbp graphs are in fact fan-planar. However, a fan-planar graph is not necessarily 1-fbp [3].

The rest of this chapter is structured as follows: In Sect. 8.2, we present examples of dense fan-planar and 1-fbp graphs. In Sect. 8.3, we present known relationships between the class of fan-planar graphs and other classes of beyond-planar graphs. In Sect. 8.4, several results for the edge density of fan-planar and 1-fbp graphs are described. Known complexity and algorithmic results are discussed in Sects. 8.5 and 8.6, respectively. We conclude in Sect. 8.7 with a list of open problems.

8.2   Examples of Dense Fan-Planar Graphs

As already mentioned, 1-planar graphs are by definition fan-planar. On the other hand, a fan-planar drawing cannot contain three mutually crossing edges (by Configuration I). Thus, by definition fan-planar graphs are 3-quasiplanar [1]. Hence:

$$1\text{-planarity} \subseteq \text{fan-planarity} \subseteq 3\text{-quasiplanarity}$$

This relationship immediately implies that the maximum number of edges of a fan-planar graph with n vertices is linear and ranges between $4n - 8$ and $6.5n - 20$ due to [1, 23], respectively. Kaufmann and Ueckerdt [22] revised both bounds by showing that a fan-planar graph with n vertices has at most $5n - 10$ edges and that this bound is tight for infinitely many values of n; refer to Sect. 8.4 for more details.

In fact, it is not difficult to construct fan-planar graphs with n vertices and exactly $5n - 10$ edges, if one observes that the maximum edge density of a fan-planar graph coincides with the edge density of optimal *2-planar graphs* (i.e., graphs with n vertices and exactly $5n - 10$ edges that admit drawings in which no edge is crossed more than twice). These graphs have been completely characterized [10] as the graphs obtained by drawing a pentagram in the interior of each face of a *pentagulation*, i.e., of a planar graph whose faces are all of length five; see Fig. 8.2a, for example. Hence, optimal 2-planar graphs are fan-planar by definition. In addition, if the pentagulation has no vertices of degree two, then the obtained graphs will have no parallel edges. Since, by Euler's formula for planar graphs, a pentagulation with n vertices has exactly $5(n - 2)/3$ edges and $2(n - 2)/3$ faces, it follows that the graphs obtained by the aforementioned procedure have exactly $5(n - 2)/3 + 5 \cdot 2(n - 2)/3 = 5n - 10$ edges, as desired. The reader is also referred to the work by Hasheminezhad et al. [19], who describe how one can generate all pentagulations with n vertices by means of eight different operations.

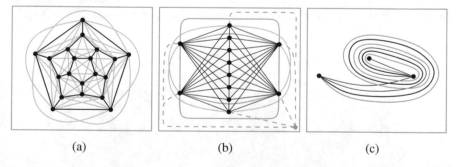

(a)	(b)	(c)

Fig. 8.2 Illustration of: **a** a fan-planar graph with $n = 20$ vertices and $5n - 10 = 90$ edges obtained by adding a pentagram in the interior of each face of the dodecahedral graph; **b** a fan-planar multi-graph with $n \geq 7$ vertices and $5n - 10$ edges, which does not contain homotopic parallel edges; **c** a non-simple fan-planar multi-graph with three vertices and infinitely many non-homotopic parallel edges

Angelini et al. [3] employed a similar construction for obtaining 1-fbp graphs of maximum density, which places the pattern of Fig. 8.1e (instead of the pentagram) at each face of the pentagulation. Thus, the resulting graphs have n vertices and $(13n - 26)/3$ edges, which is the maximum possible [3]; see also Sect. 8.4.

Another construction of n-vertex fan-planar graphs with exactly $5n - 10$ edges is the following; refer to Fig. 8.2b. Start with $K_{4,n-4}$ and connect into a path the edges of the $(n - 4)$-partition. Up to this point, the constructed graph has $4(n - 4) + n - 5 = 5n - 21$ edges. However, one can add ten more edges to this graph and a new vertex with six incident edges without violating fan-planarity and without introducing homotopic parallel edges (refer to the gray-colored edges of Fig. 8.2b). Thus, the final graph has $n + 1$ vertices and exactly $5(n + 1) - 10$ edges, as desired.

Finally, it is worth mentioning that if simplicity is relaxed in the resulting drawings, then one can construct non-simple fan-planar graphs with three vertices and infinitely many non-homotopic parallel edges, as observed by Kaufmann and Ueckerdt [22]; for an illustration refer to Fig. 8.2c.

8.3   Relationships with Other Classes of Beyond-Planar Graphs

In the following, we are discussing known inclusion relationships between the class of fan-planar graphs and other classes of beyond-planar graphs; for a summary refer to Fig. 8.3.

The inclusion relationship PLANARITY $\subset$ 1- PLANARITY $\subset$ 1- FUN- BUNDLE- PLANARITY $\subset$ FAN- PLANARITY $\subset$ 3- QUASIPLANARITY directly follows from the definitions of the corresponding graphs classes, while the fact that all the previous inclusions are strict follows from their maximum edge density; recall that n-vertex planar, 1-planar, 1-fbp, fan-planar, and 3-quasiplanar graphs have at most $3n - 6, 4n - 8, (13n - 26)/3, 5n - 10$, and $6.5n - 20$ edges, respectively, and that these bounds are tight for infinitely many values of n. The inclusion relationship 2-

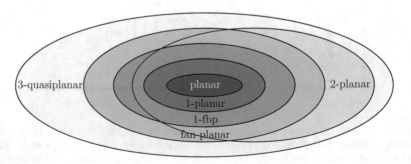

Fig. 8.3 Relationships between the class of fan-planar graphs and other classes of beyond-planar graphs

PLANARITY $\subset$ 3- QUASIPLANARITY is due to Hoffmann and Tóth [20], who described an algorithm to transform a 2-planar drawing into a 3-quasiplanar drawing by appropriately rerouting edges that violate 3-quasiplanarity. Note that this relationship is more general, as it is known that every *k-planar graph* (i.e., a graph admitting a drawing in which no edge is crossed more than k times) is *(k+1)-quasiplanar* (i.e., it admits a drawing containing no $k + 1$ mutually crossing edges) for every $k \geq 2$; for details refer to [2, 20]. Again, the fact that the classes of 2-planar and 3-quasiplanar graphs do not coincide follows from their maximum edge density.

Since optimal 2-planar graphs are fan-planar [10] and therefore lie in the intersection of 2- PLANARITY and FAN- PLANARITY, in order to complete the description of the relationships of Fig. 8.3, it remains to discuss the fact that fan-planar and 2-planar graphs form two incomparable classes [12], and that the same holds for the classes of 1-fbp and 2-planar graphs [3].

Binucci et al. [12] showed that there is a 2-planar graph which is not fan-planar by means of a more general argument: the tripartite graph $K_{1,3,4k+2}$, which can be easily seen to be fan-planar (see, e.g., Fig. 8.4a), is not k-planar for every $k \geq 1$. The reason is that $K_{1,3,4k+2}$ has crossing number $2\lfloor \frac{4k+2}{2} \rfloor \lfloor \frac{4k+1}{2} \rfloor + \lfloor \frac{4k+2}{2} \rfloor = 8k^2 + 6k + 1$ [5], while if $K_{1,3,4k+2}$ were k-planar, it would have at most $k(16k + 11)/2 = 8k^2 + 11k/2$ crossings (as it has $16k + 11$ edges); a contradiction. Angelini et al. [2] further strengthened this result by showing that there is a 1-fbp graph which is not 2-planar by a similar (but more complicated in its proof) argument. First, they showed that for every $k \geq 0$, the bipartite graph $K_{3,4k+3}$ is not k-planar. This implies that $K_{3,11}$, which can be easily seen to be 1-fbp (see Fig. 8.4b), is not 2-planar. Note that this result completely separates the two classes, as the existence of a 2-planar graph that is not 1-fbp follows from the maximum edge density of the two classes.

The proof that there is a 2-planar graph that is not fan-planar is based on a technical property of the complete graph K_7 (which itself is 2-planar and fan-planar; see Fig. 8.4c): Any pair of vertices in any fan-planar drawing of K_7 are joined by a sequence of adjacent *fragments* [12], where a fragment is defined as the portion of an edge either between two consecutive crossings or between an endvertex and the first crossing point encountered while moving towards the edge's other endvertex; see Fig. 8.4c. By arranging 13 copies of K_7 (light and dark gray in Fig. 8.4d) which pairwise share at most one vertex (refer to vertices $v_1, \ldots, v_{10}$ in Fig. 8.4d) and by introducing four additional edges (denoted by (v_1, v_7), (v_2, v_6), (v_3, v_9), and (v_4, v_8) in Fig. 8.4d), Binucci et al. [12] conclude that the derived graph is 2-planar (as each copy of K_7 is 2-planar and the additional edges define a 2-planar crossing pattern as in Fig. 8.4d) but not fan-planar. The reason is that the fragments of each copy of the light gray-colored copies of K_7 define a cyclic structure, which can be crossed neither by the fragments of the dark-gray copies of K_7 nor by the edges (v_1, v_7), (v_2, v_6), (v_3, v_9), and (v_4, v_8). This implies that the dark gray-colored copies of K_7 and the edges (v_1, v_7), (v_2, v_6), (v_3, v_9), and (v_4, v_8) must be on opposite sides of this cyclic structure. Hence, each of the edges (v_1, v_7), (v_2, v_6), (v_3, v_9), and (v_4, v_8) is crossed by two independent edges.

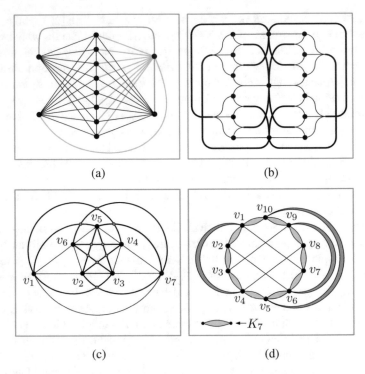

Fig. 8.4 Illustration of: **a** a fan planar drawing of the tripartite graph $K_{1,3,h}$, $h \geq 1$; **b** a 1-fbp drawing of $K_{3,12}$; **c** a drawing of the complete graph K_7 that is both 2-planar and fan-planar; the fragments of this drawing are drawn thicker; and **d** a graph consisting of 13 copies of K_7, that is not fan-planar

Table 8.1 Tight bounds on the number of edges of fan-planar and of 1-fbp graphs; the one marked with an asterisk ($*$) is tight up to a small additive constant

Model	2-layer		Outer		Bipartite		General	
	Bound	Ref.	Bound	Ref.	Bound	Ref.	Bound	Ref.
Fan-planar:	$2n - 4$	[12]	$3n - 5$	[12]	$4n - 12$ *	[4]	$5n - 10$	[22]
1-fbp:	$\frac{5n-7}{3}$	[3]	$\frac{8n-13}{3}$	[3]	–		$\frac{13n-26}{3}$	[3]

8.4 Density Results

There exist several results for the edge density of fan-planar and of 1-fbp graphs, as well as of variants of them. We provide a summary in Table 8.1, where the columns with labels "2-layer" and "outer" correspond to two common drawing models, according to which the vertices of the graph to be drawn must lie on two parallel lines (typically called *layers*) and on the outer boundary of the drawing, respectively.

In Sect. 8.2, we have already explained how to construct fan-planar graphs with $5n - 10$ edges, we now show that $5n - 10$ is also an upper bound on the number of edges of a fan-planar graph; the proof is a sketched version of the one given by Kaufmann and Ueckerdt [22].

Let $G = (V, E)$ be a fan-planar graph with $n = |V| \geq 3$ vertices and $m = |E|$ edges, and let Γ be a fan-planar drawing of G. We show that $m \leq 5n - 10$. W.l.o.g., we may assume that: (i) the graph G is *maximal fan-planar*, i.e., no edge can be added to G without destroying its fan-planarity; (ii) the drawing Γ has the maximum number of uncrossed edges among all the fan-planar drawings of G.

The proof strategy is to first partition the edge set E in an appropriate way, exploiting general considerations on the structure of Γ, and then to bound the number of edges in each subset of this partition. A first natural partitioning of E is into two subsets, based on whether an edge is crossed or uncrossed in Γ. Formally, we may write $E = E_p \dot\cup E_\chi$, where E_p is the subset of edges of G that are uncrossed in Γ, while $E_\chi = E \setminus E_p$ is the set of crossed edges. Consider now the planar subdrawing Γ_p of Γ that results from Γ after the (non-iterative) removal of all the crossed edges. By definition, Γ_p contains all the vertices of G, but differently from Γ it might not be connected, and it could even contain isolated vertices. Moreover, as every planar drawing, Γ_p divides the plane into a set F of topologically connected regions, the *faces* of Γ_p, and every crossed edge (i.e., every edge in E_χ) is drawn within exactly one face of F. Therefore, the faces of Γ_p induce a partitioning of E_χ; namely, $E_\chi = \dot\bigcup_{f \in F} E_\chi(f)$, where $E_\chi(f) \subset E_\chi$ is the subset of crossed edges that are drawn within f.

We now give a high-level description on how to bound $|E_\chi(f)|$, i.e., the number of crossed edges in a face of Γ_p. Let ∂f be the boundary of f. Due to the (possible) lack of connectivity of Γ_p, the boundary of a face is in general disconnected. It follows that ∂f is described (in general) by a disjoint collection of *facial walks* $\mathbb{W}_f = \{W_1, W_2, \ldots, W_{k_f}\}$, where $k_f \geq 1$, and $k_f = 1$ if and only if the boundary of f is connected. We recall that a *facial walk* W is a closed sequence of vertices and edges that are encountered when walking along the boundary of a face, leaving the interior of the face on the right-hand side (see, e.g., Fig. 8.5). We also recall that W may consist of a single vertex and that a same vertex and a same edge may appear more than once in W. Of course, every facial walk is uniquely described by the closed sequence of its vertices. For example, the facial walk W_4 in Fig. 8.5 is described by the closed sequence v_0-v_1-v_2-v_3-v_4-v_5-v_6-v_3-v_7-v_8-v_7-v_9-(v_0), where the vertices v_3 and v_7, as well as the edge $e = (v_7, v_8) = (v_8, v_7)$, appear twice.

The upper bound on $|E_\chi(f)|$ given in [22] is expressed in terms of the *lengths* of the facial walks in $\mathbb{W}_f$, where the *length* $l(W)$ of a facial walk W is the total number of edges in W, counting each edge with its multiplicity. More specifically, Kaufmann and Ueckerdt [22] shows (with a rather technical proof) that

$$|E_\chi(f)| \leq 2|f| + 5(k_f - 2), \tag{8.1}$$

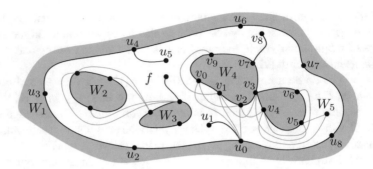

Fig. 8.5 Illustration of a face f of Γ_p (white background), whose boundary ∂f is disconnected. The boundary ∂f is described by a disjoint collection $\mathbb{W}_f = \{W_1, W_2, W_3, W_4, W_5\}$ of five facial walks (black vertices and black edges), where W_5 consists of a single vertex. The figure also includes some crossed edges of $E_\chi(f)$, which are depicted in gray

where $|f|$ denotes the overall length of the facial walks in $\mathbb{W}_f$, i.e., $|f| = \sum_{i=1}^{k_f} l(W_i)$. They came up with this result by further partitioning the crossed edges in f into two subsets: *intra facial* and *inter facial*. An *intra facial* edge of f is a crossed edge whose endvertices belong to a same facial walk, while an *inter facial* edge joins two vertices lying in distinct facial walks. For example, by referring to Fig. 8.5, the crossed edge (v_1, v_9) is an intra facial edge, because both v_1 and v_9 belong to W_4, while (u_0, v_1) is an inter facial edge, since u_0, differently from v_1, lies along W_1. The upper bound given by Inequality (8.1) was obtained by combining a bound on the number of inter facial edges with a bound on the number of intra facial edges; we omit this part of the proof as it is very technical. Combining Inequality (8.1) with the aforementioned strategies for partitioning E, we obtain

$$|E| = |E_p| + |E_\chi| = |E_p| + \sum_{f \in F} |E_\chi(f)| \le |E_p| + \sum_{f \in F} (2|f| + 5(k_f - 2)).$$

Taking into account that $\sum_{f \in F} |f| = 2|E_p|$ and that $\sum_{f \in F} (k_f - 1) = ncc(\Gamma_p) - 1$, where $ncc(\Gamma_p)$ is the number of connected components of Γ_p, we get that

$$|E| \le 5|E_p| + 5(ncc(\Gamma_p) - 1) - 5|F|.$$

Finally, by applying the Euler's formula (for disconnected plane graphs) to Γ_p, we may write $|V| + |F| - |E_p| = 1 + ncc(\Gamma_p)$, from which it follows that

$$m = |E| \le 5|V| - 10 = 5n - 10.$$

We summarize this bound in the following theorem.

Theorem 1 (Kaufmann and Ueckerdt [22]) *A fan-planar graph with $n \ge 3$ vertices has at most $5n - 10$ edges, which is a tight bound for infinitely many values of n.*

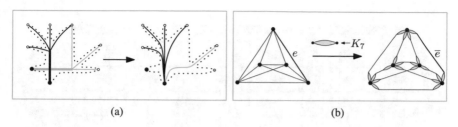

Fig. 8.6 Illustration of: **a** the transformation used to obtain the bound on the edge density of 1-fbp graphs, **b** the reduction to prove that fan-planarity is NP-hard in the general setting by Binucci et al. [12]

Angelini et al. [3] derived the bound of $(13n - 26)/3$ on the edge density of an n-vertex 1-fbp graph G by applying a simple transformation at each pair of crossing bundles of a 1-fbp drawing Γ of G (see Fig. 8.6a for an illustration), assuming without loss of generality that Γ is maximally dense; the latter assumption actually guarantees that the "boundary edges" of the crossing bundles (dotted drawn in Fig. 8.6a) exist. The result is a planar multi-graph without homotopic parallel edges on the same vertex set; thus, it contains at most $2n - 4$ faces and at most $3n - 6$ edges. To estimate the total number of edges of the original graph G, one has to observe that a single transformation involves at least three faces of the derived graph and leads to a reduction by at most two edges, if the crossing bundles contain more than two edges. Otherwise, it involves exactly two faces of the derived graph and leads to a reduction by at exactly one edge. So, if one denotes by f and ϕ the total number of such faces of the derived graph, then $f + \phi \leq 2n - 4$. Hence, the original graph has at most $3n - 6 + 2\lfloor f/3 \rfloor + \phi/2 \leq 3n - 6 + 2\lfloor (2n - 4)/3 \rfloor \leq (13n - 26)/3$. The following theorem summarizes this bound.

Theorem 2 (Angelini et al. [3]) *A 1-fbp graph with $n \geq 3$ vertices has at most $(13n - 26)/3$ edges, which is a tight bound for infinitely many values of n.*

8.5 NP-Hardness Results

Testing whether a given graph is fan-planar has been shown to be NP-complete in general and in the fixed *rotation system* setting [8, 12]; recall that a rotation system specifies the order of the edges around each vertex. Membership in NP can be proved as for the crossing number problem [16]; a non-deterministic algorithm needs (i) to guess the number of crossings of the drawing (ii) to guess the possible pairs of edges that cross (together with the order of the crossings along each edge), and (iii) to non-deterministically generate all possible planar embeddings (that also conform with the rotation system, in the fixed rotation system setting) of the graph obtained by replacing each crossing with a dummy vertex. If there is one embedding that does

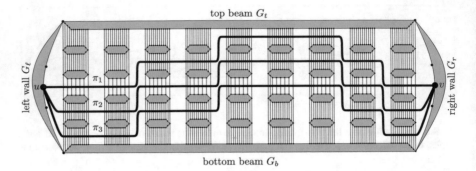

Fig. 8.7 Illustration of the reduction from 3-Partition, where $m = 3$, $B = 24$ and $A = \{7, 7, 7, 8, 8, 8, 8, 9, 10\}$. The transversal paths are routed according to the following solution: $A_1 = \{7, 7, 10\}$, $A_2 = \{7, 8, 9\}$ and $A_3 = \{8, 8, 8\}$

not contain Configurations I and II, then the instance is indeed fan-planar. Otherwise, the instance is reported as non-fan-planar.

To prove NP-hardness in the general setting, Binucci et al. [12] employed a reduction from the NP-complete problem of testing whether a graph is 1-planar [17]. In their reduction, they used the property of K_7 mentioned in Sect. 8.3, that any pair of vertices in any fan-planar drawing of K_7 are joined by a sequence of adjacent fragments (which cannot be further crossed). Thus, given an instance G of the 1-planarity testing problem, the instance G' of the fan-planarity testing problem, obtained by replacing each edge e of G with a gadget consisting of two copies of K_7 joined by an edge $\bar{e}$ (see Fig. 8.6b) has the following property: Graph G admits a 1-planar drawing Γ if and only if G' admits a fan-planar drawing Γ', such that if e and e' cross in Γ, then their corresponding edges $\bar{e}$ and $\bar{e}'$ cross in Γ'. Thus, Theorem 3 follows.

Theorem 3 (Binucci et al. [12]) *Fan-planarity testing is NP-complete.*

To prove the NP-hardness in the fixed rotation system setting, Bekos et al. [8] employed a reduction from the well known 3-Partition problem. Recall that an instance of 3-Partition is a multi-set $A = \{a_1, a_2, \ldots, a_{3m}\}$ of $3m$ positive integers in the range $(B/4, B/2)$, where B is an integer such that $\sum_{i=1}^{3m} a_i = mB$. The problem asks whether A can be partitioned into m subsets $A_1, A_2, \ldots, A_m$, each of cardinality 3, such that the sum of the numbers in each subset is B. Since 3-Partition is *strongly* NP-hard, it is not restrictive to assume that B is bounded by a polynomial in m.

Central in the reduction is the so-called *barrier gadget*, which is a graph consisting of a cycle with $n \geq 5$ vertices, called *boundary cycle*, plus all its chords connecting vertices at distance two along this cycle, called *2-hops*. The barrier gadgets are used to constraint the routes of some specific paths in the constructed graph. Indeed, if all 2-hops of a barrier gadget are embedded in the interior of its boundary cycle, then by fan-planarity no path can enter inside the boundary cycle and cross a 2-hop. If a path enters the boundary cycle of a barrier gadget without crossing any 2-hop, then it must immediately exit the boundary cycle forming a fan-crossing.

Based on an instance A of 3-Partition, an instance $\langle G, R \rangle$ of the fan-planarity testing problem is constructed as follows. Let $M = \lceil B/2 \rceil + 1$. Graph G contains as a subgraph a *global ring barrier*, which is constructed by attaching four barrier gadgets G_t, G_r, G_b and G_ℓ as depicted in Fig. 8.7. Graphs G_t and G_b are called *top* and *bottom beams*, respectively; each contains exactly $3mM$ vertices. Graphs G_ℓ and G_r are called *left* and *right walls*, respectively; each has only five vertices. Observe that G_t, G_r, G_b and G_ℓ can be embedded so that all their vertices are linkable to points within the closed region delimited by the global ring barrier. The top and bottom beams are connected by a set of $3m$ *columns*; see Fig. 8.7 for an illustration of the case $m = 3$. Each *column* consists of a stack of $2m - 1$ *cells*; a *cell* consists of a set of pairwise disjoint edges, called the *vertical edges* of that cell. In particular, there are $m - 1$ *bottommost cells*, one *central cell* and $m - 1$ *topmost cells*. Cells of the same column are separated by $2m - 2$ barrier gadgets, called *floors*. The central cells, which are $3m$ in total, have a number of vertical edges depending on the elements of A. The central cell C_i of the i-th column contains a_i vertical edges connecting its delimiting floors, $i \in \{1, 2, ..., 3m\}$. Each of the remaining cells has M vertical edges. Hence, a non-central cell contains more edges than any central cell. Further, the number of vertices of a floor is given by the number of its incident vertical edges minus two. Let u and v be the "central" vertices of the left and right walls, respectively (see also Fig. 8.7). The construction of graph G is concluded by connecting vertices u and v with m pairwise internally disjoint paths, called the *transversal paths* of G; each transversal path has exactly $(3m - 3)M + B$ edges.

The rotation system R defines a cyclic order of the edges around each vertex that is compatible with the one of Fig. 8.7, such that the 2-hops of each barrier gadget are embedded within its boundary cycle. Hence, $\langle G, R \rangle$ can be constructed in time polynomial in m.

Suppose that $\{A_1, A_2, \ldots, A_m\}$ is a partition of A. Clearly, graph G admits a fan-planar drawing preserving R, if one omits all the transversal paths; it is essentially a drawing like that one depicted in Fig. 8.7, where the columns are one next to the other within the closed region delimited by the global ring barrier. By exploiting the partition $\{A_1, A_2, \ldots, A_m\}$ of A, the transversal paths can be embedded in this partial drawing (without violating its fan-planarity) in such a way that: (R.1) they do not cross each other; (R.2) they do not cross any barrier; (R.3) each path passes through exactly 3 central cells and $3m - 3$ non-central cells; (R.4) each cell is traversed by at most one path. Eventually, each transversal path crosses exactly $(3m - 3)M + B$ vertical edges, which equals the number of its edges. Hence, these paths can be drawn such that each of their edges crosses exactly one vertical edge, which preserves fan-planarity. This implies that G admits a fan-planar drawing preserving R.

Suppose now that G admits a fan-planar drawing Γ preserving R. First, observe that the top beam and the bottom beam are disjoint, as otherwise there would be at least a 2-hop edge in one beam that is crossed by another edge of the other beam, thus violating the fan-planarity of Γ. Note, however, that the columns can partially cross each other. Indeed, an edge e of a column L might cross an edge e' of another column L', only if e is incident to a vertex in the rightmost (leftmost) side of L, e' is a leftmost (rightmost) vertical edge of L', and L and L' are two consecutive columns. With a similar argument, it is easy to see that vertices u and v must be separated by all the columns. Therefore, every transversal path satisfies conditions R.1, R.2 and it must pass through at least three central cells (if not, it would cross a number of pairwise disjoint edges that is greater than the number of its edges; hence, Γ would not be fan-planar). On the other hand, because of condition R.4, which is obviously satisfied, there cannot be any transversal path passing through more than three central cells. Otherwise, there would be some other transversal path that traverses a number of central cells that is strictly less than three. Hence, also condition R.3 is satisfied. In conclusion, every transversal path π_j, $j \in \{1, 2, \ldots, m\}$, crosses $(3m - 3)M + B$ vertical edges and traverses exactly three central cells C_j^1, C_j^2 and C_j^3. If $m(C_j^1)$, $m(C_j^2)$ and $m(C_j^3)$ denote the number of edges of these cells, then $m(C_j^1) + m(C_j^2) + m(C_j^3) = B$, because each non-central cell has M edges. Therefore, the partitioning of A defined by $A_1, A_2, \ldots, A_m$, where $A_j = \{m(C_j^1), m(C_j^2), m(C_j^3)\}$, is a solution of 3-Partition for the input instance A. Hence, Theorem 4 follows.

Theorem 4 (Bekos et al. [8]) *Fan-planarity testing with fixed rotation system is NP-complete.*

Note that by appropriately adjusting the barrier gadget, the following problems have also been shown to be NP-hard: (i) the problem of testing whether a graph is 1-planar RAC [9], (ii) the problem of testing whether a graph is gap-planar [7], and (iii) the problem of testing whether a graph with a fixed rotation system is 1-fpb [3].

8.6 Algorithmic Results

Since the fan-planarity testing problem is NP-complete, polynomial time recognition and drawing algorithms are known only for special cases of two meaningful subclasses of fan-planar graphs (at the time of writing of this chapter): (i) 2-layer fan-planar graphs, and (ii) outer-fan-planar graphs. Recall that the former are graphs that admit fan-planar drawings in which each vertex is drawn along one of two distinct horizontal layers and each edge is drawn as a straight-line segment that connects vertices of different layers, while the latter are graphs that admit fan-planar drawings, in which the vertices are incident to the unbounded face of the drawing.

Before we proceed with the description of the algorithm by Binucci et al. [11] for the class of 2-layer fan-planar graphs, we first need to introduce some definitions. An n-vertex *ladder* is a bipartite outerplane graph consisting of two paths $u_1, u_2, \ldots, u_{n/2}$ and $v_1, v_2, \ldots, v_{n/2}$ of the same length and the edges (u_1, v_1), $(u_2, v_2), \ldots (u_{n/2}, v_{n/2})$. A *snake* graph is obtained by embedding an arbitrary number of paths of length two, inside each bounded face of a ladder, that connect a pair of non-adjacent vertices of the face (for an illustration refer to Fig. 8.8a and b).

Towards a polynomial-time recognition and drawing algorithm, Binucci et al. [11] first gave the following complete characterization of the biconnected 2-layer fan-planar graphs: a graph is biconnected 2-layer fan-planar if and only if it is a spanning

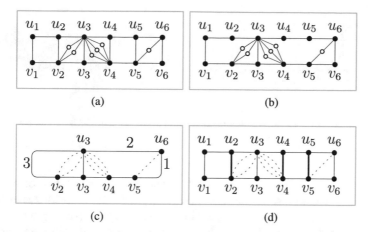

Fig. 8.8 Illustration of: **a** a snake graph with 12 vertices, whose black vertices define the underlying ladder; **b** a biconnected subgraph of it; **c** the weighted contracted (multi-)graph, whose dotted drawn edges have unit weight; and **d** the graph obtained by expanding the outer edges with a weight greater than two along with the (bold) edges added in the last step of the algorithm by Binucci et al. [11]

subgraph of a snake graph. To check whether a given graph G is 2-layer fan-planar, the algorithm by Binucci et al. [11] first *contracts* each maximal chain $\langle u, w_1, \ldots w_k, v \rangle$, with $deg(w_i) = 2$ for every $i = 1, 2, \ldots, k$, into a single edge (u, v) of weight k (see Fig. 8.8c). By the previous characterization, the weighted contracted (multi-)graph obtained this way must be outerplanar, and additionally it must have all edges with a weight greater than two on its outer face; both conditions can be checked in linear time. If one of them does not hold, then the instance is reported as negative. Otherwise, an outerplanar embedding with all edges with a weight greater than two on the outer face has been computed and the algorithm proceeds by expanding these edges (i.e., by reverting the corresponding contraction operations). What it remains to be checked in the obtained (multi-)graph is whether one can add a suitable set of internal edges (chords) connecting vertices of the outer face, such that the resulting graph is still outerplane and becomes a ladder, if one subsequently removes the internal edges of unit weight (see Fig. 8.8d). This check can be clearly done in quadratic time, but it can be also done in linear time with a little effort [11]. Graph G is 2-layer fan-planar, only if the final check is also positive, in which case a corresponding straight-line drawing can be obtained by placing (i) the vertices of each of the two paths of the ladder alternatively along two distinct horizontal lines, and (ii) each of the remaining (degree-2) vertices between their corresponding endvertices but on the opposite layer (see Fig. 8.9a). Hence, Theorem 5 follows.

Theorem 5 (Binucci et al. [11]) *Let G be a bipartite biconnected graph with n vertices. There exists an $O(n)$-time algorithm that tests whether G is 2-layer fan-planar, and that computes a 2-layer fan-planar straight-line drawing of G in the positive case.*

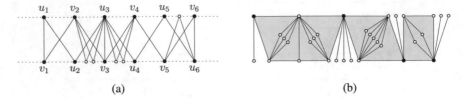

Fig. 8.9 Illustration of: **a** a 2-layer fan-planar drawing of the graph of Fig. 8.8b; and **b** a stegosaurus containing three snakes (gray shaded)

A graph is called *stegosaurus* if and only if it is either a star or a chain of snakes that are connected at common cutvertices, such that (i) a common cutvertex is incident to at most two snakes, plus a set of degree-1 vertices, called *legs* (see Fig. 8.9b), and (ii) is not adjacent to degree-2 (i.e., non-ladder) vertices. With this definition, Binucci et al. [11] gave the following characterization for simply connected 2-layer fan-planar graphs.

Theorem 6 (Binucci et al. [11]) *A graph is 2-layer fan-planar if and only if it is a subgraph of a stegosaurus.*

Note that analogous results are also known by Angelini et al. [3] for 1-fbp graphs.

We conclude this section by summarizing the linear-time algorithm, given in [8], for testing whether a given graph G is maximal outer-fan-planar, and for computing an outer-fan-planar embedding of G, in the affirmative case; recall that an outer-fan-planar graph is *maximal*, if adding any edge to it yields a graph that is no longer outer-fan-planar. Bekos et al. [8] exposed some general considerations about the structure of a maximal outer-fan-planar drawing Γ, which translate into important combinatorial properties of G: (i) the outer boundary of Γ must be a simple cycle, (ii) any two consecutive vertices along the outer boundary of Γ must be joined by a crossing-free edge, and (iii) this edge is entirely contained in the outer boundary of Γ. Hence, G must be Hamiltonian, and biconnected. As all pairs of crossing edges in Γ are determined by the cyclic order of the vertices along the outer boundary (called *outer-fan-planar embedding* of Γ), one can assume w.l.o.g. that Γ is a straight-line drawing and the vertices lie along a prescribed circle C.

As often happens when dealing with biconnected graphs, Bekos et al. [8] first designed a linear-time recognition algorithm assuming that G is 3-connected, and then, using SPQR-trees [15, 18], extended this algorithm to the biconnected case, while preserving the time complexity. Here, we will only describe the 3-connected case. For the biconnected case, we refer the reader to [8].

So, in the following we will assume that G is an n-vertex 3-connected graph. Clearly, the test whether G is maximal outer-fan-planar is trivial if $n \leq 5$, as every complete graph with up to five vertices is maximal outer-fan-planar (having at most twelve outer-fan-planar embeddings [8]).

Suppose now that $n \geq 6$. Graph G is called *complete 2-hop* if and only if it admits an outer-fan-planar drawing Γ on a circle C, which contains all the outer edges and all the 2-hop edges (i.e., all edges that are at distance at most two along C; see Fig. 8.10a). If G is a complete 2-hop, then it is clearly maximal outer-fan-planar. Testing whether G is complete 2-hop can be easily done in linear time (using the fact that G must be 4-regular). In addition, all its outer-fan-planar embeddings can also be generated in linear time (if $n = 6$, then they are at most four, while for $n \geq 7$ there exists only one outer-fan-planar embedding).

If G is not a complete 2-hop graph, then it must contain at least one *long* edge, i.e., an edge whose endvertices are at distance at least three along C. Indeed, it is shown (with a quite technical proof) that the existence of a long edge in Γ implies the presence of exactly two vertices of degree three in G; see, e.g., the white-colored vertices in Fig. 8.10b. Moreover, every vertex v of degree three is attached to a triangle, i.e., v and its three neighbors induce a K_4 in G, and the vertices of this K_4 appear consecutively along C, with v in one of the two innermost positions. By exploiting this property (called *innermost position property*), Bekos et al. [8] observed that if G is maximal outer-fan-planar, then it can be iteratively decomposed into subgraphs $G_{n-1}, \ldots, G_3$ by removing a sequence $v_n, \ldots, v_4$ of $n - 3$ vertices of degree three (which together with its three neighbors induce a K_4), until only a 3-cycle is left, that is, G_3 is a 3-cycle; see Fig. 8.10c–i.

If this decomposition is not possible, then instance G is reported as negative. Otherwise, G is rebuilt by inserting the deleted vertices in reverse order. In doing so, all the outer-fan-planar embeddings of the current subgraph of G are maintained. It can be seen that the number of these embeddings does not grow exponentially, but it is bounded by a constant; in particular, there are at most six outer-fan-planar embeddings. Moreover, by exploiting the innermost position property it can be quickly verified whether outer-fan-planarity is preserved. If some vertex v_i ($i \geq 6$) cannot be reattached to G_{i-1}, preserving outer-fan-planarity, then instance G is reported as negative. Otherwise, the test is positive and G is indeed maximal outer-fan-planar. We note that the algorithm returns all the outer-fan-planar embeddings of G and that it requires linear time. Summarizing, we obtain the following theorem.

Lemma 1 (Bekos et al. [8]) *Let G be a 3-connected graph with n vertices. There exists an $O(n)$-time algorithm that tests whether G is maximal outer-fan-planar, and that computes all outer-fan-planar straight-line drawings of G on a circle C in the positive case.*

As already mentioned, the linear-time recognition algorithm just discussed can be extended to biconnected graphs by using SPQR-trees [15, 18], while preserving the time complexity. We omit the description of this extension, by referring the interested reader to [8]. As a non-biconnected graph cannot be maximal outer-fan-planar, we obtain the following result.

Theorem 7 (Bekos et al. [8]) *Let G be a graph with n vertices. There exists an $O(n)$-time algorithm that tests whether G is maximal outer-fan-planar, and that computes an outer-fan-planar straight-line drawing of G on a circle C in the positive case.*

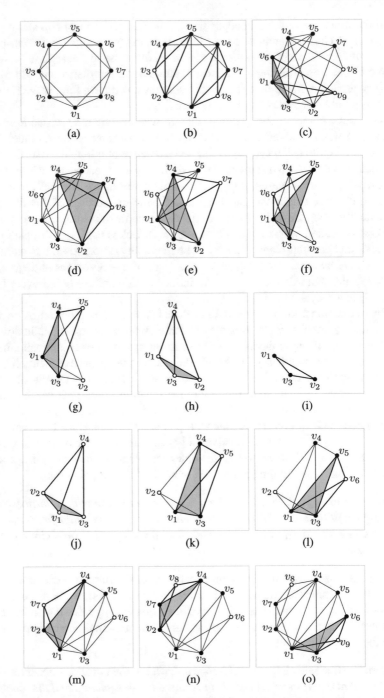

Fig. 8.10 **a** A complete 2-hop with eight vertices. **b** A maximal outer-fan-planar drawing with long edges. **c–o** A running example of the algorithm for the 3-connected case

8.7 Open Problems

In this section, we give a list of questions related to fan-planarity that still remain unanswered:

Q.1: Is it possible to characterize (or recognize in polynomial time) the *optimal* fan-planar graphs, i.e., the n-vertex fan-planar graphs with exactly $5n - 10$ edges? Note that both NP-hardness proofs of Sect. 8.5 are for fan-planar graphs that are relatively sparse.

Q.2: What is the edge density of fan-planar multi-graphs that do not contain homotopic parallel edges? In particular, is there an n-vertex such graph with more than $5n - 10$ edges?

Q.3: What is the edge density of fan-planar graphs if it is required to be drawn with straight-line edges? Note that Kaufmann and Ueckerdt [22] describe an infinite family of graphs with n vertices and $5n - 11$ edges that admit straight-line fan-planar drawings. Hence, the answer to this question is either $5n - 11$ or $5n - 10$.

Q.4: What is the edge density of adjacency-crossing graphs, which forbid Configuration I but allow Configuration II? What is the complexity of the corresponding recognition problem for this class of graphs?

Q.5: Is it possible to recognize in polynomial time whether a general graph is 2-layer fan-planar or outer-fan-planar (i.e., when maximality is relaxed)?

Q.6: What is the least amount of edges an n-vertex edge-maximal fan-planar graph can have? Density results of this kind have been established, e.g., for maximal 1- and 2-planar graphs [6, 14].

Acknowledgements The authors' research was supported in part by the DFG grant KA812/18-1 and by the scientific project: "Algoritmi e sistemi di analisi visuale di reti complesse e di grandi dimensioni" - Ricerca di Base 2017, Dipartimento di Ingegneria dell'Università degli Studi di Perugia.

References

1. Ackerman, E., Tardos, G.: On the maximum number of edges in quasi-planar graphs. J. Comb. Theory Ser. A **114**(3), 563–571 (2007)
2. Angelini, P., Bekos, M.A., Brandenburg, F.J., Da Lozzo, G., Di Battista, G., Didimo, W., Liotta, G., Montecchiani, F., Rutter, I.: On the relationship between k-planar and k-quasi-planar graphs. In: Bodlaender, H.L., Woeginger, G.J. (eds.) WG, LNCS, vol. 10520, pp. 59–74. Springer, Berlin (2017)
3. Angelini, P., Bekos, M.A., Kaufmann, M., Kindermann, P., Schneck, T.: 1-fan-bundle-planar drawings of graphs. Theor. Comput. Sci. **723**, 23–50 (2018)
4. Angelini, P., Bekos, M.A., Kaufmann, M., Pfister, M., Ueckerdt, T.: Beyond-planarity: Turán-type results for non-planar bipartite graphs. In: Hsu, W.-L., Lee, D.-T., Liao, C.-S. (eds.) ISAAC, LIPIcs, vol. 123, pp. 28:1–28:13. Schloss Dagstuhl - Leibniz-Zentrum fuer Informatik (2018)
5. Asano, K.: The crossing number of $K_{1,3,n}$ and $K_{2,3,n}$. J. Graph Theory **10**(1), 1–8 (1986)

6. Auer, C., Brandenburg, F.-J., Gleißner, A., Hanauer, K.: On sparse maximal 2-planar graphs. In: Didimo, W., Patrignani, K. (eds.) Graph Drawing, LNCS, vol. 7704, pp. 555–556. Springer, Berlin (2012)

7. Won Bae, S., Baffier, J.-F., Chun, J., Eades, P., Eickmeyer, K., Grilli, L., Hong, S.-H., Korman, M., Montecchiani, F., Rutter, I., Tóth, C.D.: Gap-planar graphs. Theor. Comput. Sci. 745, 36–52 (2018)

8. Bekos, M.A., Cornelsen, S., Grilli, L., Hong, S.-H., Kaufmann, M.: On the recognition of fan-planar and maximal outer-fan-planar graphs. Algorithmica 79(2), 401–427 (2017)

9. Bekos, M.A.: Didimo, W., Liotta, G., Mehrabi, S., Montecchiani, F.: On RAC drawings of 1-planar graphs. Theor. Comput. Sci. 689, 48–57 (2017)

10. Bekos, M.A., Kaufmann, M., Raftopoulou, C.N.: On optimal 2- and 3-planar graphs. In: Aronov, B., Katz, M.J. (eds.) SoCG, LIPIcs, vol. 77, pp. 16:1–16:16. Schloss Dagstuhl - Leibniz-Zentrum fuer Informatik (2017)

11. Binucci, C., Chimani, M., Didimo, W., Gronemann, M., Klein, K., Kratochvíl, J., Montecchiani, F., Tollis, I.G.: Algorithms and characterizations for 2-layer fan-planarity: From caterpillar to stegosaurus. J. Graph Algorithms Appl. 21(1), 81–102 (2017)

12. Binucci, C., Di Giacomo, E., Didimo, W., Montecchiani, F., Patrignani, M., Symvonis, A., Tollis, I.G.: Fan-planarity: Properties and complexity. Theor. Comput. Sci. 589, 76–86 (2015)

13. Brandenburg, F.J.: A first order logic definition of beyond-planar graphs. J. Graph Algorithms Appl. (2018)

14. Brandenburg, F.-J., Eppstein, D., Gleißner, A., Goodrich, M.T., Hanauer, K., Reislhuber, J.: On the density of maximal 1-planar graphs. In: Didimo, W., Patrignani, M. (eds.) Graph Drawing, LNCS, vol. 7704, pp. 327–338. Springer, Berlin (2012)

15. Battista, G.D., Eades, P., Tamassia, R., Tollis, I.G.: Graph Drawing: Algorithms for the Visualization of Graphs. Prentice-Hall, Upper Saddle River (1999)

16. Garey, M.R., Johnson, D.S.: Crossing number is NP-complete. SIAM J. Algebr. Discrete Methods 4(3), 312–316 (1983)

17. Grigoriev, A., Bodlaender, H.L.: Algorithms for graphs embeddable with few crossings per edge. Algorithmica 49(1), 1–11 (2007)

18. Gutwenger, C., Mutzel, P.: A linear time implementation of SPQR-trees. In: Marks, J. (ed.) Graph Drawing, LNCS, vol. 1984, pp. 77–90. Springer, Berlin (2000)

19. Hasheminezhad, M., McKay, B.D., Reeves, T.: Recursive generation of simple planar 5-regular graphs and pentagulations. J. Graph Algorithms Appl. 15(3), 417–436 (2011)

20. Hoffmann, M., Tóth, C.D.: Two-planar graphs are quasiplanar. In: Larsen, K.G., Bodlaender, H.L., Raskin, J.-F. (eds.) MFCS, LIPIcs, vol. 83, pp. 47:1–47:14. Schloss Dagstuhl - Leibniz-Zentrum fuer Informatik (2017)

21. Holten, D.: Hierarchical edge bundles: visualization of adjacency relations in hierarchical data. IEEE Trans. Vis. Comput. Graphics 12(5), 741–748 (2006)

22. Kaufmann, M., Ueckerdt, T.: The density of fan-planar graphs (2014). arXiv:1403.6184

23. Ringel, G.: Ein Sechsfarbenproblem auf der Kugel. Abh. Math. Sem. Univ. Hamb. 29, 107–117 (1965)

24. Telea, A., Ersoy, O.: Image-based edge bundles: simplified visualization of large graphs. Comput. Graph. Forum 29(3), 843–852 (2010)

Chapter 9
Right Angle Crossing Drawings of Graphs

Walter Didimo

Abstract In a *RAC drawing* of a graph, every two crossing edges form $\frac{\pi}{2}$ angles at their crossing point. The theoretical study of this type of drawings started in 2009, motivated by cognitive experiments showing that crossings with large angles do not affect too much the readability of a graph layout. Since then, the RAC drawing convention has been widely studied, both from the combinatorial and from the algorithmic point of view. RAC drawings can be also regarded as a generalization of the well-known *orthogonal drawing convention*, in which every edge is a polyline composed of horizontal and vertical segments only. In a RAC drawing there is no restriction on the slope of the edge segments, hence a vertex of any degree can be represented as a geometric point (planar orthogonal drawings with vertices drawn as points necessarily require vertices of degree at most four). In this chapter, we survey the rich literature on RAC drawings and we briefly illustrate the ideas behind some of the most interesting results.

9.1 Introduction

In a *Right Angle Crossing drawing* (*RAC drawing* for short) of a graph any two crossing edges are orthogonal, i.e., they form $\frac{\pi}{2}$ angles at their crossing point. The study of this type of drawings started in 2009 [34, 36], motivated by cognitive experiments [48, 49] that suggest a positive correlation between large-angle crossings and human understanding of graph layouts (see also [50]). Such a correlation is further witnessed by the common use of large-angle crossings in hand-drawn metro maps [58, 60] and other types of real-world diagrams; for instance, in the guidelines of the CCITT (Comité Consultatif International Téléphonique et Télégraphique) for drawing Petri nets the following requirement is reported: "There should be no acute angles where arcs cross" [21]. Examples of RAC drawings with straight-line edges are in Fig. 9.1a, b. A RAC drawing with edge bends is in Fig. 9.1c.

W. Didimo (✉)
Università degli Studi di Perugia, Perugia, Italy
e-mail: walter.didimo@unipg.it

© Springer Nature Singapore Pte Ltd. 2020
S.-H. Hong and T. Tokuyama (eds.), *Beyond Planar Graphs*,
https://doi.org/10.1007/978-981-15-6533-5_9

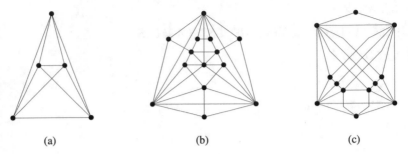

Fig. 9.1 **a** Straight-line RAC drawing of the complete graph K_5. **b** Straight-line RAC drawing of a graph with 14 vertices. **c** RAC drawing with 12 vertices and at most one bend per edge

Since their introduction, RAC drawings have been widely studied from several perspectives. This chapter surveys the main results on RAC drawings, grouping them into the following research topics:

Edge density. The edge density of a graph is defined as the ratio between its number of edges and its number of vertices. It is well known that n-vertex planar graphs have at most $3n - 6$ edges, thus their edge density is strictly less than three and asymptotically tends to this value for increasing values of n. The study of the edge density of RAC drawings aims to establish the maximum number of edges that a RAC drawing can have for any fixed number of vertices, both in the general case and in restricted scenarios in which the vertices or the edges are subject to additional drawing constraints. This question pertains the following more general *Turán-type* problem [8, 13, 53]: "What is the maximum number of edges that a drawing of a graph can have without containing a forbidden configuration of a certain type?". The study of this problem, for different types of forbidden configurations, has a long tradition in graph theory and represents one of the core research topics in the literature on graph drawing beyond planarity. For RAC drawings, the forbidden configuration consists of two edges that cross at an angle smaller than $\frac{\pi}{2}$. Section 9.3 discusses the results on the edge density of RAC drawings.

Testing and drawing algorithms. This research topic focuses on the algorithmic aspects of computing RAC drawings. Several papers study the complexity of testing whether a graph G is *RAC drawable*, i.e., whether it admits a RAC drawing. This question can be posed assuming that the embedding of the graph can be freely chosen or assuming that it is totally or partially fixed. If we know that G is RAC drawable, a natural problem is to compute a RAC drawing by using an effective layout algorithm, which possibly optimizes some well-known readability metrics, such as the number of edge bends or the drawing area. Section 9.4 surveys the main results in this research direction.

Inclusion relationships. This topic aims to establish inclusion relationships between different beyond-planar graph families. For RAC drawable graphs, the existing literature mainly concentrates on the relationships with 1-planar graphs, i.e., those graphs that can be drawn with at most one crossing per edge. Although straight-line RAC drawable graphs and 1-planar graphs are in general incomparable, some interesting inclusion relationships can be established for specific subsets of these families or when bends along the edges are allowed. This topic is explored in Sect. 9.5.

Each of Sects. 9.3–9.5 ends with a short list of interesting open problems. Section 9.2 recalls the basic terminology used in this chapter and some fundamental properties of RAC drawable graphs. Concluding remarks are in Sect. 9.6.

9.2 Basic Terminology and Properties

If not otherwise specified, in this chapter we assume that a graph $G = (V, E)$ is always *simple*, i.e., it contains neither multiple edges nor self-loops. A *drawing* Γ of G maps each vertex $v \in V$ to a distinct point p_v of the plane and each edge $(u, v) \in E$ to a simple Jordan arc with endpoints p_u and p_v. To simplify the notation and terminology, we make no distinction between a vertex of G and its corresponding point in Γ and between an edge of G and its corresponding arc in Γ.

Two edges of Γ *cross* if they have a point p in common other than their endpoints; p is called a *crossing*. We assume that an edge does not contain a vertex distinct from its endpoints, no two edges meet tangentially, and no three edges share a crossing.

A drawing Γ of G divides the plane into topologically connected regions, called *faces*. The infinite region is the *external face*; the other regions are the *internal faces*. The boundary of a face may contain both vertices and crossings. An *embedding* of G is an equivalence class of drawings of G under homeomorphism of the plane, i.e., is a class of drawings of G that define the same set of (external and internal) faces. A graph together with a fixed embedding is an *embedded graph*. A drawing without crossings is *planar*. A *planar graph* is a graph that admits a planar drawing. A *planar embedding* is the embedding of a planar drawing. A planar graph with a fixed planar embedding is an *embedded planar graph*, or briefly a *plane graph*. Note that an embedding of a graph G fixes, for each vertex v and for each crossing c, the clockwise circular order of the edges incident to v and to c.

A drawing where all the edges are straight-line segments is a *straight-line drawing*. A *polyline drawing* Γ is a drawing where the edges are mapped to chains of segments; a *bend* in Γ is a point where two segments of the same edge meet. A *k-bend drawing* is a polyline drawing with at most k bends per edge (a 0-bend drawing is a straight-line drawing). A *RAC drawing* is a polyline drawing in which any two crossing segments are orthogonal. The drawings in Fig. 9.1a, c are a straight-line RAC drawing and a 1-bend RAC drawing, respectively.

A straight-line RAC drawing has some fundamental properties, which can be easily proved [36].

Property 9.1 *If Γ is a straight-line RAC drawing, any two segments that cross the third one are parallel (see Fig. 9.2a).*

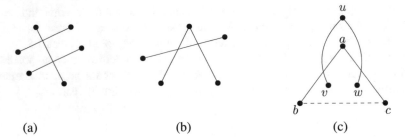

(a)   (b)   (c)

Fig. 9.2 **a** Illustration of Property 9.1. **b** The forbidden configuration of Property 9.2. **c** The forbidden configuration of Property 9.3

A *fan* in a drawing Γ is a pair of edge segments incident to the same vertex. Property 9.1 immediately implies the absence of fan crossings, as expressed by the following property (see also Fig. 9.2b).

Property 9.2 *In a straight-line RAC drawing, no edge crosses a fan.*

A third fundamental forbidden configuration in a straight-line RAC drawing Γ is expressed by Property 9.3 and illustrated in Fig. 9.2c.

Property 9.3 *Let Γ be a straight-line RAC drawing and let $T = (a, b, c)$ be a triangle such that (a, b) and (a, c) are edges of Γ. There cannot exist in Γ two adjacent edges (u, v) and (u, w) such that u is properly outside T, v and w are properly inside T, and (u, v) and (u, w) cross (a, b) and (a, c), respectively.*

Properties 9.2 and 9.3 together imply that, more in general, in a triangle T like the one of Fig. 9.2c, it is not possible that each of the two edges (u, v) and (u, w) crosses an edge of T if u is outside T and v, w are inside T.

Another fundamental property is concerned with the *crossing graph* $G^*(\Gamma)$ of a straight-line RAC drawing Γ: The vertices of $G^*(\Gamma)$ correspond to the edges of Γ and two vertices of $G^*(\Gamma)$ are connected by an edge if their corresponding edges in Γ cross. It is not difficult to see that $G^*(\Gamma)$ cannot contain odd cycles, hence the following holds.

Property 9.4 *The crossing graph of a straight-line RAC drawing is bipartite.*

Figure 9.3a depicts a straight-line RAC drawing Γ (with vertices as circles and solid edges) and its corresponding crossing graph $G^*(\Gamma)$ (with vertices as squares and dashed edges). The vertices of $G^*(\Gamma)$ are colored with three colors: red, blue, and green. A vertex is red if it is an isolated vertex in $G^*(\Gamma)$, while each blue (resp. green) vertex is adjacent to green (resp. blue) vertices only. An edge of Γ corresponding to a red, a blue, or a green vertex of $G^*(\Gamma)$ is called a *red edge*, a *blue edge*, or a *green edge*, respectively. Red edges do not cross in Γ, while a blue edge always crosses some (and only) green edges and vice versa. Such a coloring of the edges of Γ defines two straight-line sub-drawings of Γ, denoted as Γ_{rb} and Γ_{rg}, which are

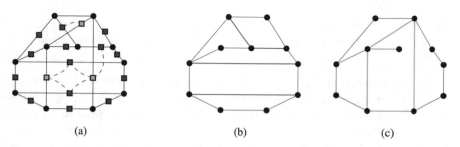

Fig. 9.3 a A straight-line RAC drawing Γ (circle-vertices and solid edges) and its crossing graph (square-vertices and dashed edges). **b** The red-blue drawing Γ_{rb}. **c** The red-green drawing Γ_{rg}

called the *red-blue* and the *green-blue* drawings. They both have the same vertex set as Γ, but Γ_{rb} consists of the red and blue edges only, while Γ_{rg} consists of the red and green edges only. Figure 9.3b, c show the red-blue and the red-green drawings of the drawing of Fig. 9.3a.

The red-blue and the red-green drawings represent a tool that is extensively used in the proof of several results on straight-line RAC drawings. In particular, observe that they are both planar graphs, which immediately implies that a straight-line RAC drawing has at most $6n - 12$ edges. A finer bound on the maximum number of edges of straight-line RAC drawings is discussed in Sect. 9.3.

9.3 Edge Density

It is known that every graph admits a RAC drawing with at most 3-bends per edge [36]. We will discuss in Sect. 9.4 efficient algorithms that compute 3-bend and 4-bend RAC drawings in polynomial area. In contrast, k-bend RAC drawings for $k \in \{0, 1, 2\}$ have a number of edges that is linear in the number of vertices, as for planar graphs and for most of the beyond-planar graph families. In this section, we survey the main bounds on the number of edges of RAC drawings. We adopt the following definitions of maximal, maximally dense, and optimal graphs, commonly used in the literature on graph drawing beyond planarity.

Given a nonnegative integer $k \le 2$, denote by $\mathscr{C}_k$ the class of graphs that admit a k-bend RAC drawing and let G be an n-vertex graph in $\mathscr{C}_k$. We say that G is *maximal* if adding any edge to it leads to a graph that is not in $\mathscr{C}_k$. G is *maximally dense* if it has the maximum number of edges over all n-vertex graphs in $\mathscr{C}_k$. G is *optimal* if it has the maximum density over all graphs of $\mathscr{C}_k$. By definition, an optimal graph is also maximally dense, while the converse may not be true. Analogously, a maximally dense graph is maximal, but not necessarily vice versa.

For straight-line drawings (i.e., for $k = 0$), the following result establishes a tight bound on the maximum number of edges of RAC drawings.

Theorem 9.1 ([36]) *Every n-vertex graph in $\mathscr{C}_0$ has at most $4n - 10$ edges. Also, for any $h \geq 3$ there exists an optimal graph in $\mathscr{C}_0$ with $n = 3h - 5$ vertices and $4n - 10$ edges.*

The upper bound of Theorem 9.1 is based on the following key observation (a detailed proof is given in [36]). Let G be a maximal graph $G \in \mathscr{C}_0$, Γ be a straight-line RAC drawing of G, and let Γ_{rb} and Γ_{rg} be a red-blue and a red-green sub-drawing of Γ, respectively. In each of the two planar drawings Γ_{rb} and Γ_{rg}, every internal face has at most two red edges and the external face consists of red edges only. This fact along with suitable applications of Euler's formula to Γ_{rb} and Γ_{rg} lead to the given bound. The second part of the theorem claims that the bound is tight. For each $h \geq 3$, an optimal graph $G_h \in \mathscr{C}_0$ is constructed as follows. Start from a maximal plane graph with h vertices and add to it the nodes and the edges of its dual plane graph, except for the node associated with the external face. Then, each face-node of the dual graph is connected to the three vertices of the corresponding face. It can be proved that G_h is straight-line RAC drawable as a consequence of a disk-packing theorem of Brightwell and Scheinerman [17]. As an example, Fig. 9.4a shows a straight-line RAC drawing of an optimal graph G_5 (having 10 vertices and 30 edges); in the figure, the vertices of the primal graph are black, while those of the dual graph are white.

We remark that an alternative proof for the upper bound given in Theorem 9.1 is given by Dujmović et al. [37]. Their proof exploits charging techniques instead of the red-blue-green coloring. The basic idea of such a technique is to assign suitably defined *charges* to the faces and vertices of the planar subdivision determined by a straight-line RAC drawing. These charges may be redistributed in order to satisfy desired properties, but leaving the total sum unchanged. Euler's formula is then used to relate the number of vertices to the sum of the charges, and the number of edges to the number of vertices.

For 1-bend and 2-bend RAC drawable graphs, the following bounds are proven by Arikushi et al. [6], but in this case the bounds are not known to be tight.

Theorem 9.2 ([6]) *Every n-vertex graph in $\mathscr{C}_1$ has at most $6.5n - 13$ edges.*

Theorem 9.3 ([6]) *Every n-vertex graph in $\mathscr{C}_2$ has at most $74.2n$ edges.*

While the proof of Theorem 9.2 still relies on charging techniques, Theorem 9.3 uses a stronger version of the popular crossing lemma due to Pach et al. [56]. The next theorem, proven by Angelini et al. [3], provides a better upper bound to the edge density of 1-bend RAC drawable graphs, which is almost tight.

Theorem 9.4 ([3]) *Every n-vertex graph in $\mathscr{C}_1$ has at most $5.5n - O(1)$ edges; also, there exist infinitely many graphs in $\mathscr{C}_1$ with $5n - O(1)$ edges.*

We finally mention that the edge density of straight-line RAC drawings has also been studied in a constrained scenario in which the graph is bipartite and the vertices of the two partition sets lie on two distinct parallel lines, called *layers*. Drawings of this type are called 2-*layer RAC drawings*. The densest n-vertex bipartite graph that

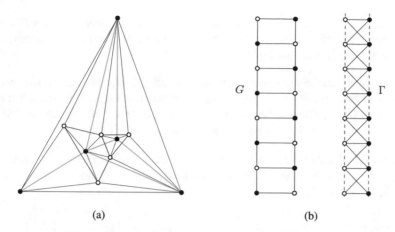

(a)   (b)

Fig. 9.4 a An optimal straight-line RAC drawing. **b** A ladder G with 16 vertices and a 2-layer RAC drawing Γ of G. The colors of the vertices denote the two partition sets

Table 9.1 Summary of results about the edge density of RAC drawings. In the third column, the symbol ✓ denotes that the bound is tight, while a symbol × means that the tightness of the given bound has not been proven

Type of RAC drawing	Maximum number of edges	Tightness	References
0-bend	$4n - 10$	✓	[36, 37]
1-bend	$5.5n - O(1)$	×	[3]
2-bend	$74.2n$	×	[6]
3-bend	$n(n-1)/2$	✓	[36]
2-layer	$1.5n - 2$	✓	[27]

admits a 2-layer RAC drawing consists of $1.5n - 2$ edges [27]. It is a biconnected graph called *ladder*, and consists of two paths of the same length $\langle u_1, u_2, \ldots, u_{\frac{n}{2}} \rangle$ and $\langle v_1, v_2, \ldots, v_{\frac{n}{2}} \rangle$, plus the edges (u_i, v_i) for each $i = 1, 2, \ldots, \frac{n}{2}$. A ladder G with 16 vertices and a 2-layer RAC drawing Γ of it are shown in Fig. 9.4b, where white and black vertices correspond to the vertices u_i and v_i, respectively. We recall that the study of the 2-layer drawing convention for bipartite graphs has a long tradition in graph drawing, as it clearly conveys the structure of a bipartite graph and, at the same time, it represents a building block for the popular Sugiyama-style framework, adopted for visualizing general graphs on several layers [59]. The algorithmic aspects concerned with the computation of 2-layer RAC drawings are discussed in Sect. 9.4.

Table 9.1 summarizes the bounds on the maximum number of edges for the different types of RAC drawings illustrated in this section. We conclude by highlighting two open problems for the edge density of RAC drawings. The first problem naturally arises from the bounds summarized in Table 9.1 (see also [32]).

Problem 9.1 *Improve the upper bounds on the maximum number of edges of 1-bend and 2-bend RAC drawable graphs given in Theorems 9.4 and 9.3, respectively, or prove that these bounds are tight.*

The second problem is concerned with providing lower bounds on the number of edges of maximal RAC drawable graphs. While this kind of question has been studied for other families of beyond-planar graphs, such as 1-planar graphs [14] and 2-planar graphs [7], it is apparently unexplored for RAC drawings.

Problem 9.2 *What is the minimum number of edges that a maximal k-bend RAC drawable graph can have, i.e., what is the density of the sparsest maximal graph $G \in \mathscr{C}_k$, for $k \in \{0, 1, 2\}$?*

9.4 Testing and Drawing Algorithms

We first discuss the known algorithms for computing RAC drawings with edge bends (Sect. 9.4.1), and then we recall the main time-complexity results about recognizing straight-line RAC drawable graphs (Sect. 9.4.2). Finally, we survey the literature about RAC drawings of planar graphs (Sect. 9.4.3).

9.4.1 RAC Drawings with Bends

As already mentioned in the previous section, it is not difficult to realize that every graph G admits a 3-bend RAC drawing [36]. There are at least two efficient algorithms that can be used to compute such a drawing Γ of G. The simplest algorithm is to place all the vertices of G on a horizontal line (in any order), using at most three bends per edge so that any two crossing segments have slopes 1 and -1, respectively (i.e., they form 45° and $-45°$ angles with the horizontal line). Figure 9.5a shows an example of a 3-bend RAC drawing of the complete graph K_6, computed with this approach. Another algorithm is based on computing an orthogonal drawing with box-vertices and at most one bend per edge, applying a technique by Papakostas and Tollis [57], and then on transforming this drawing into a RAC drawing with point-vertices and at most three bends per edge (refer to [36]). The transformation consists in replacing each box-vertex with a grid point inside the box, using at most one bend to connect this point to the boundaries of the box (which may yield up to two extra bends per edge). Both these drawing algorithms work in $O(n + m)$ time and yield 3-bend RAC drawings in $O(n^4)$ area, where n and m are the number of vertices and edges of the graph, respectively. In particular, the computed drawings have integer coordinates on a grid of size $O(n^2) \times O(n^2)$.

Di Giacomo et al. [26] provided a different trade-off between the number of bends and area requirement of RAC drawings, showing how to easily construct a 4-bend

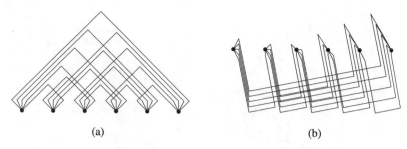

Fig. 9.5 Two RAC drawings of K_6: **a** 3-bend RAC drawing; **b** 4-bend RAC drawing

RAC drawing of any graph G in $O(n^3)$ area, where all vertices are still placed on the same horizontal line. More precisely, the drawing fits into an integer grid of size $O(n^2) \times O(n)$. An example of such a drawing for K_6 is depicted in Fig. 9.5b. The algorithm still works in $O(n + m)$ time. We summarize the aforementioned results with the following.

Theorem 9.5 ([26, 36]) *Let G be any graph with n vertices and m edges. There exist: (a) an $O(n + m)$-time algorithm that computes a 3-bend RAC drawing of G on an integer grid of size $O(n^2) \times O(n^2)$; (b) an $O(n + m)$-time algorithm that computes a 4-bend RAC drawing of G on an integer grid of size $O(n^2) \times O(n)$.*

Efficient algorithms are also known for computing 1-bend and 2-bend RAC drawings of graphs with small vertex degree. Recall that k-bend RAC drawable graphs, for $k \le 2$, are sparse, as stated by Theorems 9.1–9.3. Angelini et al. [1] prove the following result.

Theorem 9.6 ([1]) *Every n-vertex graph with vertex-degree at most $\Delta \in \{3, 6\}$ admits a $\frac{4}{3}$-bend RAC drawing in $O(n^2)$ area, which can be computed in $O(n)$ time.*

In other words, Theorem 9.6 claims that if a graph G has vertex-degree at most six, it always admits a 2-bend RAC drawing in quadratic area, while if G has vertex-degree at most three, it always admits a 1-bend RAC drawing with the same area requirement. The algorithms that construct such RAC drawings exploit the decomposition of a regular directed (multi)graph into directed 2-*factors*. For an undirected graph G, a 2-*factor* of G is a spanning subgraph consisting of a forest of cycles (i.e., vertex-disjoint cycles). Similarly, if G is directed, a *directed* 2-*factor* of G is a spanning subgraph consisting of a forest of directed cycles.

Given an undirected graph G of vertex-degree at most Δ, Eades, Symvonis, and Whitesides [40] show how to construct a Δ-regular digraph G' with the same vertex set as G such that: (i) each vertex of G' has both indegree and outdegree $d = \lceil \frac{\Delta}{2} \rceil$; ($ii$) G is a subgraph of the undirected underlying graph of G'; (iii) the edges of G' can be partitioned into d (edge-disjoint) directed 2-factors. If $\Delta = 6$, G' is partitioned into three directed 2-factors F_1, F_2, F_3. In this case, the algorithm of Theorem 9.6

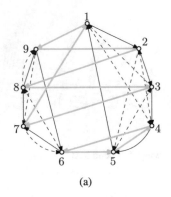
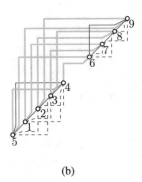

(a)   (b)

Fig. 9.6 Illustration of the technique to compute a 2-bend RAC drawing of a graph with vertex-degree $\Delta = 6$. **a** A Δ-regular digraph G' such that each vertex has both indegree and outdegree three. The set of edges is partitioned into three directed 2-factors (normal black edges, thick gray edges, dashed black edges). **b** A 2-bend RAC drawing of G'; to avoid a clutter visualization, edge directions (arrows) are not shown

Table 9.2 Summary of the algorithmic results about RAC drawings with bends. Δ is an upper bound to the maximum vertex-degree of the input graph. The table reports the number of bends per edge and the area requirement of the computed drawing, as well as the time complexity of the drawing algorithm; n and m denote the number of vertices and the number of edges of the input graph, respectively

Input graph	Bends per edge	Area requirement	Drawing time	References
Any	4	$O(n^3)$	$O(n+m)$	[26]
Any	3	$O(n^4)$	$O(n+m)$	[36]
$\Delta = 6$	2	$O(n^2)$	$O(n)$	[1]
$\Delta = 3$	1	$O(n^2)$	$O(n)$	[1]

constructs a 2-bend RAC drawing of G', placing all vertices on a line ℓ of slope 1 and such that: For each cycle C of F_1, all the edges of C except one are straight segments drawn along ℓ, while the remaining edge of C is above ℓ; each edge of F_2 is drawn above ℓ with one or two bends; each edge of F_3 is drawn below ℓ with one or two bends. Removing from the drawing the edges of G' that are not in G, we get a 2-bend RAC drawing of G. A similar technique is used when $\Delta = 3$. An illustration of the drawing technique for $\Delta = 6$ is depicted in Fig. 9.6.

Table 9.2 summarizes the main algorithmic results discussed in this section. In addition to these results, we mention that Fink et al. [42, 43] studied the problem of computing RAC drawings of graphs with few bends per edge, assuming that the vertices are mapped to a predefined set of points (locations) in the plane. A drawing of a graph where the vertices are mapped to a given set of points in the plane is known as a *point-set embedding*. Computing point-set embeddings of graphs is a well-studied problem, both when the one-to-one correspondence between vertices and points is given as part of the input (see, e.g., [55]) and when such a correspondence is not given (see, e.g., [51]). In particular, Fink et al. [42, 43] prove that every graph with n vertices and m edges can be mapped to any given set of points of an $n \times n$ integer grid in $O((n + m)^2)$ area and with at most three bends per edge, even when the one-to-one correspondence between vertices and points is prescribed (they guarantee that also the edge bends are placed at grid coordinates). For dense graphs, this result leads to the same area bound and edge-bend complexity given in [36]. Two open questions that naturally arise from the results discussed above are the following.

Problem 9.3 *Is it possible to realize any graph as a k-bend RAC drawing in quadratic area, for some value of $k \geq 3$?*

The aforementioned result by Fink et al. [42] implies that sparse graphs always admit a 3-bend RAC drawing in quadratic area.

Problem 9.4 *For $k \in \{1, 2\}$, what is the complexity of testing whether a given graph G admits a k-bend RAC drawing?*

Problem 9.4 is also mentioned in [32]. Recall that every graph admits a 3-bend RAC drawing and, as it will be discussed in Sect. 9.4.2, recognizing 0-bend RAC drawable graphs is NP-hard.

9.4.2   Straight-Line RAC Drawings

Concerning the recognition of straight-line RAC drawable graphs, the following hardness result has been proven by Argyriou, Bekos, and Symvonis [4].

Theorem 9.7 ([4]) *Given a graph G, it is NP-hard to decide whether G admits a straight-line RAC drawing.*

The proof of Theorem 9.7 exploits a reduction from the well-known 3-SAT problem. The building block gadget for this reduction is a straight-line RAC drawable graph of nine vertices, called *augmented square antiprism*. Its different straight-line RAC drawings can induce only two different embeddings (see Fig. 9.7).

Testing the existence of a straight-line RAC drawing is also NP-hard in the point-set embedding scenario, i.e., when each vertex of the graph has to be drawn at a fixed point of the plane [42].

On the positive side, polynomial-time testing algorithms for the straight-line RAC drawability are described for bipartite graphs. More precisely, the class of complete bipartite graphs that admit a straight-line RAC drawing is characterized by the following result, which immediately leads to a trivial $O(1)$-time testing algorithm for this class.

Theorem 9.8 ([35]) *A complete bipartite graph K_{n_1,n_2} ($n_1 \leq n_2$) admits a straight-line RAC drawing if and only if either $n_1 \leq 2$ or $n_1 = 3$ and $n_2 \leq 4$.*

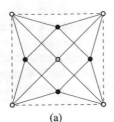
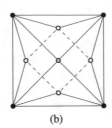
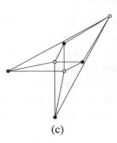

(a)   (b)   (c)

Fig. 9.7 a–b Two straight-line RAC drawings of the augmented square antiprism having different combinatorial embeddings. **c** A straight-line RAC drawing of graph $K_{3,4}$

In particular, every $K_{2,n}$ graph can be easily drawn without crossing edges while a straight-line RAC drawing of $K_{3,4}$ is depicted in Fig. 9.7c. The proof of Theorem 9.8 is then completed by showing that neither $K_{3,5}$ nor $K_{4,4}$ admit a straight-line RAC drawing. This is done by proving that there is no way to add an extra vertex into a straight-line RAC drawing of $K_{3,4}$ so to get a new straight-line RAC drawing. The proof is based on a case analysis that extensively uses the forbidden configurations of Properties 9.2 and 9.3.

Straight-line RAC drawings of bipartite graphs are also studied within the 2-layer drawing model. While edge density results about 2-layer RAC drawable graphs have been already discussed in Sect. 9.3, we recall here that such graphs can be efficiently recognized, as stated by the following result.

Theorem 9.9 ([27]) *Given a bipartite graph G, there exists a linear-time algorithm that tests whether G admits a 2-layer RAC drawing, and that computes such a drawing if it exists.*

It is important to remark that the edge crossings in a 2-layer drawing depends only on the linear ordering of the vertices on each layer; we call such an ordering a 2-*layer embedding*. The design of the linear-time testing and drawing algorithm in Theorem 9.9 is based on first showing that a connected graph is 2-layer RAC drawable if and only if it has a 2-layer embedding with no *fan crossings*, i.e., where a fan of edges is never crossed by another edge. Such an embedding is called a 2-*layer RAC embedding*. Then, the algorithm exploits a characterization of the class of graphs that admit a 2-layer RAC embedding. Roughly speaking, with reference to Fig. 9.8a, each graph in this class can consist of some (possibly none) nontrivial biconnected components, each being a spanning subgraph of a ladder (B_1 and B_2 in the figure), plus tree-like components with specific properties (formed by the dashed edges in the figure). A 2-layer RAC embedding of the graph in Fig. 9.8a is depicted in Fig. 9.8b.

Di Giacomo et al. [27] also prove that if G is an n-vertex 2-layer RAC drawable graph, it is possible to compute a 2-layer RAC embedding of G with a minimum number of edge crossings in $O(n^2 \log n)$ time. It is worth remarking that, without the restriction of having right angle crossings, the problem of computing drawings

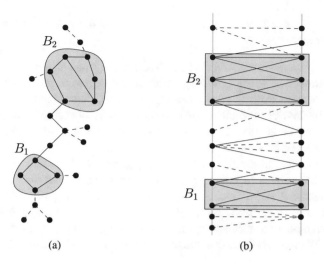

Fig. 9.8   a A bipartite graph G that is 2-layer RAC drawable. **b** A 2-layer RAC embedding of G; a RAC drawing with this embedding can be easily constructed

Table 9.3   Summary of the algorithmic results about straight-line RAC drawings

Input graph	Testing time	Drawing time	Drawing type	References
Any	NP-hard	–	–	[4]
Complete bipartite	$O(1)$	$O(n)$	Planar or Fig. 9.7c	[35]
Bipartite	$O(n)$	$O(n)$	2-layer	[27]
Bipartite	$O(n)$	$O(n^2 \log n)$	2-layer min-cross.	[27]

of graphs with minimum number of crossings is well-known to be NP-hard [45], and it remains hard for 2-layer drawings, even when the ordering of the vertices on one of the two layers is fixed and cannot be changed [39].

Table 9.3 summarizes the main algorithmic results discussed so far for straight-line RAC drawings. Additional results on 2-layered RAC drawings are concerned with the so-called *maximum* 2-*layer subgraph problem*. Namely, if G is a bipartite graph that does not admit a 2-layer RAC drawing, what is the time complexity of extracting a straight-line RAC drawable subgraph of G with the maximum number of edges? Unfortunately, this problem has been shown to be NP-hard [27], even if the order of the vertices on one of the two layers is fixed [28]. To attack the problem, different polynomial-time heuristics and a 3-approximation algorithm have been designed, implemented, and experimentally compared [28].

An interesting and still unanswered question about straight-line RAC drawings is the following (see also [32]).

Problem 9.5 *Does every graph with vertex-degree at most $\Delta = 3$ admit a straight-line RAC drawing?*

Recall that an n-vertex straight-line RAC drawing has at most $4n - 10$ edges, while the number of edges of a graph with vertex-degree at most three is at most $1.5n$. Thus, at least from an edge density perspective, there is margin for a positive answer. If the answer is negative, then the complexity of recognizing those graphs with vertex-degree at most three that are straight-line RAC drawable becomes a problem worthy of being studied.

From a practical point of view, given any graph G, there is a general lack of algorithmic techniques to compute a RAC drawing of G that also satisfies some important aesthetic requirements. In particular, since testing whether G admits a straight-line RAC drawing is NP-hard, while three bends per edge always suffice, one can design heuristics that compute RAC drawings of G with a small number of bent edges. We suggest to investigate the following research direction.

Problem 9.6 *Let G be any graph. Design polynomial-time heuristics that compute a RAC drawing of G with a small number of bent edges, or with a small number of bends in total, in polynomial area.*

9.4.3   RAC Drawings of Planar Graphs

A general research question addressed by different papers is whether one can draw a planar graph G allowing right angle crossings (i.e., relaxing the drawing planarity constraint) in order to improve some important drawing aesthetics with respect to every planar drawing of G. Some positive answers to this question are contained in a work by van Kreveld [52]. He proves that there exists an infinite family of planar graphs for which every straight-line planar drawing takes quadratic area, while a RAC drawing in linear area always exists. He also shows similar existential results for other drawing aesthetics, namely, edge-length ratio and vertex angular resolution.

Despite the above positive results, Angelini et al. [1] have shown that in general the area requirement of straight-line RAC drawings for planar graphs is the same as for planar drawings. More precisely, they prove that there exist infinitely many planar graphs for which every straight-line RAC drawing takes quadratic area. A family of planar graphs with this property is constructed as follows. Let G be a *nested triangles* graph on n vertices, that is, a 3-connected graph with $\frac{n}{3}$ 3-cycles nested one into the other, as in Fig. 9.9a. It is known that any straight-line planar drawing of G takes $\Omega(n^2)$ area [44]. Consider the graph G' obtained from G by replacing each edge e of G with the complete graph K_4, as shown in Fig. 9.9b (edge e is identified with an edge of K_4). Clearly, G is a subgraph of G', and G' has $O(n)$ vertices and edges. Angelini et al. [1] prove that in any straight-line RAC drawing of G' no two edges of G can cross, which implies that also any straight-line RAC drawing takes $\Omega(n^2)$ area.

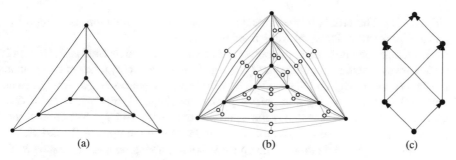

Fig. 9.9 a A nested triangles graph G. **b** The planar graph G' obtained from G by replacing each edge with a K_4 graph; the added vertices are white and the added edges are gray. **c** An upward RAC drawing of a non-upward planar digraph

If we allow bends along the edges, every planar graph with maximum vertex-degree Δ admits a 4-bend RAC drawing in area $O(n\sqrt{\Delta n})$, namely, on an integer grid of size $O(n) \times O(\sqrt{\Delta n})$ [2]. This implies that, if Δ is a sublinear function of n, this result leads to 4-bend RAC drawings in subquadratic area. Observe that, as already mentioned in Sect. 9.4.1, the result of Fink et al. [42] implies that every planar graph always admits a 3-bend RAC drawing in $O(n^2)$ area, regardless of its maximum vertex degree.

RAC drawings of planar graphs have also been investigated in combination with two popular graph drawing conventions, the *upward planar drawing* convention for directed graphs and the *simultaneous embedding* convention. We recall that a drawing of a directed graph (digraph for short) is *upward* if every edge is drawn as a curve monotonically increasing in the vertical direction, according to its orientation (see., e.g., Fig. 9.9c). A digraph is *upward planar* if it admits an upward planar drawing, i.e., an upward drawing with no edge crossing. It is not difficult to see that the digraph in Fig. 9.9c is not upward planar. A classical problem in graph drawing is testing whether a digraph is upward planar. Unfortunately, this problem is NP-hard in the general case [46], although polynomial-time testing algorithms are described for specific subfamilies of planar graphs, such as outerplanar, series-parallel, or 3-connected planar graphs (see, e.g., [20, 31, 33] for surveys and algorithmic comparisons on the subject). Moreover, although every upward planar digraph has a straight-line upward planar drawing [23], the area of such a drawing may require exponential area [24]. The introduction of *upward RAC drawings*, i.e., of drawings that are upward and RAC at the same time, has been proposed as a relaxation of the upward planar drawing convention, with the aim of overcoming the abovementioned limits in terms of recognition and area requirement. Namely, one can ask whether every planar acyclic digraph admits an upward RAC drawing and if every digraph that has an upward RAC drawing admits one with straight-line edges in polynomial area. For example, the non-upward planar digraph in Fig. 9.9c is drawn straight-line RAC in polynomial area. Unfortunately, in the general case both these questions have a negative answer, and recognizing upward RAC drawable planar digraphs remains

NP-hard [1]. The proofs of these results are based on augmentation techniques for planar digraphs similar to that illustrated in Fig. 9.9b.

The simultaneous embedding problem has been introduced by Braß et al. [16], and widely investigated in the literature (see, e.g., [12] for a survey). Given two planar graphs $G_1 = (V, E_1)$ and $G_2 = (V, E_2)$ with the same vertex set, a simultaneous embedding of G_1 and G_2 is a pair of planar drawings, Γ_1 of G_1 and Γ_2 of G_2, such that each vertex $v \in V$ has the same position in Γ_1 and Γ_2. Notice that, the edges of Γ_1 are allowed to cross the edges of Γ_2. While the upward RAC drawing convention is a generalization of the upward planar drawing one, *simultaneous RAC embeddings* of two planar graphs G_1 and G_2 restrict the classical simultaneous embedding convention, adding the constraint that the edges of Γ_1 can cross those of Γ_2 only at right angles. If we require that the edges are drawn as straight-line segments, recognizing those pairs of planar graphs that admit a simultaneous RAC embedding is an NP-hard problem [47], and constructive algorithms are known only for restricted pairs of planar graphs (even a wheel graph and a cycle might not admit a straight-line simultaneous RAC embedding) [5]. However, Bekos et al. [10] prove that every pair of planar graphs has a simultaneous RAC embedding with at most six bends per edge. We recall that three bends per edge suffice in the classical simultaneous embedding convention [41]; Di Giacomo et al. [25] prove in fact that the number of bends per edge can be easily reduced from three to two and also give smaller bounds for specific subfamilies of planar graphs, such as outerplanar graphs, paths, and cycles.

As a natural problem, one can think of studying a variant of the above described simultaneous RAC embedding convention, which generalizes the classical simultaneous embedding one, instead of restricting it. Namely, given two graphs $G_1 = (V, E_1)$ and $G_2 = (V, E_2)$, we can look for a simultaneous embedding $\langle \Gamma_1, \Gamma_2 \rangle$ of G_1 and G_2, such that each of the two drawings Γ_1 and Γ_2 is RAC, while there is no restriction on the union of the two drawings. This generalization, which we call here *simultaneous independent RAC embedding* of G_1 and G_2, may enlarge the set of graphs pairs that admit a simultaneous drawing with respect to the case in which Γ_1 and Γ_2 are required to be planar.

Problem 9.7 *What pairs of planar graphs $\langle G_1, G_2 \rangle$ admit a simultaneous independent RAC embedding $\langle \Gamma_1, \Gamma_2 \rangle$?*

We remark that a question similar to Problem 9.7 has been studied by Di Giacomo et al. [29], with the requirement that each of the two drawings Γ_1 and Γ_2 is *quasiplanar*, i.e., it does not contain three mutually crossing edges.

9.5 Inclusion Relationships with 1-Planar Graphs

We recall that n-vertex 1-planar graphs have at most $4n - 8$ edges [54], which is a tight bound. If we require that the edges are drawn as straight-line segments then 1-planar graphs have at most $4n - 9$ edges [30], which is also a tight bound. Since straight-line RAC drawings have at most $4n - 10$ edges, we immediately conclude that there are 1-planar graphs that do not admit a RAC drawing without bends. Eades and Liotta [38] prove in fact much stronger results about the relationships between the two classes. They show that: (i) There exist infinitely many 1-planar graphs with $4n - 10$ edges that cannot be drawn RAC with straight-line edges (e.g., graph G in Fig. 9.10); (ii) for values of $n \geq 85$, there exist n-vertex straight-line RAC drawable graphs that are not 1-planar (see, e.g., the graph G' in Fig. 9.10); (iii) every optimal straight-line RAC drawable graph (i.e., with $4n - 10$ edges) is 1-planar.

More recently Brandenburg et al. [15] have studied the relationship between *IC-planar* graphs and straight-line RAC drawable graphs. An IC-planar graph is a 1-planar graph that admits a 1-planar drawing in which any two pairs of crossing edges have no common endvertex (IC stands for *Independent Crossings*). They show that every IC-planar graph is straight-line RAC drawable. We remark that, the class of *NIC-planar graphs*, another subfamily of 1-planar graphs that properly includes IC-planar graphs, is not included in the family of straight-line RAC drawable graphs. A NIC-planar drawing is a 1-planar drawing in which any two pairs of crossing edges share at most one endvertex [9]. Dehkordi and Eades [22] show that also every *outer 1-planar graph* admits a straight-line RAC drawing in which every vertex appears on the external boundary. An outer 1-planar graph is a graph that admits a 1-planar drawing with all vertices on the external face (this class generalizes outerplanar graphs). More in general, Dehkordi and Eades characterize the graphs that admit a straight-line RAC drawing with all vertices on the external face.

Figure 9.10 summarizes the known inclusion relationships between 1-planar graphs and straight-line RAC drawable graphs.

Recent papers study drawings that are 1-planar and RAC at the same time. Bekos et al. [11] show that it is NP-hard to recognize 1-planar RAC drawable graphs with straight-line edges, but they also show that every 1-planar graph is 1-bend RAC drawable. Their proof is based on a constructive technique, which starts from an embedded 1-planar graph and recursively computes a 1-bend RAC drawing, by possibly changing the initial embedding. This technique does not guarantee polynomial-time area for the computed drawing. The existential results in [11] have been further strengthen by Chaplick, Lipp, Wolff, and Zink [18]. They prove that every 1-planar graph with a given embedding actually admits an embedding preserving 1-bend RAC drawing, and polynomial area can be achieved for 2-bend RAC drawings. They also prove that every NIC-plane graph admits an embedding preserving 1-bend RAC drawing in quadratic area. Concerning the area requirement of 1-planar RAC drawings, we also remark that for a simple family of embedded 1-planar graphs, called *kite-triangulations*, a cubic area lower bound for embedding preserving straight-line RAC drawings is proved by Angelini et al. [1]. A kite-triangulation is obtained by augmenting a plane triangulation with edges inside pairs of adjacent faces.

In Sect. 9.2, we have observed that straight-line RAC drawings do not contain an edge that crosses a fan (see Property 9.2). Drawings that do not have such a forbidden configuration are called *fan-crossing free* drawings [19], and a graph that has such a drawing is a *fan-crossing free* graph. Hence, the family of straight-line RAC drawable graphs is included in the class of fan-crossing free graphs, as well as

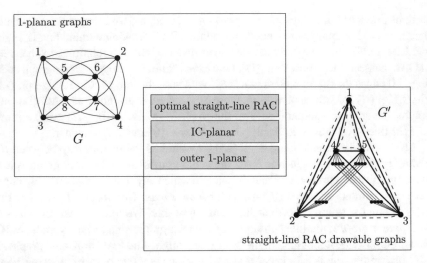

Fig. 9.10 Summary of the inclusion relationships between straight-line RAC drawable graphs and 1-planar graphs. The graph G has $4n - 10$ edges and it cannot be drawn RAC without bends. The graph G' is a schematic example of a straight-line RAC drawable graph that is not 1-planar; each dashed edge (u, v) represents four paths of three edges between u and v, like those connecting vertices $\{3, 4\}$ and vertices $\{2, 5\}$

the class of 1-planar graphs, which are fan-crossing free by definition. Concerning the edge density, as for 1-planar graphs, fan-crossing free graphs have at most $4n - 8$ edges if we allow edge bends and at most $4n - 9$ edges if we require that the edges are drawn as straight-line segments. Inspired by the results about 1-planar graphs that are not RAC drawable, we suggest to study the following open question.

Problem 9.8 *Are there n-vertex fan-crossing free graphs with at most* $4n - 10$ *edges that are not* 1*-planar and that are not straight-line RAC drawable?*

9.6   Concluding Remarks

The RAC drawing convention has inspired many papers that are not mentioned in this chapter. In particular, variants of RAC drawings have been proposed and studied, such as drawings in which all crossing edges form an angle that is either larger than a given value α or exactly α, for $\alpha \in (0, \frac{\pi}{2})$. For these variants, both edge density and algorithmic results have been described. For references about these topics, see the survey "The Crossing-Angle Resolution in Graph Drawing" [32] and the chapter "Angular Resolution: Around Vertices and Crossings" of this book, which also considers large vertex angles in addition to large crossing angles.

Acknowledgements The author's research was supported in part by the scientific project: "Algoritmi e sistemi di analisi visuale di reti complesse e di grandi dimensioni" - Ricerca di Base 2017 and 2018, Dipartimento di Ingegneria dell'Università degli Studi di Perugia.

References

1. Angelini, P., Cittadini, L., Didimo, W., Frati, F., Di Battista, G., Kaufmann, M., Symvonis, A.: On the perspectives opened by right angle crossing drawings. J. Graph Algorithms Appl. **15**(1), 53–78 (2011)
2. Angelini, P., Di Battista, G., Didimo, W., Frati, F., Hong, S., Kaufmann, M., Liotta, G., Lubiw, A.: Large angle crossing drawings of planar graphs in subquadratic area. In: Márquez, A., Ramos, P., Urrutia, J. (eds.) Computational Geometry - XIV Spanish Meeting on Computational Geometry, EGC 2011, Dedicated to Ferran Hurtado on the Occasion of His 60th Birthday, Alcalá de Henares, Spain, 27–30 June 2011, Revised Selected Papers. Lecture Notes in Computer Science, vol. 7579, pp. 200–209. Springer, Berlin (2011)
3. Angelini, P., Bekos, M.A., Förster, H., Kaufmann, M.: On RAC drawings of graphs with one bend per edge. Graph Drawing. Lecture Notes in Computer Science, vol. 11282, pp. 123–136. Springer, Berlin (2018)
4. Argyriou, E.N., Bekos, M.A., Symvonis, A.: The straight-line RAC drawing problem is NP-hard. J. Graph Algorithms Appl. **16**(2), 569–597 (2012)
5. Argyriou, E.N., Bekos, M.A., Kaufmann, M., Symvonis, A.: Geometric RAC simultaneous drawings of graphs. J. Graph Algorithms Appl. **17**(1), 11–34 (2013)
6. Arikushi, K., Fulek, R., Keszegh, B., Moric, F., Tóth, C.D.: Graphs that admit right angle crossing drawings. Comput. Geom.: Theory Appl. **45**(4), 169–177 (2012)
7. Auer, C., Brandenburg, F., Gleißner, A., Hanauer, K.: On sparse maximal 2 planar graphs. In: Didimo and Patrignani [37], pp. 555–556
8. Avital, S., Hanani, H.: Graphs. Gilyonot Lematematika **3**, 2–8 (1966)
9. Bachmaier, C., Brandenburg, F.J., Hanauer, K., Neuwirth, D., Reislhuber, J.: NIC-planar graphs. Discret. Appl. Math. **232**, 23–40 (2017)
10. Bekos, M.A., van Dijk, T.C., Kindermann, P., Wolff, A.: Simultaneous drawing of planar graphs with right-angle crossings and few bends. J. Graph Algorithms Appl. **20**(1), 133–158 (2016)
11. Bekos, M.A., Didimo, W., Liotta, G., Mehrabi, S., Montecchiani, F.: On RAC drawings of 1-planar graphs. Theor. Comput. Sci. **689**, 48–57 (2017)
12. Bläsius, T., Kobourov, S.G., Rutter, I.: Simultaneous embedding of planar graphs. In: Tamassia, R. (ed.) Handbook on Graph Drawing and Visualization, pp. 349–381. Chapman and Hall/CRC, London/Boca Raton (2013)
13. Bollobás, B.: Extremal Graph Theory. Academic, New York (1978)
14. Brandenburg, F., Eppstein, D., Gleißner, A., Goodrich, M.T., Hanauer, K., Reislhuber, J.: On the density of maximal 1-planar graphs. In: Didimo and Patrignani [37], pp. 327–338
15. Brandenburg, F.J., Didimo, W., Evans, W.S., Kindermann, P., Liotta, G., Montecchiani, F.: Recognizing and drawing IC-planar graphs. Theor. Comput. Sci. **636**, 1–16 (2016)
16. Braß, P., Cenek, E., Duncan, C.A., Efrat, A., Erten, C., Ismailescu, D., Kobourov, S.G., Lubiw, A., Mitchell, J.S.B.: On simultaneous planar graph embeddings. Comput. Geom.: Theory Appl. **36**(2), 117–130 (2007)
17. Brightwell, G., Scheinerman, E.R.: Representations of planar graphs. SIAM J. Discret. Math. **6**(2), 214–229 (1993)
18. Chaplick, S., Lipp, F., Wolff, A., Zink, J.: Compact drawings of 1-planar graphs with right-angle crossings and few bends. In: Biedl, T.C., Kerren, A. (eds.) Graph Drawing and Network Visualization - 26th International Symposium, GD 2018, Barcelona, Spain, 26–28 September 2018, Proceedings. Lecture Notes in Computer Science, vol. 11282, pp. 137–151. Springer (2018). https://doi.org/10.1007/978-3-030-04414-5_10

19. Cheong, O., Har-Peled, S., Kim, H., Kim, H.: On the number of edges of fan-crossing free graphs. Algorithmica **73**(4), 673–695 (2015)
20. Chimani, M., Zeranski, R.: Upward planarity testing in practice: SAT formulations and comparative study. ACM J. Exp. Algorithmics **20**, 1.2:1.1–1.2:1.27 (2015)
21. Comité Consultatif International Téléphonique et Télégraphique: Definition of numerical Petri nets - graphical representation, CCITT standards document, committee X (1985)
22. Dehkordi, H.R., Eades, P.: Every outer-1-plane graph has a right angle crossing drawing. Int. J. Comput. Geom. Appl. **22**(6), 543–558 (2012)
23. Di Battista, G., Tamassia, R.: Algorithms for plane representations of acyclic digraphs. Theor. Comput. Sci. **61**, 175–198 (1988)
24. Di Battista, G., Tamassia, R., Tollis, I.G.: Area requirement and symmetry display of planar upward drawings. Discret. Comput. Geom. **7**, 381–401 (1992)
25. Di Giacomo, E., Liotta, G.: Simultaneous embedding of outerplanar graphs, paths, and cycles. Int. J. Comput. Geom. Appl. **17**(2), 139–160 (2007)
26. Di Giacomo, E., Didimo, W., Liotta, G., Meijer, H.: Area, curve complexity, and crossing resolution of non-planar graph drawings. Theory Comput. Syst. **49**(3), 565–575 (2011)
27. Di Giacomo, E., Didimo, W., Eades, P., Liotta, G.: 2-layer right angle crossing drawings. Algorithmica **68**(4), 954–997 (2014)
28. Di Giacomo, E., Didimo, W., Grilli, L., Liotta, G., Romeo, S.A.: Heuristics for the maximum 2-layer RAC subgraph problem. Comput. J. **58**(5), 1085–1098 (2015)
29. Di Giacomo, E., Didimo, W., Liotta, G., Meijer, H., Wismath, S.K.: Planar and quasi-planar simultaneous geometric embedding. Comput. J. **58**(11), 3126–3140 (2015)
30. Didimo, W.: Density of straight-line 1-planar graph drawings. Inf. Process. Lett. **113**(7), 236–240 (2013)
31. Didimo, W.: Upward graph drawing. Encyclopedia of Algorithms, pp. 2308–2312. Springer, Berlin (2016)
32. Didimo, W., Liotta, G.: The crossing-angle resolution in graph drawing. In: Pach, J. (ed.) Thirty Essays on Geometric Graph Theory, pp. 167–184. Springer, New York (2013)
33. Didimo, W., Giordano, F., Liotta, G.: Upward spirality and upward planarity testing. SIAM J. Discret. Math. **23**(4), 1842–1899 (2009)
34. Didimo, W., Eades, P., Liotta, G.: Drawing graphs with right angle crossings. In: Dehne, F.K.H.A., Gavrilova, M.L., Sack, J., Tóth, C.D. (eds.) Algorithms and Data Structures, 11th International Symposium, WADS 2009, Banff, Canada, 21–23 August 2009. Proceedings. Lecture Notes in Computer Science, vol. 5664, pp. 206–217. Springer (2009)
35. Didimo, W., Eades, P., Liotta, G.: A characterization of complete bipartite RAC graphs. Inf. Process. Lett. **110**(16), 687–691 (2010)
36. Didimo, W., Eades, P., Liotta, G.: Drawing graphs with right angle crossings. Theor. Comput. Sci. **412**(39), 5156–5166 (2011)
37. Dujmovic, V., Gudmundsson, J., Morin, P., Wolle, T.: Notes on large angle crossing graphs. Chic. J. Theor. Comput. Sci. **2011** (2011)
38. Eades, P., Liotta, G.: Right angle crossing graphs and 1-planarity. Discret. Appl. Math. **161**(7–8), 961–969 (2013)
39. Eades, P., Wormald, N.C.: Edge crossings in drawings of bipartite graphs. Algorithmica **11**(4), 379–403 (1994)
40. Eades, P., Symvonis, A., Whitesides, S.: Three-dimensional orthogonal graph drawing algorithms. Discret. Appl. Math. **103**(1–3), 55–87 (2000)
41. Erten, C., Kobourov, S.G.: Simultaneous embedding of planar graphs with few bends. J. Graph Algorithms Appl. **9**(3), 347–364 (2005). http://jgaa.info/accepted/2005/ErtenKobourov2005.9.3.pdf
42. Fink, M., Haunert, J., Mchedlidze, T., Spoerhase, J., Wolff, A.: Drawing graphs with vertices at specified positions and crossings at large angles. In: van Kreveld, M.J., Speckmann, B. (eds.) Graph Drawing - 19th International Symposium, GD 2011, Eindhoven, The Netherlands, 21–23 September 2011, Revised Selected Papers. Lecture Notes in Computer Science, vol. 7034, pp. 441–442. Springer (2011)

43. Fink, M., Haunert, J., Mchedlidze, T., Spoerhase, J., Wolff, A.: Drawing graphs with vertices at specified positions and crossings at large angles. In: Rahman, M.S., Nakano, S. (eds.) WAL-COM: Algorithms and Computation - 6th International Workshop, WALCOM 2012, Dhaka, Bangladesh, 15–17 February 2012. Proceedings. Lecture Notes in Computer Science, vol. 7157, pp. 186–197. Springer (2012). https://doi.org/10.1007/978-3-642-28076-4_19
44. de Fraysseix, H., Pach, J., Pollack, R.: How to draw a planar graph on a grid. Combinatorica **10**(1), 41–51 (1990)
45. Garey, M.R., Johnson, D.S.: Computers and Intractability: A Guide to the Theory of NP-Completeness. W. H. Freeman, New York (1979)
46. Garg, A., Tamassia, R.: On the computational complexity of upward and rectilinear planarity testing. SIAM J. Comput. **31**(2), 601–625 (2001)
47. Grilli, L.: On the NP-hardness of GRacSim drawing and k-SEFE problems. J. Graph Algorithms Appl. **22**(1), 101–116 (2018)
48. Huang, W.: Using eye tracking to investigate graph layout effects. In: APVIS, pp. 97–100 (2007)
49. Huang, W., Hong, S.H., Eades, P.: Effects of crossing angles. In: PacificVis 2008, pp. 41–46 (2008)
50. Huang, W., Eades, P., Hong, S.: Larger crossing angles make graphs easier to read. J. Visual Lang. Comput. **25**(4), 452–465 (2014)
51. Kaufmann, M., Wiese, R.: Embedding vertices at points: few bends suffice for planar graphs. J. Graph Algorithms Appl. **6**(1), 115–129 (2002)
52. van Kreveld, M.J.: The quality ratio of RAC drawings and planar drawings of planar graphs. In: Brandes, U., Cornelsen, S. (eds.) Graph Drawing - 18th International Symposium, GD 2010, Konstanz, Germany, 21–24 September 2010. Revised Selected Papers. Lecture Notes in Computer Science, vol. 6502, pp. 371–376. Springer (2010)
53. Kupitz, Y.S.: Extremal Problems in Combinatorial Geometry. Lecture Notes Series. Matematisk institut, Aarhus universitet (1979)
54. Pach, J., Tóth, G.: Graphs drawn with few crossings per edge. Combinatorica **17**(3), 427–439 (1997)
55. Pach, J., Wenger, R.: Embedding planar graphs at fixed vertex locations. Graphs Comb. **17**(4), 717–728 (2001)
56. Pach, J., Radoicic, R., Tardos, G., Tóth, G.: Improving the crossing lemma by finding more crossings in sparse graphs. Discret. Comput. Geom. **36**(4), 527–552 (2006)
57. Papakostas, A., Tollis, I.G.: Efficient orthogonal drawings of high degree graphs. Algorithmica **26**(1), 100–125 (2000)
58. Roberts, M.J.: Underground Maps Unravelled: Explorations in Information Design. Maxwell J. Roberts (2012). https://books.google.it/books?id=khtYMwEACAAJ
59. Sugiyama, K., Tagawa, S., Toda, M.: Methods for visual understanding of hierarchical system structures. IEEE Trans. Syst. Man Cybern. **11**(2), 109–125 (1981)
60. Vignelli, M.: New York subway map (2008). http://secondavenuesagas.com/2008/05/02/mens-vogue-calls-on-vignelli-for-a-long-awaited-update/

Chapter 10
Angular Resolutions: Around Vertices and Crossings

Yoshio Okamoto

Abstract Angular resolution is one of the well-known esthetic criteria for graph drawing, but its theoretical properties are not well understood. For a straight-line drawing of a graph, its *vertex angular resolution* is the minimum angle formed by two consecutive edges around a vertex, and its *crossing angular resolution* is the minimum angle formed by a crossing, while the crossing angular resolution is defined to be 2π if there is no crossing. The *total angular resolution* of a straight-line drawing is the minimum of the vertex angular resolution and the crossing angular resolution. The vertex/crossing/total angular resolution of a graph is the supremum of the vertex/crossing/total angular resolution of any straight-line drawing of the graph. In this chapter, we review some of the results on angular resolution in the literature, and identify several open problems in the field.

10.1 Introduction

This chapter is concerned with straight-line drawings of undirected graphs, in which each edge is drawn as a straight-line segment joining two points that represent vertices.

Angular resolution is one of the well-known esthetic criteria for graph drawing, but its theoretical properties are not well understood. For a straight-line drawing of a graph, its *vertex angular resolution* is the minimum angle formed by two consecutive edges around a vertex, and its *crossing angular resolution* is the minimum angle formed by a crossing, while the crossing angular resolution is defined to be 2π if there is no crossing. The *total angular resolution* of a straight-line drawing is

Y. Okamoto (✉)
The University of Electro-Communications, Chofugaoka 1–5–1, Chofu, Tokyo, Japan
e-mail: okamotoy@uec.ac.jp

RIKEN Center for Advanced Intelligence Project, Nihonbashi 1-chome Mitsui Building, 15th floor, Nihonbashi 1–4–1, Chuo-ku, Tokyo, Japan

© Springer Nature Singapore Pte Ltd. 2020
S.-H. Hong and T. Tokuyama (eds.), *Beyond Planar Graphs*,
https://doi.org/10.1007/978-981-15-6533-5_10

the minimum of the vertex angular resolution and the crossing angular resolution. The vertex/crossing/total angular resolution of a graph is the supremum of the vertex/crossing/total angular resolution of any straight-line drawing of the graph. In this chapter, we review some of the results on angular resolution in the literature, and identify several open problems in the field.

This chapter is organized as follows. In Sect. 10.2, we introduce the necessary notation and look at a few examples. Section 10.3 is devoted to vertex angular resolution, where we will see two techniques to produce drawings with large vertex angular resolutions. The first one is to use a proper vertex coloring of the square of a graph, and the second one is to use a fixed set of slopes. In Sect. 10.4, we turn our attention to crossing angular resolution. There, we will see the relationship of crossing angular resolution with right-angle crossing graphs (or RAC graphs). In Sect. 10.5, we focus on total angular resolution, and, in Sect. 10.6, we conclude the chapter.

As is often the case, my view of the field is biased, and I do not even try to be comprehensive. Nevertheless, I hope this chapter still gives an introduction to the study of angular resolutions of graph drawing for interested readers.

10.2 Definitions and Examples

Let D be a straight-line drawing of an undirected graph G. In this chapter, we do not allow a vertex is placed on the relative interior of an edge, but we allow three or more edges to cross at the same point. For a vertex v in the drawing, its *angular resolution* is defined as the minimum angle formed by two edges incident to v. If the degree of v is zero, the resolution is defined to be ∞, and if the degree of v is one, the resolution is defined to be 2π. The *vertex angular resolution* of the drawing D is the minimum resolution over all vertices of D and denoted by $\mathsf{var}(D)$.

For a crossing in D, its *angular resolution* is defined as the minimum angle formed by two edges passing through the crossing. The *crossing angular resolution* of the drawing D is the minimum resolution over all crossings of D and denoted by $\mathsf{car}(D)$. If D has no crossing, then the crossing angular resolution is defined to be 2π.

The *total angular resolution* of the drawing D is the minimum of the vertex angular resolution and the crossing angular resolution, and denoted by $\mathsf{tar}(D)$.

The *vertex angular resolution of a graph* G is the supremum of the vertex angular resolutions of all straight-line drawings of G, and denoted by $\mathsf{var}(G)$. Similarly, the *crossing angular resolution of a graph* G is the supremum of the crossing angular resolutions of all straight-line drawings of G, and denoted by $\mathsf{car}(G)$. The *total angular resolution of a graph* G is the supremum of the total angular resolutions of all straight-line drawings of G, and denoted by $\mathsf{tar}(G)$.

We note that vertex (crossing, total) angular resolution is also called vertex (crossing, total) angle resolution in the literature, respectively.

Now, we will see a couple of examples. Consider K_4, a complete graph with four vertices. Refer to Fig. 10.1. In the left drawing of Fig. 10.1, the vertex angular

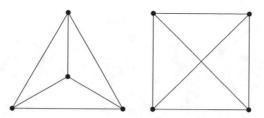

Fig. 10.1 Angular resolutions of K_4, a complete graph with four vertices. In the left drawing, the vertex angular resolution is $\pi/6$, the crossing angular resolution is 2π, and thus the total angular resolution is $\pi/6$. In the right drawing, the vertex angular resolution is $\pi/4$, the crossing angular resolution is $\pi/2$, and thus the total angular resolution is $\pi/4$

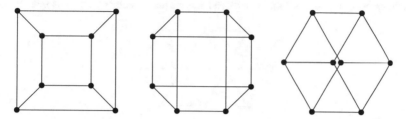

Fig. 10.2 Angular resolutions of Q_3, a three-dimensional cube. In the left drawing, the vertex angular resolution is $\pi/4$, the crossing angular resolution is 2π, and thus the total angular resolution is $\pi/4$. In the middle drawing, the vertex angular resolution is $\pi/4$, the crossing angular resolution is $\pi/2$, and thus the total angular resolution is $\pi/4$. In the right drawing, the vertex angular resolution is $\pi/3$, the crossing angular resolution is $\pi/3$, and thus the total angular resolution is $\pi/3$

resolution is $\pi/6$, the crossing angular resolution is 2π, and thus the total angular resolution is $\pi/6$. On the other hand, in the right drawing of Fig. 10.1, the vertex angular resolution is $\pi/4$, the crossing angular resolution is $\pi/2$, and thus the total angular resolution is $\pi/4$. A lesson learned is that introducing a crossing may increase the total angular resolution of the drawing. It turns out that the total angular resolution of K_4 is $\pi/4$. A proof will be given in Sect. 10.5.

Next, consider a three-dimensional cube Q_3. It has 8 vertices and 12 edges. Refer to Fig. 10.2. In the left drawing, the vertex angular resolution is $\pi/4$, the crossing angular resolution is 2π, and thus the total angular resolution is $\pi/4$. In the middle drawing, the vertex angular resolution is $\pi/4$, the crossing angular resolution is $\pi/2$, and thus the total angular resolution is $\pi/4$. In the right drawing, the vertex angular resolution is $\pi/3$, the crossing angular resolution is $\pi/3$, and thus the total angular resolution is $\pi/3$. It turns out that the total angular resolution of Q_3 is $\pi/3$. A proof will be given in Sect. 10.5.

Next, consider the Petersen graph. Refer to Fig. 10.3. In the left drawing, the vertex angular resolution is $\pi/5$, the crossing angular resolution is $2\pi/5$, and thus the total angular resolution is $\pi/5$. In the middle drawing, the vertex angular resolution is $\pi/6$, the crossing angular resolution is $\pi/3$, and thus the total angular resolution is $\pi/6$. In the right drawing, the vertex angular resolution is $\pi/3$, the crossing angular

Fig. 10.3 Angular resolutions of the Petersen graph. In the left drawing, the vertex angular resolution is $\pi/5$, the crossing angular resolution is $2\pi/5$, and thus the total angular resolution is $\pi/5$. In the middle drawing, the vertex angular resolution is $\pi/6$, the crossing angular resolution is $\pi/3$, and thus the total angular resolution is $\pi/6$. In the right drawing, the vertex angular resolution is $\pi/3$, the crossing angular resolution is $\pi/3$, and thus the total angular resolution is $\pi/3$

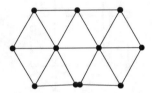

Fig. 10.4 An example in which the vertex angular resolution is never attained [13]. The vertex angular resolution of the drawing is $\pi/3 - \varepsilon$, where $\varepsilon > 0$ depends on the small gap between two vertices at the bottom

resolution is $\pi/3$, and thus the total angular resolution is $\pi/3$. It turns out that the total angular resolution of the Petersen graph is $\pi/3$. A proof will be given in Sect. 10.5.

One may wonder why the definitions of angular resolutions of a graph use "supremum" rather than "maximum." This is because the maximum is not necessarily attained. Refer to Fig. 10.4. The vertex angular resolution of the drawing is $\pi/3 - \varepsilon$, where ε depends on the small gap between two vertices at the bottom. This gap cannot be zero, as otherwise the drawing is degenerated (i.e., the map from the vertex set to the plane is not injective). Thus, the vertex angular resolution of this graph is $\pi/3$, which is never attained.

10.3 Vertex Angular Resolution

Vertex angular resolution was first investigated by Formann, Hagerup, Haralambides, Kaufmann, Leighton, Simvonis, Welzl, and Woeginger [13], but under the name of "resolution." As observed by them, the following easy upper bound for the vertex angular resolution can be obtained. Namely, for every undirected graph G with at least one edge, $\mathsf{var}(G) \le 2\pi/\Delta(G)$, where $\Delta(G)$ is the maximum degree of G. This is because a vertex v of maximum degree is incident to $\Delta(G)$ edges and those $\Delta(G)$ edges partition the degree of 2π. In other words, we have

Fig. 10.5 Proof of $\mathrm{var}(G) = \Omega(1/\Delta(G)^2)$ by Formann et al. [13]. The left figure shows a given graph G, and the middle figure is the square G^2, together with a proper 5-coloring. The right figure shows how to place the vertices of G around the corner of a regular pentagon

$$\mathrm{var}(G) = O(1/\Delta(G)) \tag{10.1}$$

for every undirected graph G.

As a general lower bound for the vertex angular resolution, Formann et al. [13] proved that

$$\mathrm{var}(G) = \Omega(1/\Delta(G)^2) \tag{10.2}$$

for every undirected graph G. This implies that the vertex angular resolution of a bounded-degree graph is constant (i.e., does not depend on the number of vertices). Since their argument is nice and short, we reproduce the proof here. In the proof, we explicitly create a straight-line drawing D of G such that $\mathrm{var}(D) = \Omega(1/\Delta(G)^2)$.

The idea is to use a proper vertex coloring of the square G^2. The *square* G^2 of a graph G is defined as follows. The vertex set of G^2 is the same as that of G, and two vertices u and v are adjacent in G^2 if and only if they are within a distance of two in G.

The construction goes as follows. Refer to Fig. 10.5. First, we find a proper vertex coloring of the square G^2. It is a basic fact that the greedy algorithm always finds a proper vertex coloring with $\Delta(H) + 1$ colors for every undirected graph H.[1] Therefore, the vertices of the square G^2 can be properly colored with $\Delta(G^2) + 1 = O(\Delta(G)^2)$ colors. Let $\chi = O(\Delta(G)^2)$ be the number of colors used in this coloring. Then, consider a regular χ-gon, and associate each of the χ colors with a corner of the χ-gon. For every vertex v of the graph, we place v around the corner of the χ-gon associated with the color of v. We draw every edge as it runs in G. This completes the construction. Note that the construction can be done in quadratic time.

We can observe that the vertex angular resolution of the constructed drawing is $\Omega(1/\chi)$ as follows. Let v be a vertex of G, and u_1, u_2 be two neighbors of v. By the construction of G^2, in a proper vertex coloring of G^2, the three vertices v, u_1, and

---

[1]This can be seen as follows. We will color the vertices of H with colors in $\{1, 2, \ldots, \Delta(H) + 1\}$. When we color a vertex v, the number of vertices that have already been colored is at most the degree of v, which is at most $\Delta(H)$. Thus, there is still a color that remains unused and this color can be used for v.

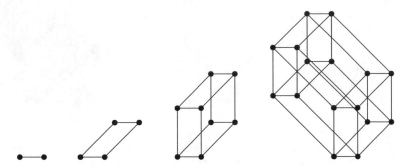

Fig. 10.6 The drawing of a d-dimensional cube proposed by Formann et al. [13], when $d = 4$

u_2 receive different colors. Therefore, the angle formed by two edges u_1v and u_2v in the constructed drawing is at least π/χ approximately, which is $\Omega(1/\chi)$. Since $\chi = O(\Delta(G)^2)$, this gives a straight-line drawing D with $\mathsf{var}(D) = \Omega(1/\Delta(G)^2)$.

This coloring argument was used to show that $\mathsf{var}(G) = \Omega(1/\Delta(G))$ for every planar graph G. This is true because G^2 has a proper vertex coloring with $O(\Delta(G))$ colors when G is planar. A short and simple argument for this bound can be found in [16]. The determination of a tight upper bound for the chromatic number of the square of a planar graph has a long history in graph theory. See [15] for the current best bound of $\frac{3}{2}\Delta(G) + o(\Delta(G))$.

Consider the case when G is a d-dimensional cube. Formann et al. [13] proved that $\mathsf{var}(G) \geq \pi/d$. To this end, they gave the following algorithm. We first fix the following set of d slopes of line segments used for drawing edges. The set is $\{0, \pi/d, 2\pi/d, \ldots, (d-1)\pi/d\}$. The drawing is constructed iteratively. The one-dimensional cube (i.e., two vertices and one edge) is drawn horizontally. Suppose that, for each k, $1 \leq k < d$, the k-dimensional cube is drawn using the first k slopes in our slope set. Then, to draw a $(k + 1)$-dimensional cube, we create two copies of the drawing of a k-dimensional cube, and translate one of them along the $(k + 1)$th slope in the slope set. Then, the missing edges between two copies are drawn using the $(k + 1)$th slope. Figure 10.6 shows an example. Since the slopes in the drawing are restricted to our set, the vertex angular resolution of the constructed drawing is π/d.

This drawing for d-dimensional cubes is an example of *the method of a set of fixed slopes*. The method recurs for crossing angular resolution and total angular resolution in the subsequent sections.

The lower bound in Eq. (10.2) is almost tight. Formann et al. [13] proved by a probabilistic argument that there exists a Δ-regular graph G such that $\mathsf{var}(G) = O((\log \Delta)/\Delta^2)$ for any Δ.

On the computational side, Formann et al. [13] proved that it is NP-hard to decide if the vertex angular resolution of a given undirected graph G is $2\pi/\Delta(G)$ or not, even when $\Delta(G) = 4$. This implies that the computation of the vertex angular resolution of a given undirected graph is NP-hard.

Fig. 10.7 Planar graphs G used by Garg and Tamassia [14] for which every straight-line plane drawing D has $\mathsf{var}(D) = O(\sqrt{(\log \Delta(G))/\Delta(G)^3})$

Straight-line drawings obtained by the results in Formann et al. [13] may produce crossings. Therefore, people started to look at the vertex angular resolutions of straight-line *plane* drawings, i.e., straight-line drawings without edge crossings. Of course, in this case, graphs under investigation must be planar.

For lower bounds, Malitz and Papakostas [21] proved that every planar graph G has a straight-line plane drawing D with $\mathsf{var}(D) = \Omega(1/\alpha^{\Delta(G)})$ with $\alpha = 1/(3 + 2\sqrt{3}) \approx 0.15$. They also proved that every outerplanar graph G has a straight-line plane drawing D with $\mathsf{var}(D) = \Omega(1/\Delta(G))$. Their proof for planar graphs used the coin representation of a planar graph [19] (i.e., every planar graph can be represented by a system of touching disks) together with an appropriate Möbius transformation. This idea was also used for the planar slope number problem by Keszegh, Pach, and Pálvölgyi [18]. The *planar slope number* of a planar graph G is the minimum number of slopes formed by edges in all straight-line plane drawings of G. Keszegh et al. [18] proved that the planar slope number of a planar graph only depends on the maximum degree.

For an upper bound, Garg and Tamassia [14] constructed a planar graph G such that every straight-line plane drawing D of G has $\mathsf{var}(D) = O(\sqrt{(\log \Delta(G))/\Delta(G)^3})$. Their recursive construction is illustrated in Fig. 10.7. This gives rise to the following open problem.

Problem 10.1 Determine the asymptotically tight bound for the maximum vertex angular resolution over all straight-line plane drawings of a planar graph with maximum degree Δ.

10.4   Crossing Angular Resolution

As far as the author knows, the explicit introduction of crossing angular resolution was first made by Di Giacomo, Didimo, Eades, Hong and Liotta [8], who simply called the crossing angular resolution the *crossing resolution*. They proved that the crossing angular resolution of a complete graph K_n with n vertices is at least $\pi/\lceil n/3 \rceil$, and at most $\pi/(\bar{\theta}(K_n) - 1)$, where $\bar{\theta}(G)$ is the geometric thickness of a graph G. The *geometric thickness* of an undirected graph $G = (V, E)$ is defined as the minimum number k of such that E is the disjoint union of k sets $E_1 \cup E_2 \cup \cdots \cup E_k$ and there exists a straight-line drawing of G in which no pair of edges in E_i crosses for any $i \in$

$\{1, 2, \ldots, k\}$. Dillencourt, Eppstein, and Hirschberg [11] proved that the geometric thickness of a complete graph with n vertices is at least $\lceil (n/5.646) + 0.342 \rceil$ when $n \geq 12$. Therefore, we obtain the following bounds:

$$\frac{3\pi}{n}(1 + o(1)) \leq \mathsf{car}(K_n) \leq \frac{5.646\pi}{n}(1 + o(1)).$$

This gives rise to the following problem.

Problem 10.2 Determine (the leading term of) the crossing angular resolution of a complete graph with n vertices.

Indeed, the upper bound holds in general: Di Giacomo et al. [8] also proved that

$$\mathsf{car}(G) \leq \frac{\pi}{\bar{\theta}(G) - 1} \tag{10.3}$$

for every undirected non-planar graph G. Equation (10.3) has several implications. Among others, we observe that the crossing angular resolution of bounded-degree graphs can be arbitrarily small. To see this, we use the following fact proved by Barát, Matoušek, and Wood [5]: for all $\Delta \geq 9$ and $\varepsilon > 0$, there exists a Δ-regular n-vertex graph H such that $\bar{\theta}(H) = \Omega(\sqrt{\Delta}n^{1/2-4/\Delta-\varepsilon})$. Their proof is based on a counting argument, and thus non-constructive. Then, by Eq. (10.3), for that graph H

$$\mathsf{car}(H) = O(1/\sqrt{\Delta}n^{1/2-4/\Delta-\varepsilon}). \tag{10.4}$$

On the other hand, this bound only holds when $\Delta \geq 9$ and it is not clear if the crossing angular resolution of an undirected graph G can be bounded by a function of the maximum degree $\Delta(G)$ when $\Delta(G) \leq 8$. When $\Delta(G) = 3$, this is possible, as we will see in Sect. 10.5. However, the following is still open.

Problem 10.3 Can the crossing angular resolution be bounded by a constant from below if the maximum degree is between 4 and 8?

As one might have already guessed, the crossing angular resolution is closely related to the so-called right-angle crossing drawings of graphs. A straight-line drawing of an undirected graph is a *right-angle crossing drawing* (or a *RAC drawing*) if the crossing angular resolution is at least $\pi/2$. An undirected graph is a *right-angle crossing graph* (or a *RAC graph*) if it admits a straight-line RAC drawing.

The concepts of RAC drawings and RAC graphs were first introduced by Didimo, Eades, and Liotta [9]. They proved that every RAC graph with $n \geq 4$ vertices can have at most $4n - 10$ edges, and this is tight. Argyriou, Bekos, and Symvonis [3] proved that it is NP-hard to decide if a given undirected graph is a RAC graph. This implies that the computation of the crossing angular resolution of a given undirected graph is NP-hard.

Motivated by RAC graphs, Dujmovic, Gudmundsson, Morin, and Wolle [12] studied the large-angle crossing graphs. Let α be a real number such that $0 < \alpha <$

$\pi/2$. A straight-line drawing of an undirected graph is an α-*angle crossing drawing* (or an αAC *drawing*) if the crossing angular resolution is at least α. An undirected graph is an α-*angle crossing graph* (or an αAC *graph*) if it admits a straight-line αAC drawing. Dujmovic et al. [12] proved that every αAC graph with $n \geq 3$ vertices has at most $(\pi/\alpha)(3n - 6)$ edges. This implies the following: if an n-vertex graph G has more than $(\pi/\alpha)(3n - 6)$ edges, then $\mathsf{car}(G) \leq \alpha$. For example, if the graph G has $\Theta(n^2)$ edges, then $\mathsf{car}(G) = O(1/n)$. Similarly, if we denote the minimum degree of G by $\delta(G)$, then the number of edges is at least $\delta(G)n/2$, and consequently

$$\mathsf{car}(G) = O(1/\delta(G)) \tag{10.5}$$

for every graph G.

So far, we mainly looked at upper bounds for the crossing angular resolution of a graph. Now, we move our attention to lower bounds.

When a graph G is planar, we know there exists a straight-line plane drawing of G by Fáry's theorem. By definition, the crossing angular resolution of a straight-line plane drawing is 2π. Therefore, the crossing angular resolution of every planar graph is 2π.

As we have already seen in this section, for a complete graph K_n, we have $\mathsf{car}(K_n) = \Theta(1/n)$. For a complete bipartite graph $K_{m,n}$, we have $\mathsf{car}(K_{m,n}) = \Omega(1/\max\{m, n\})$. This is a consequence from its total angular resolution; see Sect. 10.5.

As a useful method to give a straight-line drawing with large crossing angular resolution, we employ the method of a set of fixed slopes. We used this method for the drawing of the d-dimensional cube with large vertex angular resolution in Sect. 10.3, and we will revisit this method for total angular resolution in Sect. 10.5. Here, we concentrate on bounding crossing angular resolution. For example, if each edge is drawn horizontally or vertically, then the crossing angular resolution will be $\pi/2$.

As an example, consider a four-dimensional cube Q_4. Refer to Fig. 10.8. It has two disjoint copies of a three-dimensional cube Q_3. Since Q_3 is planar, it has a straight-line plane drawing. Now we squeeze those copies so that they are almost flat, and all edges are almost horizontal. We place each of them beside parallel lines, and complete our drawing of Q_4 by drawing the remaining edges vertically between the two subcubes. Then, the crossing angular resolution of the resulting drawing is $\pi/2 - \varepsilon$, where ε depends on the "flatness" of the drawing of Q_3. Therefore, $\mathsf{car}(Q_4) \geq \pi/2$.

The same method shows that $\mathsf{car}(G) \geq \pi/2$ when G is a three-dimensional grid (Fig. 10.9) and a six-dimensional cube Q_6 (Fig. 10.10 (left)). If we use three slopes $0, \pi/3, 2\pi/3$, then we can obtain the crossing angular resolution $\pi/3$. This is illustrated with a nine-dimensional cube Q_9 in Fig. 10.10 (right). In general, the crossing angular resolution of a d-dimensional cube Q_d is at least $\pi/\lceil d/3 \rceil$, which can be observed by using $\lceil d/3 \rceil$ slopes. Note that Q_d has $dn/2$ edges, and thus $\mathsf{car}(Q_d) < 6\pi/d$ by the result on α-angle crossing drawings mentioned above [12]. This motivates the following problem.

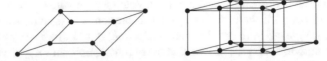

Fig. 10.8 The crossing angular resolution of a four-dimensional cube. The left figure shows a straight-line plane drawing of a three-dimensional cube. In the right figure, this drawing of a three-dimensional cube is made almost flat and duplicated. Adding the vertical edges results in a straight-line drawing with crossing angular resolution approximately $\pi/2$

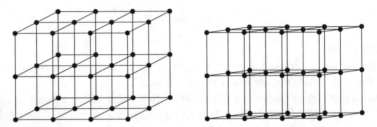

Fig. 10.9 The crossing angular resolution of a three-dimensional grid is $\pi/2$. The left figure is a "usual" straight-line drawing, and the right figure is a straight-line drawing with crossing angular resolution approximately $\pi/2$

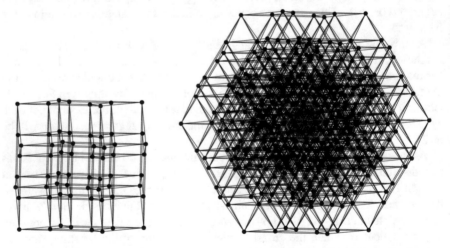

Fig. 10.10 The crossing angular resolution of a six-dimensional cube is $\pi/2$ (left), and the crossing angular resolution of a nine-dimensional cube is at least $\pi/3$ (right)

Problem 10.4 Determine the crossing angular resolution of a d-dimensional cube.

Another example was given by Didimo, Kaufmann, Liotta, Okamoto, and Spill-ner [10]. They studied a leveled drawing of a leveled tree. In a *leveled tree*, distinct real numbers are associated with the vertices, and in a *leveled drawing* of a leveled tree, the y-coordinate of each vertex v must be identical to the real number associated

Fig. 10.11 A leveled
drawing of a leveled tree
(left), and another leveled
drawing of the same leveled
tree with crossing angular
resolution $\pi/2 - \varepsilon$ (right).
The arbitrarily selected root
is depicted by white

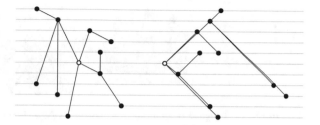

to v. Didimo et al. [10] proved that every leveled tree has a straight-line leveled drawing of crossing angular resolution $\pi/2 - \varepsilon$ for every $\varepsilon > 0$. They used the slopes of
$\pi/4$ and $-\pi/4$ in their drawings. Refer to Fig. 10.11. First, fix an arbitrary vertex as
a root. Then, each edge is drawn with the following rule. For an edge $\{u, v\}$ of the
tree, suppose that u is the parent of v. The edge $\{u, v\}$ is drawn with slope $\pi/4$ if the
y-coordinate of v is larger than the y-coordinate of u; otherwise, the edge $\{u, v\}$ is
drawn with slope $-\pi/4$. We may need to perturb the vertices a little to avoid degener-
acy. The resulting drawing has the crossing angular resolution $\pi/2 - \varepsilon$, where $\varepsilon > 0$
is an artifact of the perturbation.

Recent results on crossing angular resolutions can be found in [6].

10.5   Total Angular Resolution

The total angular resolution was first introduced by Argyriou, Bekos, and Symvonis [4], where the total angular resolution was simply called the *total resolution*.
They proved that $\mathsf{tar}(K_n) = \Theta(1/n)$ for a complete graph K_n, and $\mathsf{tar}(K_{m,n}) =
\Theta(1/\max\{m, n\})$ for a complete bipartite graph $K_{m,n}$. Their bounds can be observed
in the following way.

For upper bounds, by combining the upper bound for the vertex angular resolution
in Eq. (10.1) and the upper bound for the crossing angular resolution in Eq. (10.3),
we obtain

$$\mathsf{tar}(G) = O(\min\{1/\Delta(G), 1/(\bar{\theta}(G) - 1)\}) \tag{10.6}$$

for every non-planar graph G. Since $\Delta(K_n) = n - 1$ and $\Delta(K_{m,n}) = \max\{m, n\}$, we
readily obtain $\mathsf{tar}(K_n) = O(1/n)$ and $\mathsf{tar}(K_{m,n}) = O(1/\max\{m, n\})$.

To prove the lower bounds, Argyriou et al. [4] studied particular straight-line
drawings of K_n and $K_{m,n}$. For K_n, they placed the vertices on the corners of a regular
n-gon. For $K_{m,n}$, they considered a square, and placed the vertices of one partite set on
the top side and the vertices of the other partite set on the bottom side, respectively.
For concrete constructions, see Fig. 10.12. In the figure, $K_{4,5}$ is drawn, where the
partite set A with four vertices lies on the bottom side and the partite set B with five
vertices lies on the top side. On the top side, one vertex of B is placed at the top-left
corner of the square, and four rays emanate from that vertex to the bottom side. The

 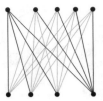

Fig. 10.12 Total angular resolutions of complete graphs and complete bipartite graphs: the construction by Argyriou et al. [4]

angle formed by each consecutive pair of those four rays is equal, and the rightmost ray goes through the bottom-right corner of the square. Then, the four vertices in A are placed at the intersections of those rays and the bottom side. Symmetrically, five rays emanate from the bottom-right vertex of the square to the top side, and the intersections with the top side determine the positions of five vertices in B. Argyriou et al. [4] proved that in this drawing, the total angular resolution is attained by the angle formed by the rightmost edge and the second rightmost edge incident to the vertex at the bottom-left corner of the square.

The method of a set of fixed slopes can be used to bound the total angular resolution of a graph from below. For example, the drawing produced by Formann et al. [13] for d-dimensional cubes Q_d proves the total angular resolution of Q_d is at least π/d. Refer to Fig. 10.6. Another example is a leveled drawing of a leveled tree. For the definition, see Sect. 10.4. Didimo et al. [10] constructed a leveled drawing of a leveled tree with total angular resolution $\pi/\Delta - \varepsilon$, where Δ is the maximum degree of the tree and $\varepsilon > 0$ is arbitrary. Refer to Fig. 10.13. In their drawing, we use the slopes in $\{i\pi/\Delta + \alpha \mid i \in \{0, 1, \ldots, \Delta - 1\}\}$, where α is any constant such that $0 < \alpha < \pi/\Delta$. We first fix an arbitrary vertex as root and then consider a proper edge-coloring with Δ colors, which always exists. Then, an edge e is drawn with the ith slope $i\pi/\Delta + \alpha$ if and only if e receives the ith color. The drawing can be obtained by traversing the tree from root to leaves. It may happen that a vertex is placed on an edge: in that case, we introduce a tiny perturbation to slide the vertex along the direction parallel to the x-axis. The resulting drawing has the total angular resolution $\pi/d - \varepsilon$, where $\varepsilon > 0$ depends on the perturbation.

The method of a set of fixed slopes can be used to draw an undirected graph with maximum degree three. To this end, we use a result by Mukkamala and Pálvölgyi [22]. They proved that an undirected graph with maximum degree three has a straight-line drawing that only uses the slopes in $\{0, \pi/4, \pi/2, 3\pi/4\}$. Examples are given in Fig. 10.14. Therefore, their drawing readily gives a straight-line drawing of total angular resolution $\pi/4$. This bound is tight as a complete graph with four vertices shows; recall Fig. 10.1.

On the other hand, the total angular resolution cannot be bounded by a constant for bounded-degree graphs because the crossing angular resolution cannot as we saw in the previous section. As with the crossing angular resolution, we do not know if this is the case already when the maximum degree is 4, 5, 6, 7, or 8.

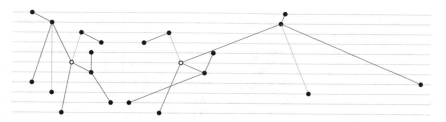

Fig. 10.13 Total angular resolution of a leveled drawing of a leveled tree, studied by Didimo et al. [10]. In this example, the maximum degree Δ is equal to four, and thus the total angular resolution is $\pi/4 - \varepsilon$ for any arbitrary small ε. The left figure shows a given leveled tree with a proper edge coloring with $\Delta = 4$ colors. The right figure shows the construction of a leveled drawing with total angular resolution $\pi/4$. You may observe that the edges with the same color share the same slope

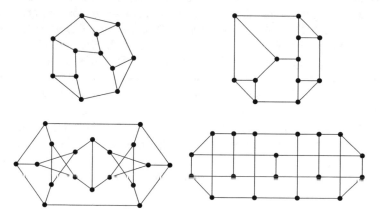

Fig. 10.14 Total angular resolutions of graphs of maximum degree three. Since they can be drawn only with slopes of $0, \pi/4, \pi/2, 3\pi/4$, the total angular resolution is at least $\pi/4$. The top row shows the Frucht graph, and the bottom row shows the second Blanuša snark. In the right figures, only the four slopes are used

Problem 10.5 Can the total angular resolution be bounded by a constant from below if the maximum degree is between 4 and 8?

Determining the total angular resolution of a given graph is NP-hard. As it was pointed out in Sect. 10.3, Formann et al. [13] proved that deciding if a given graph of maximum degree four has a straight-line drawing of vertex angular resolution $\pi/2$ is NP-hard. Their reduction indeed proves that deciding if a given graph of maximum degree four has a straight-line drawing of total angular resolution $\pi/2$ is NP-hard.

In Sect. 10.2, we saw a few examples of graphs with their total angular resolutions. Here, we give proofs of the correctness. First, consider the complete graph K_4 with four vertices. Then, the convex hull boundary of a straight-line drawing of K_4 contains at most four vertices. Therefore, the convex hull must have at least one vertex of angle at most $\pi/2$. Since three edges of the graph are incident to that vertex, the vertex angular resolution must be at most $\pi/4$, which implies that the total angular resolution must be at most $\pi/4$, too.

For the three-dimensional cube Q_3 and the Petersen graph, we proceed as follows. As a common property of those two examples, we use the fact that those two graphs are 3-regular (i.e., every vertex has degree three). Now, for the sake of contradiction, suppose that the total angular resolution is more than $\pi/3$. Then, observe that the convex hull boundary of a straight-line drawing of those graphs must contain at least seven vertices. To see this, suppose that the convex hull boundary contains at most six vertices. Then, one of the convex hull vertices has an angle of less than $2\pi/3$. Since that vertex is incident to three edges, it should create an angle smaller than $\pi/3$.

Consider a straight-line drawing D of those graphs of total angular resolution more than $\pi/3$. By the above observation, the convex hull boundary of D has at least seven vertices. Let $\tilde{D}$ be the planarization of D in the sense that we insert vertices at all crossings so that the $\tilde{D}$ is a straight-line plane drawing. See Fig. 10.15 for the construction. Let n be the number of vertices of D, and let c be the number of crossings in D, which is equal to the number of newly inserted vertices in $\tilde{D}$. Let $V(\tilde{D})$, $E(\tilde{D})$ and $F(\tilde{D})$ be the sets of vertices, edges and faces of $\tilde{D}$, respectively. Then, $|V(\tilde{D})| = n + c$ and $|E(\tilde{D})| \geq \frac{3}{2}n + 2c$. Since we have assumed that the total angular resolution of D is more than $\pi/3$, the drawing $\tilde{D}$ has no triangle face. Therefore, by counting the number of pairs of incident edges and faces in two ways, we obtain $2|E(\tilde{D})| \geq 4(|F(\tilde{D})| + 7 = 4|F(\tilde{D})| + 3$, where we use the assumption that the convex hull boundary contains at least seven vertices. Then, Euler's formula tells us that $|V(\tilde{D})| - |E(\tilde{D})| + |F(\tilde{D})| = 2$, which implies that

$$
\begin{aligned}
2 &= |V(\tilde{D})| - |E(\tilde{D})| + |F(\tilde{D})| \\
&\leq |V(\tilde{D})| - |E(\tilde{D})| + \frac{1}{2}|E(\tilde{D})| - \frac{3}{4} \\
&= |V(\tilde{D})| - \frac{1}{2}|E(\tilde{D})| - \frac{3}{4} \\
&\leq (n + c) - \frac{1}{2}\left(\frac{3}{2}n + 2c\right) - \frac{3}{4} \\
&= \frac{1}{4}n - \frac{3}{4}.
\end{aligned}
$$

Therefore, $n \geq 11$. This leads to a contradiction since Q_3 has eight vertices and the Petersen graph has ten vertices.

Recent results on total angular resolutions can be found in [1].

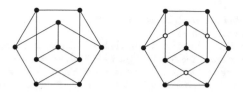

Fig. 10.15 A straight-line drawing with crossings (left) and its planarization (right). In the right drawing, white vertices are newly inserted at the crossings of the left drawing

10.6   Concluding Remarks

From the examples in this chapter, we may observe that there are trade-offs between vertex angular resolution and crossing angular resolution. Total angular resolution tries to balance those two aspects of esthetics in graph drawing. Recently, several authors have started to study the trade-offs between several esthetic criteria in graph drawing, such as area, edge lengths, the number of crossings [2, 7, 17, 20]. This will lead to studying the aspect of multi-criteria optimization in graph drawing and information visualization, which is another open area to be explored further.

Acknowledgements The author's research was partially supported by JSPS/MEXT KAKENHI Grant Numbers JP24106005 and JP15K00009, JST CREST Grant Number JPMJCR1402, and Kayamori Foundation of Informational Science Advancement. The author's thanks go to many people: Seok-Hee Hong and Takeshi Tokuyama for inviting him to NII Shonan Meeting "Algorithmics for Beyond Planar Graphs" in 2016, where a part of the material of this chapter was presented; Martin Nöllenburg for his detailed comments that improved the text considerably; Michael Bekos for pointing out that the hardness proof by Formann et al. [13] implies the hardness of computing the total angular resolution; Oswin Aichholzer, Fabian Klute, Irene Parada, Daniel Perz, and Birgit Vogtenhuber for their proof of the fact that the total angular resolutions of Q_3 and the Petersen graph are at most $\pi/3$, which is an outcome from the Japan-Austrian Bilateral Seminar: Computational Geometry Seminar with Applications to Sensor Networks in 2018.

References

1. Aichholzer, O., Korman, M., Okamoto, Y., Parada, I., Perz, D., van Renssen, A., Vogtenhuber, B.: Graphs with large total angular resolution. In: Archambault, D., Tóth, C.D. (eds.) Graph Drawing and Network Visualization - 27th International Symposium, GD 2019, Prague, Czech Republic, 17–20 September 2019, Proceedings. Lecture Notes in Computer Science, vol. 11904, pp. 193–199. Springer (2019). https://doi.org/10.1007/978-3-030-35802-0_15
2. Angelini, P., Cittadini, L., Didimo, W., Frati, F., Di Battista, G., Kaufmann, M., Symvonis, A.: On the perspectives opened by right angle crossing drawings. J. Graph Algorithms Appl. **15**(1), 53–78 (2011). https://doi.org/10.7155/jgaa.00217
3. Argyriou, E.N., Bekos, M.A., Symvonis, A.: The straight-line RAC drawing problem is NP-hard. J. Graph Algorithms Appl. **16**(2), 569–597 (2012). https://doi.org/10.7155/jgaa.00274
4. Argyriou, E.N., Bekos, M.A., Symvonis, A.: Maximizing the total resolution of graphs. Comput. J. **56**(7), 887–900 (2013). https://doi.org/10.1093/comjnl/bxs088

5. Barát, J., Matoušek, J., Wood, D.R.: Bounded-degree graphs have arbitrarily large geometric thickness. Electron. J. Comb. **13**(1) (2006)
6. Bekos, M.A., Förster, H., Geckeler, C., Holländer, L., Kaufmann, M., Spallek, A.M., Splett, J.: A heuristic approach towards drawings of graphs with high crossing resolution. In: Biedl, T.C., Kerren, A. (eds.) Graph Drawing and Network Visualization - 26th International Symposium, GD 2018, Barcelona, Spain, 26–28 September 2018, Proceedings. Lecture Notes in Computer Science, vol. 11282, pp. 271–285. Springer (2018). https://doi.org/10.1007/978-3-030-04414-5_19
7. Di Giacomo, E., Didimo, W., Liotta, G., Meijer, H.: Area, curve complexity, and crossing resolution of non-planar graph drawings. Theory Comput. Syst. **49**(3), 565–575 (2011). https://doi.org/10.1007/s00224-010-9275-6
8. Di Giacomo, E., Didimo, W., Eades, P., Hong, S., Liotta, G.: Bounds on the crossing resolution of complete geometric graphs. Discret. Appl. Math. **160**(1–2), 132–139 (2012). https://doi.org/10.1016/j.dam.2011.09.016
9. Didimo, W., Eades, P., Liotta, G.: Drawing graphs with right angle crossings. Theor. Comput. Sci. **412**(39), 5156–5166 (2011). https://doi.org/10.1016/j.tcs.2011.05.025
10. Didimo, W., Kaufmann, M., Liotta, G., Okamoto, Y., Spillner, A.: Vertex angle and crossing angle resolution of leveled tree drawings. Inf. Process. Lett. **112**(16), 630–635 (2012). https://doi.org/10.1016/j.ipl.2012.05.006
11. Dillencourt, M.B., Eppstein, D., Hirschberg, D.S.: Geometric thickness of complete graphs. J. Graph Algorithms Appl. **4**(3), 5–17 (2000). https://doi.org/10.7155/jgaa.00023
12. Dujmovic, V., Gudmundsson, J., Morin, P., Wolle, T.: Notes on large angle crossing graphs. Chic. J. Theor. Comput. Sci. **2011** (2011)
13. Formann, M., Hagerup, T., Haralambides, J., Kaufmann, M., Leighton, F.T., Symvonis, A., Welzl, E., Woeginger, G.J.: Drawing graphs in the plane with high resolution. SIAM J. Comput. **22**(5), 1035–1052 (1993). https://doi.org/10.1137/0222063
14. Garg, A., Tamassia, R.: Planar drawings and angular resolution: algorithms and bounds (extended abstract). In: van Leeuwen, J. (ed.) Algorithms - ESA '94, Second Annual European Symposium, Utrecht, The Netherlands, 26–28 September 1994, Proceedings. Lecture Notes in Computer Science, vol. 855, pp. 12–23. Springer (1994). https://doi.org/10.1007/BFb0049393
15. Havet, F., van den Heuvel, J., McDiarmid, C., Reed, B.: List colouring squares of planar graphs (2008). arXiv:0807.3233
16. van den Heuvel, J., McGuinness, S.: Coloring the square of a planar graph. J. Graph Theory **42**(2), 110–124 (2003). https://doi.org/10.1002/jgt.10077
17. Hoffmann, M., van Kreveld, M.J., Kusters, V., Rote, G.: Quality ratios of measures for graph drawing styles. In: Proceedings of the 26th Canadian Conference on Computational Geometry, CCCG 2014, Halifax, Nova Scotia, Canada, 2014. Carleton University, Ottawa, Canada (2014). http://www.cccg.ca/proceedings/2014/papers/paper05.pdf
18. Keszegh, B., Pach, J., Pálvölgyi, D.: Drawing planar graphs of bounded degree with few slopes. SIAM J. Discret. Math. **27**(2), 1171–1183 (2013). https://doi.org/10.1137/100815001
19. Koebe, P.: Kontaktprobleme der konformen Abbildung. Ber. Sächs. Akad. Wiss. Leipzig Math.-Phys. Kl. **88**, 141–164 (1936)
20. van Kreveld, M.J.: The quality ratio of RAC drawings and planar drawings of planar graphs. In: Brandes, U., Cornelsen, S. (eds.) Graph Drawing - 18th International Symposium, GD 2010, Konstanz, Germany, 21–24 September 2010. Revised Selected Papers. Lecture Notes in Computer Science, vol. 6502, pp. 371–376. Springer (2010). https://doi.org/10.1007/978-3-642-18469-7_34
21. Malitz, S.M., Papakostas, A.: On the angular resolution of planar graphs. SIAM J. Discret. Math. **7**(2), 172–183 (1994). https://doi.org/10.1137/S0895480193242931
22. Mukkamala, P., Pálvölgyi, D.: Drawing cubic graphs with the four basic slopes. In: van Kreveld, M.J., Speckmann, B. (eds.) Graph Drawing - 19th International Symposium, GD 2011, Eindhoven, The Netherlands, 21–23 September 2011, Revised Selected Papers. Lecture Notes in Computer Science, vol. 7034, pp. 254–265. Springer (2011). https://doi.org/10.1007/978-3-642-25878-7_25

Chapter 11
Crossing Layout in Non-planar Graph Drawings

Martin Nöllenburg

Abstract Edge crossings are a major obstruction for the readability of graph layouts as has been shown in several empirical studies. Yet, non-planar graphs are abundant in network visualization applications. Therefore, graph layout techniques are needed that optimize readability and comprehensibility of graph drawings in the presence of edge crossings. This chapter deals with aesthetic ideas for improving the appearance of crossings and presents alternative layout styles and algorithmic results that go beyond solely optimizing the crossing-number metric. In particular, we review edge casing in geometric graphs as a way to represent crossings, the slanted layout of crossings in orthogonal graph layouts, and minimizing bundled rather than individual crossings. Further, we look at concepts such as confluent graph layout and partial edge drawings, which both have no visible crossings.

11.1 Introduction

The readability of a graph drawn as a typical node-link diagram, that is, vertices as points and edges as simple (straight-line) curves, is affected by various aesthetic criteria as has been confirmed in several empirical user studies testing a variety of graph reading tasks [45, 53–55, 62]. One of the main readability criteria is the number of edge crossings, where, generally speaking, graph layouts with fewer crossings are more readable. Clearly, planar graphs that have crossing-free layouts should thus be drawn without crossings, which is not only a natural representation but also optimal in terms of the crossing aesthetic. There is a large body of work on drawing planar graphs [52, 61]. Yet most of the graphs that are studied in practical applications are usually non-planar, be it social networks, collections of webpages/documents with hyperlinks/references, communication networks, or networks in the life sciences. Since the beginning of the field, the question of minimizing crossings in non-planar graph layouts has been a focus of interest in graph drawing research. The problem is well known to be NP-hard [40], and consequently, several heuristics and exact algo-

M. Nöllenburg (✉)
Algorithms and Complexity Group, TU Wien, Vienna, Austria
e-mail: noellenburg@ac.tuwien.ac.at

© Springer Nature Singapore Pte Ltd. 2020
S.-H. Hong and T. Tokuyama (eds.), *Beyond Planar Graphs*,
https://doi.org/10.1007/978-981-15-6533-5_11

rithms have been developed in graph drawing that explicitly or implicitly minimize crossings [22].

While aiming for graph layouts with few crossings is certainly an important goal, graph readability is not only affected by the number of crossings but also by their actual appearance. Huang et al. [44] studied the effect of crossing angles on graph readability and showed empirically that crossings with large angles between 70 and 90 degrees lead to better graph reading performance than crossings with very acute angles. Subsequently, right-angle crossing (RAC) graphs [27] and large-angle crossing (LAC) graphs [28] have been investigated in the literature. Chapter 10 in this book discusses the angular resolution of straight-line graph drawings in detail, hence we do not consider specific aspects of optimizing crossing angles in this chapter.

Instead, our focus is on further visual techniques that aim to reduce the negative effects of edge crossings on the readability of graph layouts. When considering the crossing layout in graph drawings we distinguish two different input settings. The first type of crossing layout problems deals with a given geometric input layout, for example, obtained from a suitable graph layout algorithm, where the task is to change the rendering of the drawing around the edge crossings to improve readability. In the second type of crossing layout problems, the input graph does not come with a fully specified geometry and the task is rather to compute a complete layout of the graph, including the final positioning of vertices and crossings and the routing of the edges. There is a gradual transition between the two extremes, with layout problems, where some combinatorial restrictions of the output layout may be specified in the input, such as an embedding of the planarized graph (planar graph obtained by replacing crossings by dummy vertices) or a certain vertex ordering for a circular graph layout that must be respected.

A second dimension along which crossing layout techniques can be distinguished is whether crossings remain clearly visible, but with a less disruptive appearance, or whether crossings become more or less invisible. An intermediate technique between these two extremes tries to bundle crossings of locally parallel edges so that a whole group of crossings appears as one single crossing, at least from some distance. Table 11.1 shows the classification of the covered crossing layout techniques according to these two dimensions.

Table 11.1 Different crossing layout techniques with their typical input graph specifications and visibility of crossings in the output layout

Technique	Graph input			Output crossings	
	Drawing	Embedding	Graph only	Visible	Invisible
Edge casing	×			×	
Slanted orthogonal		×		×	
Bundled crossings		×	×	×	
Confluent drawings			×		×
Partial edge drawings	×		×		×

This chapter is organized as follows. In Sect. 11.2 we give basic definitions and notation. In Sect. 11.3 we consider three techniques (edge casing, slanted orthogonal drawings, and bundled crossings) that improve the appearance of crossings, whereas Sect. 11.4 deals with two techniques (confluent drawings and partial edge drawings) that eliminate all visible crossings. We report on the state of the art of the aforementioned techniques from an algorithmic perspective and collect the main open questions from the literature.

11.2 Basic Definitions

Let $G = (V, E)$ be an undirected, simple, connected graph with n vertices and m edges. A drawing Γ of G maps each vertex $v \in V$ to a point $\Gamma(v) \in \mathbb{R}^2$ and each edge $e = (u, v) \in E$ to a simple open curve $\Gamma(e)$ with endpoints $\Gamma(u)$ and $\Gamma(v)$. Unless stated otherwise we will, for the sake of simplicity, identify a vertex v and its image $\Gamma(v)$ as well as an edge e and its image $\Gamma(e)$. Two edges cross in Γ if they intersect in an interior point x, in which their relative order changes. We make some standard regularity assumptions for Γ. In particular, no three edges intersect in the same point, two edges cross at most once, and edges incident to the same vertex do not cross at all. Further, no edge passes through a non-incident vertex. For a straight-line drawing we have that each edge $e = (u, v)$ is drawn as the line segment $\overline{uv}$ and for a polyline drawing we have that e is drawn as a polygonal path $p(e) = (u = p_0, p_1, \ldots, p_{l-1}, p_l = v)$ with $l \geq 1$ segments between the vertices u and v. We use $|uv|$ to denote the Euclidean distance between u and v.

For a planar graph G, a *planar embedding*, *combinatorial embedding*, or simply *embedding* is an equivalence class of planar drawings of G. A particular embedding comprises all planar drawings of G that share the same face structure (including the external face) and the same cyclic orderings of the incident edges of each vertex (also called a *rotation scheme*). An embedding is completely described by its rotation scheme and the external face.

For a non-planar graph G, the notion of a (combinatorial) embedding can be extended by considering planarizations of G. A *planarization* of G is the embedded planar graph obtained by replacing each crossing in a (non-planar) drawing Γ of G with a dummy vertex of degree 4. Then a *(non-planar) embedding* of G is an embedding of a planarization of G. In particular, an embedding of a non-planar graph G specifies, via the embedding of a planarization of G, which pairs of edges of G cross and in which sequence the crossings occur along each edge. We call a non-planar embedding *k-planar* if no edge has more than k crossings.

11.3 Improving the Appearance of Crossings

In this section, we present three approaches for improving the visual appearance of crossings by introducing visual cues that indicate an above–below relationship for crossing edges (Sect. 11.3.1), by requiring distinctive angular crossing patterns in orthogonal drawings (Sect. 11.3.2), and by rerouting edges so that groups of crossings become visually more salient than individual pairwise crossings (Sect. 11.3.3).

11.3.1 Edge Casing

Edge casing is a well-known method in technical drawings, for example, circuit diagrams, that indicates an above–below ordering of two paths or wires by interrupting the lower one locally around the intersection point. In computer graphics, this concept is called *halo effect* [3]. In graph drawing, edge casing, as introduced by Eppstein et al. [33], helps to distinguish edge crossings visually from vertices. This becomes relevant especially when relatively small disks are used as vertex symbols and crossings and vertices are densely distributed. It also adds a somewhat 3-dimensional touch to the graph layout. Figure 11.1 shows an example of a non-cased and a cased drawing of the same graph.

Eppstein et al. [33] considered edge casing as a post-processing technique that can be used to enhance the visual quality of a given graph drawing Γ. In a *cased drawing*, an ordering needs to be specified for each crossing c such that the lower of the two edges gets interrupted. Technically, each edge is drawn with a background-color casing of fixed width and hence drawing the upper edge later than the lower edge yields this visual interruption. The upper edge of c is also called *bridge* and the lower one *tunnel*. The Gestalt principles of continuation and closure [56] explain that humans still perceive each edge as one object despite some short interruptions. It is assumed that crossings and vertices are spaced sufficiently far apart so that no cased crossing interferes with any other crossing or vertex.

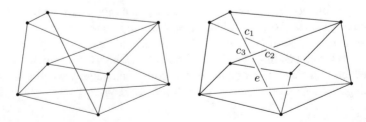

Fig. 11.1 Non-cased drawing (left) and cased drawing (right) in the weaving model. Crossings c_1, c_2, c_3 cannot be realized in the stacking model. Edge e is a bridge at c_1, a tunnel at c_3 and has three switches in total

Table 11.2 Algorithmic results and running times on different edge casing problems. Recall that $|V| = n$, $|E| = m \in \Omega(n)$, and let $k \in \Omega(m) \cap O(m^2)$ be the number of crossings

Model	Stacking	Weaving
MINTOTALSWITCHES	*Open*	$O(k^{5/2} \log^{3/2} k)$
MAXTOTALSWITCHES	*Open*	$O(k^{5/2} \log^{3/2} k)$
MINMAXSWITCHES	*Open*	*open*
MINMAXTUNNELS	$O(m \log m + k)$ exp.	$O(m^4)$
MINMAXTUNNELLENGTH	$O(m \log m + k)$ exp.	NP-hard
MAXMINTUNNELDISTANCE	$O((m + k) \log m)$ exp.	$O(m^3 \log m)$ exp.

Two models for ordering crossing edges are considered: the *stacking model*, in which a global ordering of all edges is applied and the *weaving model*, in which edges are ordered independently for each crossing. Every drawing in the stacking model also satisfies the weaving condition. Further, each edge with multiple crossings forms a sequence of tunnel and bridge crossings. Whenever two consecutive crossings along an edge are of different types, they form a *switch*.

In each of the two models a number of optimization problems can now be defined. In particular, Eppstein et al. argue that an edge or an entire drawing with many switches has a more chaotic appearance and thus becomes harder to read. Further, since for a fixed casing width the tunnel length depends on the crossing angle, they argue that in drawings with shorter tunnel lengths (and thus larger crossing angles) edges are generally easier to recognize. The six considered problems are [33]:

- MINTOTALSWITCHES
  minimize the total number of switches;
- MAXTOTALSWITCHES
  maximize the total number of switches (motivated by creating difficult puzzles);
- MINMAXSWITCHES
  minimize the maximum number of switches on any single edge;
- MINMAXTUNNELS
  minimize the maximum number of tunnels on any single edge;
- MINMAXTUNNELLENGTH
  minimize the maximum total tunnel length on any single edge;
- MAXMINTUNNELDISTANCE
  maximize the minimum distance between any two consecutive tunnels.

Each of the six problems can be considered both in the stacking and in the weaving model, giving rise to the set of open questions and results obtained by Eppstein et al. [33] as shown in Table 11.2. Some algorithms are randomized and expected running times are given. For more detailed running times using additional parameters that relate to properties of the face polygons in the input drawing, we refer to the original paper [33].

While the above results are of a more theoretical interest, Rusu et al. [56] implemented a standard force-based layout algorithm and then applied edge casing to all crossings by considering a stacking model in random order. Using the resulting cased drawings, they performed a preliminary empirical study, in which users could rate whether edge casing would increase or decrease the readability of a set of straight-line drawings. They received mixed results. Some participants gave positive feedback about introducing edge casing, while others preferred traditional drawings. A more systematic study, also using optimized rather than random edge casing, is necessary to establish reliable empirical evidence on readability of cased graph drawings.

Recently, the edge casing model led to a generalization of the notion of k-planarity. According to Bae et al. [5], a k-*gap planar* graph G is a graph that admits a cased drawing, where every edge has at most k tunnels (or *gaps* in their terminology). In their paper, the authors proved a number of density and existence results and studied the recognition complexity and relationships to other beyond-planar graph classes. In particular, a k-gap planar graph has at most $O(\sqrt{k} \cdot n)$ edges and for $k = 1$ they give a tight bound of $5n - 10$ edges. The complete graph K_n is 1-gap planar only for $n \leq 8$. Further, deciding whether a graph is 1-gap planar is NP-complete, even if a fixed rotation system is given for each vertex. Finally, every $2k$-planar graph is k-gap planar and every k-gap planar graph is $(2k + 2)$-quasiplanar. The open problems mentioned by Bae et al. [5] include finding a tight upper bound on the edge density in 2-gap planar graphs, studying the existence of 1-gap planar drawings for complete bipartite graphs such as $K_{6,6}$, studying the complexity of the recognition problem of 1-gap planar graphs with all vertices in the outer face, and answering the question whether 1-gap planar graphs admit RAC drawings with few bends.

11.3.2   Slanted Orthogonal Layout

Orthogonal drawings of graphs, where each edge is drawn as a polyline consisting of horizontal and vertical segments, are well studied in graph drawing and visualization research, both for planar and non-planar graphs [13, 29, 49, 59]. For non-planar graphs, orthogonal layouts guarantee that all edge crossings form optimal 90° angles, that is, they are RAC drawings [27]. But especially for large non-planar graphs it can get even more difficult than in general straight-line drawings to distinguish vertices and crossings, as a degree-4 vertex and an edge crossing are both locally incident to four line segments forming a +-shape.

This motivated Bekos et al. [10] to initiate the study of slanted orthogonal draw-ings as a generalization of orthogonal drawings in the following sense. A *slanted orthogonal* drawing (or *slog* drawing) Γ of a graph G with maximum degree 4 is a drawing such that

(i) each vertex is mapped to an integer grid point with four exclusive edge ports aligned with the coordinate axes,
(ii) each edge is mapped to an octilinear polyline consisting of horizontal, vertical, and 45°-diagonal segments (starting and ending on a horizontal or vertical segment),

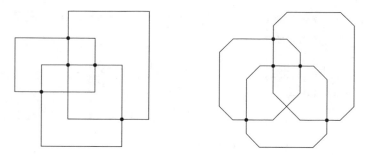

Fig. 11.2 Graph K_5 drawn as an orthogonal layout (left) and as a slanted orthogonal layout (right)

(iii) edge crossings are interior points of diagonal polyline segments,
(iv) all bend angles are 135° (so-called *half-bends*).

Figure 11.2 shows an example of a slanted orthogonal drawing in comparison to a
standard orthogonal drawing. In addition to clearly distinguishable crossings with
an ×-shape compared to vertices with their +-pattern, slanted orthogonal drawings
have a smoother general appearance due to their larger 135° bends. Yet, the number
of half-bends is at least twice the number of bends in a bend-minimal orthogonal
drawing [10], but may even be significantly higher.

Bekos et al. [10] studied bend minimization for slanted orthogonal drawings. By
modifying Tamassia's flow network for bend minimization of plane orthogonal drawings [59] in the topology-shape metrics (TSM) framework, the authors obtained a
network flow model, in which a minimum-cost flow corresponds to a slog representation with minimum number of bends for a given embedded planarization of the input
graph G. Since the resulting slog representation is only a combinatorial description
of the shape of a slog drawing, the next step is to assign actual coordinates to vertices, bends, and crossings. Here Bekos et al. described a heuristic that computes
slanted orthogonal drawings with $O(n^2)$ area for a given (bend-minimal) slog representation. However, this heuristic may introduce additional half-bends on some of
the edges. Therefore, the authors additionally presented a linear program to compute
bend-minimal slog drawings. They showed that in contrast to the heuristic drawings
with non-optimal bend number but $O(n^2)$ area, bend-minimal drawings may need
exponential area. Both drawing methods have been implemented by Bekos et al. [10]
and there is experimental evidence that actually every bend-minimal slog representation obtained from the flow network admits a corresponding slanted orthogonal
drawing. Yet there is no formal proof and thus the question remains open.

Subsequently, Bekos et al. [9, 11] extended the concept of slanted orthogonal
drawings. *Sloggy* drawings [9] allow crossings not only on diagonal edge segments,
but also on vertical and horizontal segments. While this removes to some extent
the positive effect of showing crossings exclusively as a visually distinctive ×-
pattern, the advantage of sloggy drawings is that the number of bends can be greatly
reduced, see Fig. 11.3a. In fact, Bekos et al. proved that optimal sloggy drawings use
exactly twice the number of bends of an optimal orthogonal drawing, that is, each 90°

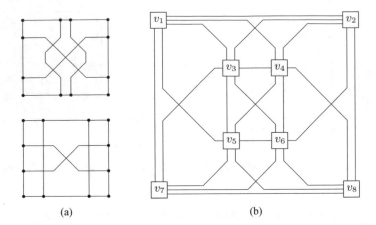

Fig. 11.3 (**a**) Comparison of a slanted orthogonal layout with twelve bends (top) and a sloggy layout with four bends (bottom). (**b**) Example of a Sloginsky drawing.creditImages courtesy of M. Bekos

bend is replaceable by two 135° bends. Unfortunately, the worst-case exponential area bound of bend-minimal slanted orthogonal drawings applies for bend-minimal sloggy drawings as well. They further implemented an integer linear program (ILP) that transforms a bend-minimal orthogonal representation into a sloggy drawing with twice the number of bends. Using a weighted objective function their ILP maximizes the number of diagonal crossings and at the same time aims for an even distribution of the bends in the drawing. Open questions in the context of sloggy drawings include whether optimal sloggy drawings can actually be computed in polynomial time (avoiding the use of ILP) and, similarly as for slanted orthogonal drawings, whether a geometric realization of a given sloggy representation can be computed efficiently.

Another disadvantage of slanted orthogonal drawings is their restriction to graphs of maximum degree 4. The Kandinsky model in orthogonal layouts removes this restriction by replacing vertices by boxes and allowing multiple edges to attach to the same side of each box [38]. However, bend minimization in the Kandinsky model is NP-hard [15]. *Sloginsky drawings* [11] combine the advantages of slanted orthogonal drawings and the Kandinsky model. The only difference to the slanted orthogonal model is that each vertex is represented as a box that may have multiple edge ports on each of its four sides, see Fig. 11.3b. In order to compute Sloginsky drawings, Bekos et al. [11] described and implemented an ILP model for minimizing bends, again in the TSM framework assuming a graph with an embedded planarization as the input. Based on the resulting slog representation, they use a modification of the LP model of Bekos et al. [10] to test whether the slog representation is realizable and assign vertex and bend coordinates if the corresponding Sloginsky drawing exists. The main open question in the Sloginsky setting is whether every graph admits a bend-minimal Sloginsky representation that is actually realizable. To achieve this, the avoidance of S-shaped edges may be necessary according to Bekos et al. [11]. From a practical point of view, the worst-case exponential area requirement of bend-minimal Sloginsky drawings (and also of slanted orthogonal and sloggy drawings) needs to be reduced, for example, by allowing a few additional bends or by allowing axis-aligned crossings as in the sloggy layout style [9].

Another interesting open question related to variants of orthogonal drawings for non-planar graphs is to study *smooth orthogonal drawings* with edge crossings, that is, drawings with edges that are composed of orthogonal (or octilinear) line segments and pieces of circular arcs [8, 12] meeting smoothly in points with a common tangent. Since it is straight-forward to replace bends by smooth arcs, the main challenge is to use as few edge segments as possible. Initially, only planar smooth orthogonal drawings had been studied. Depending on the type of crossings one is willing to accept and the graph class considered, questions are to minimize the total edge complexity or to bound the maximum edge complexity. Recently, Argyriou et al. [4] were the first to study upper and lower edge complexity bounds for 1-plane and outer-1-plane graphs in orthogonal and smooth orthogonal drawings. In particular, they showed that every 1-plane graph (of maximum degree 4) admits a polynomial-area smooth orthogonal drawing with at most three segments per edge, but it is still open whether this bound is tight. For outer-1-plane graphs, they proved a tight bound of two on the edge complexity. As further research directions, smooth orthogonal drawings of 2-plane or more general beyond-planar graphs can be considered, as well as the size of the crossing angles on the smooth arcs.

11.3.3 Bundled Crossings

Edge bundling is a popular technique in network visualization that strongly reduces the visual clutter of a dense graph drawing by pulling together edges of similar length and position [50]. A drawing with edge bundling has a much sparser appearance than a regular straight-line drawing of the same graphs. Edge bundling generally simplifies the layout and, especially in dense graphs, it also leads to fewer perceivable crossings since individual edges disappear. Yet, the disadvantage of edge bundling is that only a coarse image of the graph is given and often detailed adjacencies cannot be inferred; in that sense edge bundling is usually not *faithful* [51], that is, the underlying graph cannot be reconstructed from the image.

Fink et al. [36] proposed *bundled crossings* as a new concept in graph drawing that locally groups suitable crossings into bundles rather than having a large number of individual crossings scattered over the drawing in an unstructured way. In contrast to edge bundling, bundling only the crossings generally maintains the traceability of individual edges much better. Further, Fink et al. suggested that the number of bundled crossings may be a better measure for the quality of a graph drawing than the number of individual crossings.

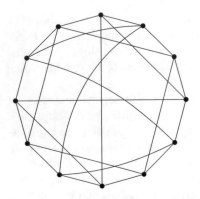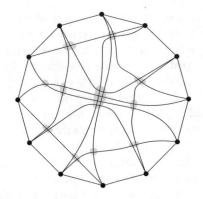

Fig. 11.4 Two circular drawings of the Chvaátal graph. A drawing with 28 individual crossings (left) and a drawing with 11 bundled crossings and a different non-planar embedding (right)

A *bundled crossing* in a drawing Γ of a graph $G = (V, E)$ is defined as a set C of pairwise edge crossings with the following properties:

- There are two disjoint sets of edges $E_1 \subset E$ and $E_2 \subset E$ such that C contains exactly the crossings between all pairs of edges (e_1, e_2) with $e_1 \in E_1$ and $e_2 \in E_2$. The sets E_1 and E_2 are called the *bundles* of C.
- The set C can be separated from all remaining crossings of Γ by a pseudodisk that does not intersect any other edge $e \notin E_1 \cup E_2$. The existence of a pseudodisk D shows that the crossings in C are visually separated from the rest of Γ.

Obviously, every individual crossing is also a bundled crossing, but the goal is to find drawings with a smaller number of bundled crossings. Figure 11.4 shows an example of a drawing with 11 bundled crossings representing 28 pairwise crossings.

Fink et al. [36] considered embedded non-planar graphs and the algorithmic problem of partitioning the crossings into a minimum number of bundled crossings without changing the embedding. For a given embedding $\mathscr{E}$ of a non-planar graph G, the *bundled crossing number* $\mathrm{bc}(\mathscr{E})$ is the minimum number of bundled crossings into which the crossings of $\mathscr{E}$ can be partitioned. They showed that minimizing the number of bundled crossings in a given embedding is NP-hard. Using a connection between minimizing bundled crossings and dissecting rectilinear polygons with holes into a minimum number of rectangles, Fink et al. designed a heuristic algorithm for minimizing bundled crossings. In fact, this algorithm yields a factor-10 approximation for graphs with a given circular embedding, that is, an embedding, where all vertices lie in the unbounded external face. It is an open question to find approximation algorithms for graphs that are not circularly embedded.

Alam et al. [1] considered the bundled crossing number for the more general case of graphs without a given embedding. First, they showed that in the unrestricted setting the edges may self-intersect and cross other edges multiple times, the bundled crossing number equals the genus of the graph, which is known to be NP-hard to compute [60]. When following the natural crossing restrictions of Sect. 11.2, the bundled crossing number is at least the genus of the graph and Chaplick et al. [24] recently proved that computing the bundled crossing number is NP-hard in that case. *Circular layouts* are graph layouts, where all vertices are placed on a circle and all edges are drawn inside the circle, see Fig. 11.4. Such layouts are also of practical interest [39] and they are topologically equivalent to 1-page book drawings. Alam et al. [1] presented an $O(m^2)$-time algorithm that computes a circular layout with a fixed cyclic vertex order that has at most $m - 1$ bundled crossings. This immediately serves as a general upper bound on the bundled crossing number. The authors could further show a lower bound of $m/16$ bundled crossings for a fixed vertex order, which implies that their algorithm computes a 16-approximation. Via a different lower bound for the case of unspecified cyclic vertex orders they obtained that applying their algorithm to an arbitrary vertex order still provides a $6c/(c - 2)$-approximation for a constant $c > 2$ such that $m \geq cn$. In the case of a fixed cyclic vertex order, they additionally showed that deciding whether a drawing with at most k bundled crossings exists is fixed-parameter tractable for the parameter k. Finally, even if arbitrary layouts are considered, Alam et al. showed that computing a circular layout with their algorithm for an arbitrary vertex order still provides a $6c/(c - 3)$-approximation for a constant $c > 3$ such that $m \geq cn$. Some upper bounds on the bundled crossing numbers are also presented. Remaining open questions include improving the approximation results, finding further fixed-parameter algorithms for bundled crossing minimization problems, and determining the algorithmic complexity of computing the bundled crossing number for circular layouts (with or without a given embedding or fixed cyclic vertex order). It would not be surprising if this is NP-hard as well, but a proof is missing. In fact, it has recently been established by Chaplick et al. [24] that deciding whether the bundled crossing number of a graph in a simple circular layouts is at most some integer k is fixed-parameter tractable in k. This leaves open the natural question whether the same is true for general simple graph layouts.

Angelini et al. [2] recently combined the concept of bundled crossings with *fan planarity*. A fan-planar graph G is a graph that admits a drawing Γ such that for any edge $e \in E$ all edges that cross e in Γ form a *fan*, that is, they are all incident to a common vertex [48]. Angelini et al. [2] introduced bundling of fans in the sense that each edge may belong to a *fan bundle* of either of its end-vertices, while the middle part of the edge is not bundled. Now a *1-fan-bundle-planar* (1-fbp) drawing is a drawing in which every fan bundle has at most one crossing with another fan bundle. The middle part of each edge has no crossings. Hence each crossing in a 1-fbp drawing is actually a bundled crossing. The results obtained by Angelini et al. are mostly of theoretical interest. For example, they proved relationships with other beyond-planar graph classes, upper and lower bounds (some of which are tight) on the edge density of 1-fbp graphs, NP-completeness of the general recognition problem, and efficient recognition algorithms for special graph classes.

11.4 Eliminating Crossings

In the previous section, we presented different approaches that help to make crossings either more distinctive to avoid confusion with vertices or re-arrange them locally so that clutter is reduced and crossings become less salient. In this section, we discuss two graph layout styles that can be used to represent non-planar graphs without actually showing any explicit edge crossings at all. Obviously, this can only be achieved by using alternative definitions of how to draw edges. In Sect. 11.4.1, we consider representing edges as smooth paths in a planar system of junctions and arcs; and in Sect. 11.4.2, we report on a layout style that takes edge casing to an extreme in the sense that the entire middle parts of edges with crossings are not drawn.

11.4.1 Confluent Drawings

One of the disadvantages of edge bundling (recall Sect. 11.3.3) is that a bundled set of edges creates the false impression of an all-to-all connection between the two sets of vertices on both ends of the edge bundle, even if the subgraph induced by these vertices is much sparser. Confluent drawings of graphs may be seen as a mathematically precise and planar version of edge bundling in the following sense.

A (planar) *confluent drawing* [26] D of a graph $G = (V, E)$ consists of a set of points to represent the vertex set V and a set J of *junction points*. For simplicity, we denote the set of vertex points by V as well. Further, there is a set A of smooth simple curves called *arcs* that start and end at vertices or junction points. No two arcs intersect, except at common endpoints. For each junction point j it is required that all arcs meeting in j share the same tangent line in j. Basically, a confluent drawing D can be seen as a planar drawing of a graph with vertex set $V \cup J$ and edges with a special tangency requirement in junction vertices. Finally, D must represent G as follows: There is an edge $(u, v) \in E$ if and only if there is a smooth, locally monotone path $p_{uv} = (u, j_1, \ldots, j_k, v)$ from u to v in D such that all $k \geq 0$ internal path vertices $j_1, \ldots, j_k$ are distinct junction points. In particular, p_{uv} must pass straight through each junction without any sharp turns and p_{uv} cannot have any self-intersections. Without loss of generality one may assume that every junction is a *binary* junction of degree 3, in which two arcs merge into one. A graph is called *confluent* if it has a confluent drawing. Due to their alternative way of reading edges, confluent drawings can in fact represent very dense non-planar graphs as a planar diagram. For example, all complete graphs or complete bipartite graphs are confluent, see Fig. 11.5.

Since the representations of multiple edges can share parts of their smooth paths, we can interpret the shared parts as an edge bundle. However, the definition of a confluent drawing ensures that an arc is part of multiple edge paths if and only if all the vertices reachable from both ends of the arc are actually mutually adjacent in G. This definition has also been compared to a system of train tracks and switches, where two vertices u and v are adjacent if and only if a train can move from u

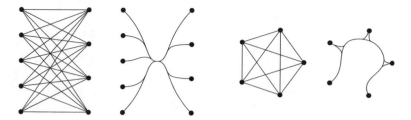

Fig. 11.5 Straight-line and confluent layouts of $K_{5,4}$ and K_5

to v without changing its direction or making sharp turns [26, 46]. The alternative definition of adjacencies and edges by reachability on smooth paths in the confluent drawing implies that confluent graphs are not closed under edge deletion; removing edges from cliques and bicliques can make a graph non-confluent.

A number of variations of confluent drawings have been introduced subsequently. Hui et al. [46] defined *strongly confluent* graphs as graphs that have a *strongly confluent* drawing, in which the smooth path representing an edge of G may self-intersect, that is, the requirement of local monotonicity is relaxed. They showed that strongly confluent graphs are a proper subclass of confluent graphs.

In the original definition by Dickerson et al. [26] the smooth path representing some edge e is not necessarily unique. Eppstein et al. [35] defined the restricted notion of *strict* confluent drawings, which have a unique smooth path for every edge of G and disallow smooth paths from a vertex to itself. Strictness of a confluent drawing D implies that there is no need for distinguishing whether D is strongly confluent or confluent because any smooth path with a self-intersection would yield a violation of strictness.

Hui et al. [46] further defined *tree-confluent* graphs as graphs that admit a confluent drawing D that is actually a drawing of a tree, that is, there are no (smooth or non-smooth) cycles in the graph $(V \cup J, A)$. Eppstein et al. [31] extended the tree-confluent graphs into Δ-*confluent* graphs, whose drawings additionally allow 3-way Δ-junctions, in which each arc connects smoothly to the other two arcs. In this terminology, the confluent representation of K_5 in Fig. 11.5 actually uses three Δ-junctions.

The main theoretical research questions on confluent graphs and layouts that have been addressed in the literature concern the characterization and recognition of the different confluent graph classes, as well as showing relationships with other graph classes. Dickerson et al. [26] showed that large classes of non-planar graphs are confluent, for example, all interval graphs, all cographs [25] and all complements of trees and cycles, but they also showed that there are non-confluent graphs, such as the Petersen graph, the 4-dimensional hypercube, and certain subdivisions of non-planar graphs.

Hui et al. [46] were the first to show that recognizing strongly confluent graphs is a problem in NP by giving a polynomial upper bound on the number of junctions and arcs needed. Yet, it remains open if recognizing strongly confluent graphs is in

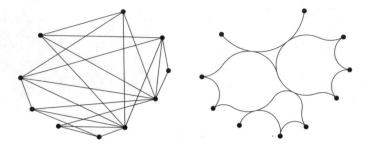

Fig. 11.6 A 10-vertex graph and a strict outerconfluent layout

fact NP-complete or not. It is also open whether confluent graphs can be recognized in NP and whether this problem is NP-hard. Eppstein et al. [35] showed that testing whether a graph has a strict confluent drawing is actually NP-complete. At least for subclasses of confluent graphs such as the tree-confluent graphs and their natural generalization to forest-confluent graphs the recognition problem is polynomial-time solvable. In fact, Hui et al. [46] showed that tree-confluent graphs are characterized by a certain vertex elimination ordering and that they are a subset of the chordal bipartite graphs. More precisely, they showed that the forest-confluent graphs are equivalent to the (6, 2)-chordal bipartite graphs, which are bipartite graphs in which every cycle of length at least six contains at least two chords. Further, the closely related class of Δ-confluent graphs is shown by Eppstein et al. [31] to be equivalent to the distance hereditary graphs, where again the characterizing elimination sequence plays an important role in the proof.

In analogy to outerplanar graphs, Hui et al. [46] defined *(strongly) outerconfluent graphs* as those graphs that have a *(strongly) outerconfluent drawing* with all vertices on the outer face. Recognizing outerconfluent graphs again remains an open problem. However, for outerconfluent bipartite graphs as a subclass (graphs that allow an outerconfluent drawing, where the two parts of the bipartition form two contiguous intervals in the cyclic order of vertices around the outer face), Hui et al. showed that these are equivalent to bipartite permutation graphs.

Eppstein et al. [35] combined their strictness condition with outerconfluency and showed that for a graph with a given cyclic vertex ordering around the outer face the existence of a strict outerconfluent drawing can be tested in quadratic time. In fact, if such a drawing exists, they also presented an algorithm to construct a drawing based on circle packings, where each arc in A consists of at most two circular arcs, see Fig. 11.6 for example. Without a prescribed cyclic vertex ordering it remains an open problem whether the strict outerconfluent graphs can be recognized in polynomial time. Several observations on subclasses and superclasses of strict confluent and strict outerconfluent graphs as well as some graph-parametric properties were recently presented by Förster et al. [37]. Table 11.3 summarizes the known results and lists remaining open problems for recognizing confluent graph classes.

While the previously mentioned results were mostly of theoretical nature concerning characterizations of graph classes and corresponding recognition problems,

Table 11.3 Results and open problems for recognition problems of confluent graphs and subclasses

Graph class	Recognition problem	Alternative characterizations	References
Confluent	*Open*	–	[26]
Strongly confluent	$\in$ NP	–	[46]
Strict confluent	NP-complete	–	[35]
Outerconfluent	*Open*	–	[46]
Outerconfluent bipartite	$\in$ P	Bipartite permutation	[46, 57]
Strict outerconfluent	*Open* Fixed vertex order $\in$ P	–	[35, 37]
Tree/forest-confluent	$\in$ P	(6, 2)-chordal bipartite	[46]
Δ-confluent	$\in$ P	Distance hereditary	[31]

there are also some constructive algorithms for creating confluent graph drawings. Already in their initial paper that defined confluent drawings, Eppstein et al. [26] presented a heuristic algorithm that iteratively replaces large cliques or bicliques by the appropriate confluent junctions. If this procedure succeeds in transforming the input graph into a planar graph, a corresponding confluent drawing can be obtained immediately. For graphs with bounded arboricity the algorithm actually runs in linear time. Hirsch et al. [43] introduced another heuristic algorithm for creating confluent drawings by finding a set of cliques and bicliques that cover all edges of the graph G. From that edge cover, they derive the auxiliary biclique edge cover graph G_b, find a drawing of G_b, and finally replace vertices in the drawing of G_b by confluent structures. If the drawing of G_b is planar, then the result will be a confluent drawing. Otherwise, it is a non-planar confluent drawing with some (non-confluent) crossings on the arcs. In an experimental comparison with the heuristic of Eppstein et al. [26], they could show that their approach has a higher success rate in finding planar confluent drawings.

Subsequently, confluent drawings have been combined with special layout styles for more restricted types of drawings. Eppstein et al. [32] took the classic layered graph drawing style [42, 58] for directed graphs and introduced upward confluent subdrawings in between two or more layers. In short, layered drawings place the vertices of a directed acyclic graph on a finite set of horizontal lines so that all edges are pointing upward; for non-acyclic directed graphs the number of upward pointing edges should be maximized. In each layer a vertex ordering is sought so that as few edges as possible cross in between layers, which is an NP-hard problem [30]. But often a large number of crossings remain after optimizing the vertex ordering using good heuristics or even exact algorithms. The idea of Eppstein et al. is to compute small biclique covers of the edges in between layers, which form a bipartite subgraph, using a reduction to the vertex coloring problem. Since both problems are known as NP-hard they used well-known efficient vertex coloring heuristics to obtain biclique covers of small cardinality in $O(m^3)$ time. Eppstein et al. further investigated and implemented algorithms for positioning junctions and control points of smooth Bézier curves to actually draw the confluent arcs nicely. Planarity of the confluent drawings is not guaranteed, but the number of crossings is strongly reduced by confluency.

Eppstein and Simons [34] employed the confluent drawing style for visualizing Hasse diagrams, which are upward drawings of transitively reduced directed acyclic graphs (DAGs) used for visually representing partially ordered sets (posets). In that sense, they focused on transitively reduced DAGs. The authors showed that planar upward confluent drawings exist if and only if the underlying poset has order dimension at most two. If the DAG satisfies this property, they provide an $O(n^2)$-time algorithm to compute a confluent drawing on an $O(n) \times O(n)$-grid with a minimum number of $O(n^2)$ confluent junctions. If the given poset is actually series parallel, the running time reduces to linear and the number of junctions is bounded by $O(n)$ as well. Eppstein and Simons implemented their algorithms and performed an experimental evaluation showing that confluent Hasse diagrams use substantially less ink (and consequently have less visual clutter) than traditional straight-line Hasse diagrams.

11.4.2   Partial Edge Drawings

A radical approach to remove clutter from dense graph visualizations was suggested by Becker et al. [7], who simply erased a large part (roughly 90%) in the middle of each edge in a traditional straight-line graph drawing leaving only two stubs at the incident vertices behind. This idea not only reduces the ink used in the drawing, but more importantly it often reduces overplotting of other vertices and it removes all edge crossings that involve the erased edge parts. While this may look like cheating on first sight, the idea of Becker et al. relies on the established closure and continuation principles in Gestalt theory, which imply that humans can still see a full line segment based only on the remaining edge stubs by filling in the missing information in our brains. User studies have confirmed that such drawings remain readable and reduce clutter significantly as long as the edge shortening is not exaggerated [23].

Bruckdorfer and Kaufmann [18] formalized the edge shortening idea of Becker et al. [7] and defined partial edge drawings. A *partial edge drawing* (PED) of a graph G is a straight-line graph drawing Γ, in which every edge is partitioned into three subsegments: two *stubs* that include the two endpoints and a middle part between the two stubs. Only the two stubs are drawn in a PED and no two edge stubs may cross, see Fig. 11.7b for example. In other words, every crossing of Γ must be located on the middle part of at least one of the crossing edges so that it "disappears" in the PED. Since this property is easily obtained by drawing infinitesimally short stubs, the obvious optimization problem is to find as long stubs as possible while remaining crossing-free. We note that for edges with a single crossing, PEDs with a short gap around the crossing point are similar to the edge casing approach (see Sect. 11.3.1).

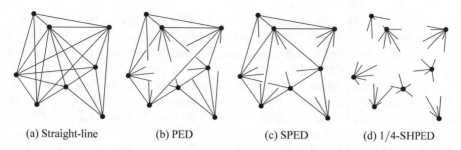

(a) Straight-line	(b) PED	(c) SPED	(d) 1/4-SHPED

Fig. 11.7 Straight-line drawing of a graph G and different partial edge drawings. The PED in **b** is neither symmetric nor homogeneous, the SPED in **c** is not homogeneous

Yet for edges with many crossings, edge casing would create multiple small gaps, whereas PEDs have a single but possibly longer gap.

Of course, partially drawn edges make it more difficult to see which two nodes are actually connected by an edge, especially if the stub of an edge $e = (u, v)$ at vertex u indicates only the direction in which to find v, but not the distance to v or the stub length at v. Therefore, more restricted variations of PEDs have been proposed by Bruckdorfer and Kaufmann [18]: *Symmetric* PEDs (SPEDs), in which both stubs of an edge must have the same length (see Fig. 11.7c), and *homogeneous* PEDs (HPEDs), in which the ratio of the stub length to the total edge length is the same for all edges (see Fig. 11.7d). Finally, *symmetric homogeneous* PEDs (SHPEDs) combine both concepts. For a parameter $0 < \delta \leq 1/2$ we define a δ-SHPED as a SHPED, where the stubs of each edge (u, v) have length $\delta \cdot |uv|$. Since $\delta = 1/2$ implies that all edges are completely drawn, a $1/2$-SHPED is actually a planar straight-line drawing. Symmetric PEDs make it easier to find the second stub of each edge as both stubs have the same length. Symmetric homogeneous PEDs additionally allow to narrow down the region, where to search for the second stub as the stub length is a constant δ-fraction of the length of the edge.

Results on PEDs fall into two groups: For graphs without input drawing, the question is to find all graphs that admit a particular type of PED, whereas for graphs with a given input drawing, the task is to maximize the stub lengths while adhering to the restrictions of the specified PED model. Bruckdorfer and Kaufmann [18] first showed that the complete graph K_n (and thus every n-vertex graph) admits a δ-SHPED for $\delta \leq 1/\sqrt{4n/\pi}$ by surrounding each vertex with an appropriately sized disk containing all its stubs and then tightly packing these disks in multiple concentric annuli. However, this bound is not tight. For example, Bruckdorfer and Kaufmann showed that K_{16} has a $1/4$-SHPED, whereas Bruckdorfer et al. [17] proved that K_{164} has no $1/4$-SHPED, or, more generally, that for any $0 < \delta < 1/2$ there is an integer N_δ such that K_n has no δ-SHPED for any $n \geq N_\delta$. For more specific graph classes, Bruckdorfer et al. [17] showed constructions proving that for a given $0 < \delta < 1/2$ complete bipartite graphs $K_{n,n}$ have a δ-SHPED if $n \leq c/\delta^2$ for some constant c. Moreover, $K_{2k,n}$ has a δ-SHPED for arbitrary integer n and $k < \log \delta / \log(1 - \delta)$. Finally, they showed that any graph of bandwidth k

has a $1/(2\sqrt{2k})$-SHPED, regardless of the number of vertices. Bruckdorfer and Kaufmann [18] also proved that the j-th power of any subgraph of a triangular tiling admits a $1/(2j)$-SHPED. Interesting open problems include finding further graph classes that always admit δ-SHPEDs for given parameter δ.

For graphs with a given non-planar drawing Γ and a fixed stub-length ratio $0 < \delta < 1/2$, it is easy to test whether Γ can be transformed into a δ-SHPED. One simply needs to test in $O(n^2)$ time whether every crossing point lies on a broken part of an edge. Bruckdorfer and Kaufmann [18] also showed that every straight-line drawing can be transformed in polynomial time into a SHPED or into an *optimal* SPED, which is a SPED that maximizes the minimum stub length.

A more challenging problem, introduced by Bruckdorfer and Kaufmann [18] again, is to maximize the total stub length (or *ink*) in a SPED for a given drawing Γ. This problem is known as maxSPED. Bruckdorfer and Kaufmann presented an ILP formulation to compute a maxSPED for any drawing Γ. In his thesis, Bruck-dorfer [16, Chap. 5.3.1] reports an NP-hardness proof for maxSPED in general and Hummel et al. [47] strengthened the NP-hardness to 3-planar input drawings. Yet a number of positive results on maxSPED are known. For 1-planar drawings a simple greedy algorithm can solve maxSPED [18]. For 2-planar drawings, the edge intersection graph consists of paths and cycles, which forms the structural basis for an $O(n \log n)$-time dynamic programming algorithm by Bruckdorfer et al. [17]. Hummel et al. [47] gave efficient dynamic programming algorithms for k-planar drawings with $k > 2$ whose edge intersection graph is a collection of trees or, more generally, edge intersection graphs of bounded treewidth. Moreover, Bruckdorfer et al. [17] considered the dual problem minSPED, where the task is to minimize the total length of the erased middle parts of the edges in a SPED of the given drawing Γ. While the optimal solutions for maxSPED and minSPED coincide, Bruckdor-fer et al. presented a 2-approximation for minSPED using a connection with the minimum-weight 2-SAT problem, which can be 2-approximated [6]. No results are known for maxSPED on graphs without a fixed drawing. For ink maximization in non-symmetric PEDs (maxPED), Hummel et al. [47] proved NP-hardness even for 4-plane input drawings. This leaves open the complexity of maxPED for 3-plane drawings. Further, Hummel et al. gave fixed-parameter maxPED algorithms for edge intersection graphs of bounded treewidth, similarly to their results on maxSPED. An open question is the complexity of deciding whether a given drawing admits a (not necessarily symmetric) δ-HPED for a given fixed ink ratio δ. In this setting only the sum of the two stub lengths of each edge is specified, but not the length of individual stubs.

Bruckdorfer et al. [19] extended the partial edge drawing idea to orthogonal draw-ings of graphs with maximum degree 4 and exactly one bend per edge. In a *1-bend orthogonal PED* (OPED) the longer of the two orthogonal segments is removed from each edge. In a *1-bend homogeneous OPED* (HOPED) half of the ink/length of each edge is removed with the shorter of the two segments always drawn entirely. Finally, in a *1-bend symmetric homogeneous OPED* (SHOPED) again half of the ink/length of each edge is removed; however, this time by erasing the two adjacent halves of both orthogonal segments. As in PEDs, the stubs must be free of crossings. Bruckdorfer et al. [19] showed that any graph that admits a 1-bend orthogonal drawing also admits a 1-bend OPED and a 1-bend HOPED. Further, all graphs with maximum degree 3 admit a 1-bend SHOPED. Among the graphs with maximum degree 4, there are some graphs that always admit a 1-bend SHOPED (all 2-circulant graphs that have 1-bend orthogonal drawings), but Bruckdorfer et al. also provided a graph with maximum degree 4 that does not have a 1-bend SHOPED. Hence the recognition problem for graphs with maximum degree 4 that admit 1-bend SHOPEDs is an interesting open question.

Finally, PEDs have not only been studied from a theoretical point of view. Bruckdorfer et al. [20] considered the particular case of 1/4-SHPEDs, which erase exactly 50% of each edge, and presented a force-based algorithm that aims to push all crossings onto the undrawn parts of the edges. Since it cannot be guaranteed that all stubs are crossing-free, those drawings are called 1/4-nearly SHPEDs. The algorithm was implemented and its performance evaluated on standard graph drawing benchmark sets. The computed drawings are substantially more effective in reducing visible crossings compared to 1/4-nearly SHPEDs obtained from erasing the central part of all edges in a standard spring-embedder layout.

A small user study [21] on the formal PED models defined by Bruckdorfer and Kaufmann [18] showed that the tested 1/4-nearly SHPEDs had slower completion time and similar or lower error rate compared to traditional straight-line drawings for tasks testing adjacencies and vertex accessibility. Yet, no statistically significant results could be obtained. A second study by Binucci et al. [14] investigated the effect of homogeneity, that is, they compared a heuristic for obtaining SPEDs (either crossing-free or optionally allowing large-angle crossings) with 1/4-nearly SHPEDs, both using the same straight-line input drawing computed with the FM3 algorithm [41]. Binucci et al. considered four tasks involving direct and indirect connectivity as well as farthest and nearest neighbor finding. Their results indicate that homogeneity is more important than gaining some additional ink or fewer crossings with the SPED heuristics, as the 1/4-nearly SHPEDs showed faster response time and lower error rate on the evaluated tasks. No empirical readability results on the orthogonal PED models are known so far.

11.5   Summary

Non-planar graphs play an important role in many practical application domains, from social science to economy, from natural sciences to engineering and technology. Yet, unlike the visual representations of vertices and edges, crossings in a graph drawing are visual entities without a structural meaning in the underlying non-geometric network data. Non-planar graphs may still be sparse, in which case a main objective in creating readable graph layouts is to avoid confusion of crossings and vertices. We have seen edge casing (Sect. 11.3.1) and slanted orthogonal drawings (Sect. 11.3.2) as visual techniques that give vertices and crossings distinctive visual properties. Partial edge drawings (Sect. 11.4.2) may also be most suitable for graphs with not too many crossings, as otherwise the remaining stubs might get too short to be readable, at least if they are required to be crossing-free. Further, we have seen two techniques that can work well even for dense graphs. Confluent drawings (Sect. 11.4.1) can easily represent very dense graphs like complete graphs and complete bipartite graphs. Yet, it is unclear in most cases whether a given graph admits a confluent drawing at all. Finally, bundled crossings (Sect. 11.3.3) can be used to reroute edges as to form fewer large crossings rather than many pairwise crossings in order to reduce visual clutter.

Many of the results mentioned in this chapter are theoretical results about properties and recognition of graphs and graph drawings with a particular crossing layout. Nonetheless, they all have some clear links to practical graph visualization, for example, by being explicitly or implicitly implemented, at least in parts, in layout algorithms or by empirical support from user studies. A better understanding of the theory of beyond-planar graph drawing, and in particular of drawings with certain crossing layout is likely to influence future improvements in practical graph layout algorithms. It can be expected that future research on drawing non-planar graphs will intensify on aspects of representing crossings in a way that minimizes their visual saliency. While past work often simply counted crossings to measure the crossing-related layout quality of non-planar graphs, recent results in the literature clearly indicate that a more detailed consideration of crossings in graph layouts is appropriate.

The individual sections of this chapter mentioned some immediate open questions related to the particular crossing layout technique. In a more general sense, it would be interesting to investigate combinations of different techniques, for example, confluent layouts allowing crossings, but representing those using edge casing, bundling, or in a slanted way. Further, many specialized graph layout styles different from standard straight-line node-link diagrams such as circular layouts, book drawings, storylines, and others may be augmented by combining them with improved crossing layout techniques. Finally, while there are individual empirical studies about the effects of, for example, crossings, crossing angles, edge bundling, and partial edge drawings, it would be very useful from a practical point of view to find suitable and empirically validated quantitative quality measures to judge the combined visual effects of the crossings in a graph layout. Such measures may improve, for instance, heuristic graph layout algorithms based on local optimization.

References

1. Alam, M.J., Fink, M., Pupyrev, S.: The bundled crossing number. In: Y. Hu, M. Nöllenburg (eds.) Graph Drawing (GD'16), *LNCS*, vol. 9801, pp. 399–412. Springer International Publishing (2016). https://doi.org/10.1007/978-3-319-50106-2_31

2. Angelini, P., Bekos, M.A., Kaufmann, M., Kindermann, P., Schneck, T.: 1-fan-bundle-planar drawings of graphs. Theor. Comput. Sci. **723**, 23–50 (2018). https://doi.org/10.1016/j.tcs.2018.03.005
3. Appel, A., Rohlf, F.J., Stein, A.J.: The haloed line effect for hidden line elimination. SIG-GRAPH Comput. Graph. **13**(2), 151–157 (1979). https://doi.org/10.1145/800249.807437
4. Argyriou, E., Cornelsen, S., Förster, H., Kaufmann, M., Nöllenburg, M., Okamoto, Y., Raftopoulou, C., Wolff, A.: Orthogonal and smooth orthogonal layouts of 1-planar graphs with low edge complexity. In: T. Biedl, A. Kerren (eds.) Graph Drawing and Network Visualization (GD'18), LNCS, vol. 11282, pp. 509–523. Springer International Publishing (2018). https://doi.org/10.1007/978-3-030-04414-5_36
5. Bae, S.W., Baffier, J.F., Chun, J., Eades, P., Eickmeyer, K., Grilli, L., Hong, S.H., Korman, M., Montecchiani, F., Rutter, I., Tóth, C.D.: Gap-planar graphs. Theor. Comput. Sci. **745**, 36–52 (2018). https://doi.org/10.1016/j.tcs.2018.05.029
6. Bar-Yehuda, R., Rawitz, D.: Efficient algorithms for integer programs with two variables per constraint. Algorithmica **29**, 595–609 (2001). https://doi.org/10.1007/s004530010075
7. Becker, R.A., Eick, S.G., Wilks, A.R.: Visualizing network data. IEEE Trans. Vis. Comput. Graph. **1**(1), 16–28 (1995). https://doi.org/10.1109/2945.468391
8. Bekos, M.A., Kaufmann, M., Kobourov, S.G., Symvonis, A.: Smooth orthogonal layouts. J. Graph Algorithms Appl. **17**(5), 575–595 (2013). https://doi.org/10.7155/jgaa.00305
9. Bekos, M.A., Kaufmann, M., Krug, R.: Sloggy drawings of graphs. In: Information, Intelligence, Systems and Applications (IISA'14) (2014). https://doi.org/10.1109/IISA.2014.6878764
10. Bekos, M.A., Kaufmann, M., Krug, R., Ludwig, T., Näher, S., Roselli, V.: Slanted orthogonal drawings: model, algorithms and evaluations. J. Graph Algorithms Appl. **18**(3), 459–489 (2014). https://doi.org/10.7155/jgaa.00332
11. Bekos, M.A., Kaufmann, M., Krug, R.: Sloginsky drawings of graphs. In: Information, Intelligence, Systems and Applications (IISA'15) (2015). https://doi.org/10.1109/IISA.2015.7388121
12. Bekos, M.A., Förster, H., Kaufmann, M.: On smooth orthogonal and octilinear drawings: relations, complexity and kandinsky drawings. Algorithmica **81**(5), 2046–2071 (2019). https://doi.org/10.1007/s00453-018-0523-5
13. Biedl, T., Kant, G.: A better heuristic for orthogonal graph drawings. Comput. Geom. Theory Appl. **9**(3), 159–180 (1998). https://doi.org/10.1016/S0925-7721(97)00026-6
14. Binucci, C., Liotta, G., Montecchiani, F., Tappini, A.: Partial edge drawing: Homogeneity is more important than crossings and ink. In: Information, Intelligence, Systems Applications (IISA'16), pp. 1–6 (2016). https://doi.org/10.1109/IISA.2016.7785427
15. Bläsius, T., Brückner, G., Rutter, I.: Complexity of higher-degree orthogonal graph embedding in the Kandinsky model. In: A.S. Schulz, D. Wagner (eds.) Algorithms (ESA'14), LNCS, vol. 8737, pp. 161–172. Springer (2014). https://doi.org/10.1007/978-3-662-44777-2_14
16. Bruckdorfer, T.: Schematics of graphs and hypergraphs. Ph.D. thesis, Universität Tübingen (2015)
17. Bruckdorfer, T., Cornelsen, S., Gutwenger, C., Kaufmann, M., Montecchiani, F., Nöllenburg, M., Wolff, A.: Progress on partial edge drawings. J. Graph Algorithms Appl. **21**(4), 757–786 (2017). https://doi.org/10.7155/jgaa.00438
18. Bruckdorfer, T., Kaufmann, M.: Mad at edge crossings? Break the edges! In: E. Kranakis, D. Krizanc, F. Luccio (eds.) Fun with Algorithms (FUN'12), LNCS, vol. 7288, pp. 40–50. Springer (2012). https://doi.org/10.1007/978-3-642-30347-0_7
19. Bruckdorfer, T., Kaufmann, M., Montecchiani, F.: 1-bend orthogonal partial edge drawings. J. Graph Algorithms Appl. **18**(1), 111–131 (2014). https://doi.org/10.7155/jgaa.00316
20. Bruckdorfer, T., Kaufmann, M., Lauer, A.: A practical approach for 1/4-SHPEDs. In: Information, Intelligence, Systems and Applications (IISA'15) (2015). https://doi.org/10.1109/IISA.2015.7387994
21. Bruckdorfer, T., Kaufmann, M., Leibßle, S.: PED user study. In: E. Di Giacomo, A. Lubiw (eds.) Graph Drawing (GD'15), LNCS, vol. 9411, pp. 551–553. Springer International Publishing (2015). https://doi.org/10.1007/978-3-319-27261-0_47

22. Buchheim, C., Chimani, M., Gutwenger, C., Jünger, M., Mutzel, P.: Crossings and planarization. In: R. Tamassia (ed.) Handbook of Graph Drawing and Visualization, chap. 2, pp. 43–85. CRC Press (2013)

23. Burch, M., Vehlow, C., Konevtsova, N., Weiskopf, D.: Evaluating partially drawn links for directed graph edges. In: M. van Kreveld, B. Speckmann (eds.) Graph Drawing (GD'11), LNCS, vol. 7034, pp. 226–237. Springer (2012). https://doi.org/10.1007/978-3-642-25878-7_22

24. Chaplick, S., van Dijk, T.C., Kryven, M., Park, J., Ravsky, A., Wolff, A.: Bundled crossings revisited. In: D. Archambault, C.D. Tóth (eds.) Graph Drawing and Network Visualization (GD'19), LNCS, vol. 11904, pp. 63–77. Springer (2019). https://doi.org/10.1007/978-3-030-35802-0_5

25. Corneil, D.G., Lerchs, H., Burlingham, L.S.: Complement reducible graphs. Discrete Appl. Math. 3(3), 163–174 (1981). https://doi.org/10.1016/0166-218X(81)90013-5

26. Dickerson, M., Eppstein, D., Goodrich, M.T., Meng, J.Y.: Confluent drawings: visualizing nonplanar diagrams in a planar way. J. Graph Algorithms Appl. 9(1), 31–52 (2005). https://doi.org/10.7155/jgaa.00099

27. Didimo, W., Eades, P., Liotta, G.: Drawing graphs with right angle crossings. Theor. Comput. Sci. 412(39), 5156–5166 (2011). https://doi.org/10.1016/j.tcs.2011.05.025

28. Dujmovi, V., Gudmundsson, J., Morin, P., Wolle, T.: Notes on large angle crossing graphs. Chicago J. Theor. Comput. Sci. 4, 1–14 (2011). https://doi.org/10.4086/cjtcs.2011.004

29. Duncan, C.A., Goodrich, M.T.: Planar orthogonal and polyline drawing algorithms. In: R. Tamassia (ed.) Handbook of Graph Drawing and Visualization, chap. 7, pp. 223–246. CRC Press (2013)

30. Eades, P., Wormald, N.C.: Edge crossings in drawings of bipartite graphs. Algorithmica 11, 379–403 (1994). https://doi.org/10.1007/BF01187020

31. Eppstein, D., Goodrich, M.T., Meng, J.Y.: Delta-confluent drawings. In: P. Healy, N. Nikolov (eds.) Graph Drawing (GD'05), LNCS, vol. 3843, pp. 165–176. Springer (2006). https://doi.org/10.1007/11618058_16

32. Eppstein, D., Goodrich, M.T., Meng, J.Y.: Confluent layered drawings. Algorithmica 47, 439–452 (2007). https://doi.org/10.1007/s00453-006-0159-8

33. Eppstein, D., van Kreveld, M., Mumford, E., Speckmann, B.: Edges and switches, tunnels and bridges. Comput. Geom. Theory Appl. 42, 790–802 (2009). https://doi.org/10.1016/j.comgeo.2008.05.005

34. Eppstein, D., Simons, J.A.: Confluent Hasse diagrams. J. Graph Algorithms Appl. 17(7), 689–710 (2013). https://doi.org/10.7155/jgaa.00312

35. Eppstein, D., Holten, D., Löffler, M., Nöllenburg, M., Speckmann, B., Verbeek, K.: Strict confluent drawing. J. Comput. Geom. 7(1), 22–46 (2016). https://doi.org/10.20382/jocg.v7i1a2

36. Fink, M., Hershberger, J., Suri, S., Verbeek, K.: Bundled crossings in embedded graphs. In: E. Kranakis, G. Navarro, E. Chávez (eds.) Theoretical Informatics (LATIN'16), LNCS, vol. 9644, pp. 454–468. Springer Berlin Heidelberg (2016). https://doi.org/10.1007/978-3-662-49529-2_34

37. Förster, H., Ganian, R., Klute, F., Nöllenburg, M.: On strict (outer-)confluent graphs. In: D. Archambault, C.D. Tóth (eds.) Graph Drawing and Network Visualization (GD'19), LNCS, vol. 11904, pp. 147–161. Springer (2019). https://doi.org/10.1007/978-3-030-35802-0_12

38. Fößmeier, U., Kaufmann, M.: Drawing high degree graphs with low bend numbers. In: F.J. Brandenburg (ed.) Graph Drawing (GD'95), LNCS, vol. 1027, pp. 254–266. Springer (1996). https://doi.org/10.1007/BFb0021809

39. Gansner, E.R., Koren, Y.: Improved circular layouts. In: M. Kaufmann, D. Wagner (eds.) Graph Drawing (GD'06), LNCS, vol. 4372, pp. 386–398. Springer Berlin Heidelberg (2006). https://doi.org/10.1007/978-3-540-70904-6_37

40. Garey, M.R., Johnson, D.S.: Crossing number is NP-complete. SIAM J. on Algebraic and Discrete Methods 4(3), 312–316 (1983). https://doi.org/10.1137/0604033

41. Hachul, S., Jünger, M.: Drawing large graphs with a potential-field-based multilevel algorithm. In: J. Pach (ed.) Graph Drawing (GD'04), LNCS, vol. 3383, pp. 285–295. Springer (2005). https://doi.org/10.1007/978-3-540-31843-9_29

42. Healy, P., Nikolov, N.S.: Hierarchical drawing algorithms. In: R. Tamassia (ed.) Handbook of Graph Drawing and Visualization, chap. 13, pp. 409–454. CRC Press (2014)
43. Hirsch, M., Meijer, H., Rappaport, D.: Biclique edge cover graphs and confluent drawings. In: Graph Drawing (GD'06), LNCS, vol. 4372, pp. 405–416. Springer (2007). https://doi.org/10.1007/978-3-540-70904-6_39
44. Huang, W., Hong, S.H., Eades, P.: Effects of crossing angles. In: Pacific Visualization Symposium (PacificVis'08), pp. 41–46. IEEE (2008). https://doi.org/10.1109/PACIFICVIS.2008.4475457
45. Huang, W., Eades, P., Hong, S.H.: A graph reading behavior: Geodesic-path tendency. In: Pacific Visualization Symposium (PacificVis'09), pp. 137–144. IEEE (2009). https://doi.org/10.1109/PACIFICVIS.2009.4906848
46. Hui, P., Pelsmajer, M.J., Schaefer, M., štefankovič, D.: Train tracks and confluent drawings. Algorithmica 47, 465–479 (2007). https://doi.org/10.1007/s00453-006-0165-x
47. Hummel, M., Klute, F., Nickel, S., Nöllenburg, M.: Maximizing ink in partial edge drawings of k-plane graphs. In: D. Archambault, C.D. Tóth (eds.) Graph Drawing and Network Visualization (GD'19), LNCS, vol. 11904, pp. 323–336. Springer (2019). https://doi.org/10.1007/978-3-030-35802-0_25
48. Kaufmann, M., Ueckerdt, T.: The density of fan-planar graphs. CoRR **abs/1403.6184** (2014). http://arxiv.org/abs/1403.6184
49. Kieffer, S., Dwyer, T., Marriott, K., Wybrow, M.: HOLA: human-like orthogonal network layout. IEEE Trans. Vis. Comput. Graph. **22**(1), 349–358 (2016). https://doi.org/10.1109/TVCG.2015.2467451
50. Lhuillier, A., Hurter, C., Telea, A.: State of the art in edge and trail bundling techniques. Comput. Graph. Forum **36**(3), 619–645 (2017). https://doi.org/10.1111/cgf.13213
51. Nguyen, Q., Eades, P., Hong, S.H.: On the faithfulness of graph visualizations. In: Pacific Visualization Symposium (PacificVis'13), pp. 209–216. IEEE (2013). https://doi.org/10.1109/PacificVis.2013.6596147
52. Nishizeki, T., Rahman, M.S.: Planar Graph Drawing. Lecture Notes Series on Computing, vol. 12. World Scientific (2004)
53. Purchase, H.: Which aesthetic has the greatest effect on human understanding? In: Graph Drawing (GD'97), LNCS, vol. 1353, pp. 248–261. Springer (1997). https://doi.org/10.1007/3-540-63938-1_67
54. Purchase, H.C., Cohen, R.F., James, M.: Validating graph drawing aesthetics. In: F.J. Brandenburg (ed.) Graph Drawing (GD'95), LNCS, vol. 1027, pp. 435–446. Springer (1996). https://doi.org/10.1007/BFb0021827
55. Purchase, H.C., Pilcher, C., Plimmer, B.: Graph drawing aesthetics created by users, not algorithms. IEEE Trans. Vis. Comput. Graph. **18**(1), 81–92 (2012). https://doi.org/10.1109/TVCG.2010.269
56. Rusu, A., Fabian, A.J., Jianu, R., Rusu, A.: Using the gestalt principle of closure to alleviate the edge crossing problem in graph drawings. In: Proceedings of 15th International Conference on Information Visualisation (IV'11), pp. 488–493. IEEE (2011). https://doi.org/10.1109/IV.2011.63
57. Spinrad, J., Brandstädt, A., Stewart, L.: Bipartite permutation graphs. Discrete Appl. Math. **18**(3), 279–292 (1987). https://doi.org/10.1016/S0166-218X(87)80003-3
58. Sugiyama, K., Tagawa, S., Toda, M.: Methods for visual understanding of hierarchical system structures. IEEE Trans. Syst. Man Cybern. **11**(2), 109–125 (1981). https://doi.org/10.1109/TSMC.1981.4308636
59. Tamassia, R.: On embedding a graph in the grid with the minimum number of bends. SIAM J. Comput. **16**(3), 421–444 (1987). https://doi.org/10.1137/0216030
60. Thomassen, C.: The graph genus problem is NP-complete. J. Algorithms **10**(4), 568–576 (1989). https://doi.org/10.1016/0196-6774(89)90006-0
61. Vismara, L.: Planar straight-line drawing algorithms. In: R. Tamassia (ed.) Handbook of Graph Drawing and Visualization, chap. 6, pp. 193–222. CRC Press (2013)
62. Ware, C., Purchase, H., Colpoys, L., McGill, M.: Cognitive measurements of graph aesthetics. Inf. Vis. **1**, 103–110 (2002). https://doi.org/10.1057/palgrave.ivs.9500013

Chapter 12
Beyond Clustered Planar Graphs

Patrizio Angelini and Giordano Da Lozzo

Abstract Many real-world networks exhibit an inherent clustering structure, which may arise from the presence of communities inside the network or from semantic affinities among nodes. Constructing effective visualizations for such networks is a crucial task that poses several practical and theoretical challenges. The standard theoretical model for readable representations of clustered graphs is the one, of c-planarity, introduced in the 90s and still a central research topic in graph drawing. The goal of this model is to realize drawings avoiding unnecessary crossings involving edges or clusters. This chapter reviews alternative models that have been proposed to enlarge the set of clustered graphs allowing for a representation that still conveys the clustering and relational information. First, we deal with a relaxed notion of c-planarity, in which some crossings are allowed. Then, we present two popular models for hybrid representations of clustered networks, namely NodeTrix and Intersection-Link representations, which combine different drawing paradigms for the inter-cluster and the intra-cluster relationships.

12.1 Introduction

The problem of displaying together multiple relationships among the same set of entities is a central subject of research in graph drawing and network visualization. Clustered graphs offer a powerful model for simultaneously describing a binary relation among pairs of entities alongside with a cluster hierarchy defined on subsets of entities, the latter grouping together entities with semantic affinities. As an example, the structure of a social network is defined by friendships between pairs of users, while communities can be independently identified based on several criteria, such as musical preferences or memberships to social groups.

P. Angelini
John Cabot University, Rome, Italy
e-mail: pangelini@johncabot.edu

G. Da Lozzo (✉)
Roma Tre University, Rome, Italy
e-mail: giordano.dalozzo@uniroma3.it

© Springer Nature Singapore Pte Ltd. 2020
S.-H. Hong and T. Tokuyama (eds.), *Beyond Planar Graphs*,
https://doi.org/10.1007/978-981-15-6533-5_12

In mathematical terms, a network equipped with a cluster hierarchy on its nodes can be modeled as a *clustered graph* $C(G, T)$ (for short c-graph), which consists of a graph G and of a rooted tree T whose leaves are the vertices of G. Such a combinatorial structure is used to enrich the vertices of the graph with the hierarchical information. In fact, each non-leaf node μ of T, called *cluster*, represents the subset V_μ of the vertices of G that are the leaves of the subtree of T rooted at μ. If $|V_\mu| = 1$, then μ is a *trivial cluster*, otherwise it is a *non-trivial cluster*. Tree T, which defines the inclusion relationships among clusters, is called *inclusion tree*, while G is the *underlying graph* of $C(G, T)$. *Intra-cluster edges* connect vertices in the same cluster, while *inter-cluster edges* connect vertices in different clusters.

In a *clustered drawing* of a c-graph $C(G, T)$, vertices and edges of G are drawn as points and Jordan arcs, respectively, and each cluster $\mu \in T$ is represented as a region R_μ homeomorphic to a closed disk containing exactly the vertices of μ. Also, if μ is a descendant of a cluster ν, then R_ν contains R_μ.

A clustered drawing can have three types of crossings; refer to Fig. 12.1. *Edge–edge crossings* (*ee-crossings*) occur between edges of G. Two kinds of crossings involve regions, instead. If an edge e intersects more than once the boundary of R_μ, with $\mu \in T$, we have *edge–region crossings* (*er-crossings*). Finally, if the boundary of R_μ intersects the boundary of R_ν, with $\mu, \nu \in T$, we have *region–region crossings* (*rr-crossings*).

A *c-planar drawing* of a c-graph is a clustered drawing without *ee-*, *er-*, and *rr-*crossings, and a *c-planar c-graph* is one that admits a c-planar drawing. A c-graph is *flat* if T has height 2, that is, no cluster different from the root contains other clusters, and it is *non-flat*, otherwise. The *clusters-adjacency graph* of a flat c-graph is the graph obtained by contracting each cluster to a single vertex, and by removing loops and multiple edges.

The problem of testing whether a c-graph is c-planar has been introduced by Feng, Cohen, Eades [31] in 1995 under the name of C-Planarity , even though previous related work by Lengauer [55] dates back to 1989. Due to its practical relevance and to its theoretical appeal, this problem has become extremely popular and has motivated a great deal of effort in the graph drawing research community in the last decades. This effort resulted in polynomial-time algorithms for instances satisfying several kinds of restrictions, such as

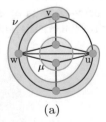
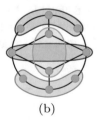
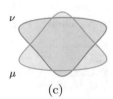

(a)   (b)   (c)

Fig. 12.1 **a** A drawing with two *er*-crossings. **b** A drawing with one *rr*-crossing. **c** Intersections between regions generating two *rr*-crossings

- Assuming that each cluster induces a small number of connected components [20, 30, 42, 43, 49, 50]. In particular, the case in which the graph is *c-connected* , that is, for each cluster $\mu \in \mathcal{T}$ the graph induced by the vertices of μ is connected, has been deeply investigated [22, 27, 31].
- Restricting to c-graphs containing a small number of clusters [1, 34, 48].
- Focusing on particular families of underlying graphs [23, 33, 52].
- Fixing the embedding of the underlying graph [19, 28, 49]. We remark that a sub-exponential algorithm for C- PLANARITY has been provided for embedded flat c-graphs with bounded face size [25].
- Assuming that the inter-cluster edges connecting each pair of clusters are grouped into pipes [1, 3, 8, 23, 35, 36].

Alternative approaches, such as algebraic methods [38, 45], ILP formulations [17, 18], and FPT techniques [12, 17, 26, 52], have also been considered.

Patrignani and Cortese recently showed that C- PLANARITY retains the same computational complexity for flat and non-flat instances [21]. Exploiting this result, Fulek and Tóth eventually settled the long-standing question regarding the complexity of the C- PLANARITY problem, by providing the first polynomial-time algorithm for general instances [37].

Finally, deep connections with the SIMULTANEOUS EMBEDDING WITH FIXED EDGES (SEFE) problem have been shown [2, 60]. In particular, the result of Fulek and Tóth combined with the one in [2] implies a polynomial-time algorithm for the SEFE problem for instances composed of two graphs whose intersection forms a connected graph.

In the next sections, we describe three different paradigms for the visualization of clustered networks that are alternative to the classical c-planar drawings, which also extend to c-graphs whose underlying graph may be non-planar. This latter extension is particularly relevant, since in several applications there is the need to represent graphs that are globally sparse but contain dense non-planar subgraphs. As an example, the Internet network is composed of strongly connected local networks, which are then interconnected via a sparser backbone structure.

In Sect. 12.2, we consider clustered drawings of c-graphs in which the c-planarity constraints are relaxed to allow certain types of crossings. In the subsequent sections, we discuss hybrid representations of c-graphs, in which the representation of the clusters and one of their connections employ different drawing models. Namely, in Sect. 12.3, we investigate *NodeTrix representations*, which combine adjacency matrices to depict clusters and node-link diagrams to show their relationships, while in Sect. 12.4 we overview *intersection-link representations*, which combine node-link diagrams with the well-established drawing paradigm of intersection representations.

12.2 Clustered Drawings with Crossings

In this section, we consider clustered drawings in which *ee*-, *er*-, and *rr*-crossings may occur. This model, introduced by Angelini et al. [4], generalizes the classical notion of c-planar drawings, where none of these crossings is allowed, and its employment is encouraged by the fact that the class of c-planar instances may be too small for some application contexts. However, by bounding the number and type of allowed crossings, these drawings may still meet the requirements of many typical graph drawing applications.

We start by formally defining how to count the different types of crossings in a clustered drawing. Following standard assumptions, we do not allow more than two Jordan arcs to cross at the same point, and we assume that any two Jordan arcs meet in finitely many points and that they alternate around each crossing point. As for *ee*-crossings, we simply count each crossing independently. If an edge e crosses the boundary of a region R_μ representing a cluster μ in k points, the number of *er*-crossings between e and R_μ is $\lfloor \frac{k}{2} \rfloor$. Note that, if e intersects the boundary of R_μ exactly once, this is not considered as an *er*-crossing, since there is no way of connecting the endpoints of e without intersecting the boundary of R_μ. In Fig. 12.1a, edge (u, w) traverses R_μ and edge (u, v) exits and enters R_v. Finally, if two regions R_μ and R_v cross (which can only happen if μ is not an ancestor of v, and vice-versa), then the number of *rr*-crossings between R_μ and R_v is equal to the number of the topologically connected regions resulting from the relative complement of R_μ in R_v (i.e., $R_\mu \setminus R_v$) minus one; see Fig. 12.1b, c.

Definition 1 An $\langle \alpha, \beta, \gamma \rangle$-*drawing* of a c-graph is clustered drawing with α *ee*-crossings, β *er*-crossings, and γ *rr*-crossings.

Note that, when $\alpha = \beta = \gamma = 0$, an $\langle \alpha, \beta, \gamma \rangle$-drawing is a c-planar drawing. Hence, the existence of an $\langle \alpha, \beta, \gamma \rangle$-drawing of a c-graph $\mathcal{C}(G, \mathcal{T})$, for some values of α, β, and γ, is a non-trivial necessary condition for the c-planarity of $\mathcal{C}(G, \mathcal{T})$, which further justifies the study of this generalization.

12.2.1 Complexity

Since the readability of drawings is negatively affected by the presence of crossings [59], the first and most natural problem for $\langle \alpha, \beta, \gamma \rangle$-drawings is the one of minimizing the number of their crossings. Formally, given a c-graph $\mathcal{C}(G, \mathcal{T})$ and an integer $k > 0$, the (α, β, γ)-CLUSTER CROSSING NUMBER $((\alpha, \beta, \gamma)$-CCN) problem asks whether $\mathcal{C}(G, \mathcal{T})$ admits an $\langle \alpha, \beta, \gamma \rangle$-drawing with $\alpha + \beta + \gamma \leq k$. The (α, β, γ)-CCN problem has been proved NP-complete [4, Theorem 15] even when G is a forest of stars, by means of a polynomial-time reduction from the CROSSING NUMBER problem, proved NP-complete by Garey and Johnson [40].

The (α, β, γ)-CCN problem has been also studied when only one out of α, β, and γ is allowed to be different from 0. The corresponding decision problems are

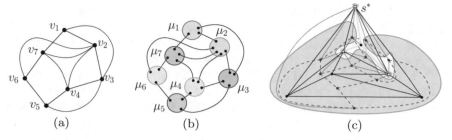

Fig. 12.2 Illustration for Theorem 1: **a** Graph G^* and **b** the c-graph $\mathcal{C}(G, \mathcal{T})$ corresponding to G^*. **c** Illustration for Theorem 2. Edges of G are solid (black); edges of H are dashed (red) and dotted (blue); edges of T^* are yellow; non-terminal vertices and terminals in G are black circles and white squares, respectively; vertices of H are red circles and white squares

called α-CCN, β-CCN, and γ-CCN, respectively. The NP-completeness proof for the general (α, β, γ)-CCN problem can be easily modified to show that all of these restricted variants are also NP-complete, for the same class of c-graphs. However, stronger NP-hardness proofs can be shown for α-CCN and β-CCN for even more restricted classes of c-graphs.

Theorem 1 ([4], Theorem 16) *The α-CCN problem is NP-complete even when the underlying graph is a matching.*

The NP-hardness of Theorem 1 can be proved by means of a polynomial-time reduction from the CROSSING NUMBER problem [40]. Namely, an instance $\langle \mathcal{C}(G, \mathcal{T}), k \rangle$ of α-CCN can be constructed starting from an instance $\langle G^*, k^* \rangle$ of CROSSING NUMBER as follows; see Fig. 12.2a, b. Initialize $G = G^*$, $k = k^*$, and $\mathcal{T}$ to a tree containing only the root cluster ρ. Subdivide each edge of G with two subdivision vertices. For each vertex v_i of G, add a cluster μ_i to $\mathcal{T}$ as a child of ρ containing all the neighbors of v_i, and remove from G vertex v_i and its incident edges. Note that graph G is a matching.

We claim that the two instances are equivalent. In fact, a drawing of G^* containing ℓ crossings can be turned into an $\langle \alpha, 0, 0 \rangle$-drawing of $\mathcal{C}(G, \mathcal{T})$ containing ℓ ee-crossings, by "splitting" each vertex v_i of G^* in the interior of a small disk d_i enclosing v_i whose boundary can be used to define region R_{μ_i}. On the other hand, an $\langle \alpha, 0, 0 \rangle$-drawing of $\mathcal{C}(G, \mathcal{T})$ containing ℓ ee-crossings can be turned into a drawing of G^* containing ℓ crossings, by "merging" all the vertices in the same cluster into the vertex of G^* they stem from. This can be done, since these vertices lie in the interior of R_{μ_i}, where μ_i is the cluster of $\mathcal{T}$ containing all such vertices, and since V_{μ_i} is an independent set.

Next, we present a strengthening of [4, Theorem 17], which states the NP-completeness of β-CCN for c-graphs whose underlying graph is a triconnected planar multigraph, overcoming the need for multiple edges.

Theorem 2 *The β-CCN problem is NP-complete even for c-connected flat c-graphs whose underlying graph is triconnected and planar.*

The NP-hardness of Theorem 2 is proved by means of a polynomial-time reduction from the NP-complete STEINER TREE problem in planar graphs (STPG) [39], defined as follows. Given a planar graph $G = (V, E)$ whose edges have weights $w : E \rightarrow \mathbb{N}$, a set $S \subset V$ of *terminals*, and an integer $k > 0$, does a tree $T^* = (V^*, E^*)$ exist such that (1) $S \subseteq V^* \subseteq V$, (2) $E^* \subseteq E$, and (3) $\sum_{e \in E^*} w(e) \leq k$? Differently from [4], we use here the variant of STPG, called UNIFORM TRIANGULATED PST (UTPST), where G is a triangulation and all edge weights are equal to 1, which is also NP-complete [5].

An instance $\langle \mathcal{C}(H, \mathcal{T}), k \rangle$ of β-CCN can be constructed starting from an instance $\langle G, S, k \rangle$ of UTPST as follows; see Fig. 12.2c. Since G is a triangulation, its unique dual graph H is triconnected, cubic, and simple. For each $s \in S$, consider the set $E_G(s)$ of the edges incident to s in G and the face f_s of H composed of the edges dual to those in $E_G(s)$. Add s to H, embed it inside f_s, and connect it to the vertices incident to f_s. Finally, for each $s \in S$, tree $\mathcal{T}$ has a cluster $\mu_s = \{s\}$. All the other vertices of H are in a cluster v.

Suppose that $\langle G, S, k \rangle$ admits a solution T^*. Consider a terminal $s^* \in S$ and construct a planar embedding of H in which s^* is incident to the outer face. We can then draw cluster v as a region R_v homeomorphic to a closed disk that encloses H, except for a small region surrounding T^*. Also, draw each cluster $\mu_s \neq v$, for each $s \in S$, as a region homeomorphic to a closed disk that surrounds s, without intersecting R_v. Since R_v intersects all and only the edges of H dual to edges in T^*, there are at most $k^* = k$ *er*-crossings.

For the other direction, let Γ be a $\langle 0, \beta, 0 \rangle$-drawing of $\mathcal{C}(H, \mathcal{T})$ with the minimum number $\beta \leq k$ of *er*-crossings. Consider the graph T^* composed of the edges that are dual to the edges of H participating in some *er*-crossing. The proof is based on showing that T^* has at least one edge incident to each terminal in S and that T^* is connected. The claim implies that T^* is a solution to the instance $\langle G, S, k \rangle$ of STPG, since T^* has at most k edges.

12.2.2 Allowing only One Type of Crossings

In this subsection, we explore the existence of drawings in which only one type of crossings is allowed. We call these drawings $\langle \infty, 0, 0 \rangle$-, $\langle 0, \infty, 0 \rangle$-, and $\langle 0, 0, \infty \rangle$-drawings, respectively. Our investigation uncovers that allowing different types of crossings has a different impact on the existence of drawings of c-graphs (see Fig. 12.3). In particular, while every c-graph admits an $\langle \infty, 0, 0 \rangle$-drawing (even if its underlying graph is not planar) and a $\langle 0, \infty, 0 \rangle$-drawing (only if its underlying graph is planar), there exist c-graphs not admitting any $\langle 0, 0, \infty \rangle$-drawing (even if the underlying graph is planar).

We first show that allowing only edge–edge crossings is sufficient to construct clustered drawings of any c-graph.

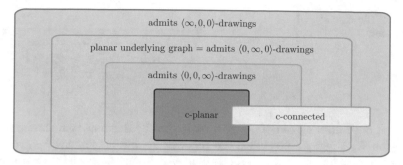

Fig. 12.3 Containment relationships among classes of c-graphs. Note that any $\langle 0, 0, \infty \rangle$-drawing of a c-connected c-graph $\mathcal{C}(G, \mathcal{T})$ can be suitably modified to obtain a c-planar drawing of $\mathcal{C}(G, \mathcal{T})$

Theorem 3 *Every c-graph $\mathcal{C}(G, \mathcal{T})$ admits an $\langle \alpha, 0, 0 \rangle$-drawing of $\mathcal{C}(G, \mathcal{T})$ with $\alpha \in O(n^4)$. If G is planar, then $\alpha \in O(n^2)$.*

Proof Let $\sigma = v_1, \ldots, v_n$ be an ordering of the vertices of G such that vertices of the same cluster are consecutive in σ. A drawing Γ of G can be constructed as follows. Place the vertices of G along a convex curve in the order they appear in σ. Draw the edges of G as straight-line segments. Since vertices belonging to the same cluster are consecutive in σ, drawing each cluster as the convex hull of the points assigned to its vertices yields a drawing without rr- and er-crossings (see Fig. 12.4). Further, since G has $O(n^2)$ edges, and since edges are drawn as straight-line segments, drawing Γ contains $O(n^4)$ ee-crossings. On the other hand, if G is planar, it has $O(n)$ edges, and thus Γ contains $O(n^2)$ ee-crossings. $\square$

We now give an algorithm to construct a $\langle 0, \beta, 0 \rangle$-drawing of any c-graph $\mathcal{C}(G, \mathcal{T})$. The algorithm consists of the following three steps.

1. A spanning tree T of the vertices of G is constructed in such a way that, for each cluster $\mu \in \mathcal{T}$, the subgraph T_μ of T induced by the vertices of μ is connected. The algorithm constructs T in two different ways, based on whether $\mathcal{C}(G, \mathcal{T})$ is c-connected or not; in particular, T is a subgraph of G if $\mathcal{C}(G, \mathcal{T})$ is c-connected, while it is not necessarily a subgraph of G if $\mathcal{C}(G, \mathcal{T})$ is not c-connected.
2. A simultaneous embedding of G and T is computed. Recall that a *simultaneous embedding* of two graphs $G_1(V, E_1)$ and $G_2(V, E_2)$, on the same set V of vertices, is a drawing of $G(V, E_1 \cup E_2)$ such that any crossing involves an edge from E_1 and an edge from E_2 [13]. If $\mathcal{C}(G, \mathcal{T})$ is c-connected, since each edge of T is also an edge of G, any planar drawing of G determines a simultaneous embedding of G and T in which no edge of G properly crosses an edge of T. Otherwise, if $\mathcal{C}(G, \mathcal{T})$ is not c-connected, we can apply the algorithm by Kammer [53] (see also [29]) to construct a simultaneous embedding of G and T in which each edge has at most two bends, which implies that each pair of edges $\langle e_1 \in G, e_2 \in T \rangle$ crosses a constant number of times.

Fig. 12.4 Illustration
for Theorem 3

3. A $\langle 0, \beta, 0 \rangle$-drawing of $\mathcal{C}(G, \mathcal{T})$ is constructed by drawing each cluster μ as a region R_μ slightly surrounding the edges of T_μ and the regions $R_{\mu_1}, \ldots, R_{\mu_k}$ representing the children $\mu_1, \ldots, \mu_k$ of μ. Hence, each crossing between an edge $e_1 \in G$ and an edge $e_2 \in T$ determines two intersections (hence one er-crossing) between e_1 and the boundary of each cluster ν such that $e_2 \in T_\nu$. Further, each edge $(u, v) \in G$ such that $(u, v) \notin T$ and u and v belong to the same cluster ν has a er-crossing with the boundary of R_ν.

Theorem 4 *Every c-graph $\mathcal{C}(G, \mathcal{T})$ such that G is planar admits a $\langle 0, \beta, 0 \rangle$-drawing with $\beta \in O(n^3)$. Further, if $\mathcal{C}(G, \mathcal{T})$ is either c-connected or flat, then $\beta \in O(n^2)$. Finally, if $\mathcal{C}(G, \mathcal{T})$ is both c-connected and flat, then $\beta \in O(n)$.*

We now turn our attention to establish bounds on the minimum value of γ in a $\langle 0, 0, \gamma \rangle$-drawing of a c-graph.

Theorem 5 *Every c-graph $\mathcal{C}(G, \mathcal{T})$ that admits a $\langle 0, 0, \infty \rangle$-drawing also admits a $\langle 0, 0, \gamma \rangle$-drawing with $\gamma \in O(n^3)$. If $\mathcal{C}(G, \mathcal{T})$ is flat, then $\gamma \in O(n^2)$.*

Proof Suppose that $\mathcal{C}(G, \mathcal{T})$ admits a $\langle 0, 0, \infty \rangle$-drawing. Then, consider the drawing Γ of the underlying graph G in any such a drawing. Visit the clusters in a bottom-up traversal of $\mathcal{T}$; for each cluster μ, place a vertex $u_{\mu, f}$ inside each face f of Γ that contains at least one vertex belonging to μ, and connect $u_{\mu, f}$ to all the vertices of μ incident to f. Note that the graph composed by the vertices of μ and by the added vertices is connected. Then, construct a spanning tree S of such a graph, so that S contains the spanning trees of the children of μ, and draw R_μ slightly surrounding S. The cubic bound on γ comes from the fact that each of the $O(n)$ clusters crosses each of the $O(n)$ other clusters a linear number of times. Finally, if $\mathcal{C}(G, \mathcal{T})$ is flat, then each of the $O(n)$ clusters crosses each of the $O(n)$ other clusters just once. $\qquad\square$

We give two examples of c-graphs not admitting any $\langle 0, 0, \infty \rangle$-drawing. Let $\mathcal{C}(G, \mathcal{T})$ be a c-graph such that G is a triconnected planar graph and has a cycle of vertices belonging to a cluster μ separating two vertices not in μ (see Fig. 12.5a). Thus, even in the presence of rr-crossings, one of the two vertices not in μ is enclosed by R_μ in any $\langle 0, 0, \infty \rangle$-drawing of $\mathcal{C}(G, \mathcal{T})$. This example exploits that G has a unique planar embedding. Next we show that even c-graphs with series-parallel underlying graph may not admit any $\langle 0, 0, \infty \rangle$-drawing. Namely, let $\mathcal{C}(G, \mathcal{T})$ be a c-graph such that G has eight vertices and is composed of parallel paths p_1, p_2, p_3, and p_4 sharing two vertices. Tree $\mathcal{T}$ is such that cluster μ_1 contains one vertex of

Fig. 12.5 Two c-graphs not admitting any $\langle 0, 0, \infty \rangle$-drawing. The underlying graph of **a** is a triconnected planar graph, while the underlying graph of **b** is a series-parallel graph

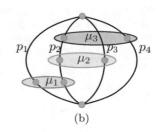

(a)   (b)

p_1 and one of p_2; cluster μ_2 contains one vertex of p_2 and one of p_3; cluster μ_3 contains one vertex of p_2 and one of p_4 (see Fig. 12.5b). In any $\langle 0, 0, \infty \rangle$-drawing of $\mathcal{C}(G, \mathcal{T})$, path p_2 should share a face with each of the other paths, which is not possible in a planar drawing of G.

In view of these negative examples, it is worth studying the complexity of testing whether a c-graph $\mathcal{C}(G, \mathcal{T})$ admits a $\langle 0, 0, \infty \rangle$-drawing. For that, we present a characterization of the planar embeddings of G that allow for the realization of a $\langle 0, 0, \infty \rangle$-drawing of $\mathcal{C}(G, \mathcal{T})$. Namely, let $\mathcal{E}$ be a planar embedding of G. For each cluster $\mu \in \mathcal{T}$, let $H_\mu(\mathcal{E})$ be an auxiliary graph, whose vertices are those of cluster μ, that contains an edge between two vertices if and only if these vertices are incident to the same face in $\mathcal{E}$.

Theorem 6 ([4], Lemma 1) *A c-graph $\mathcal{C}(G, \mathcal{T})$ admits a $\langle 0, 0, \infty \rangle$-drawing if and only if there exists a planar embedding $\mathcal{E}$ of G such that, for each cluster $\mu \in \mathcal{T}$, it holds that (i) graph $H_\mu(\mathcal{E})$ is connected and (ii) there exists in $\mathcal{E}$ no cycle composed of vertices of μ containing in its interior a vertex $v \notin \mu$.*

Proof We first prove the necessity. As for Condition (i), suppose that $H_\mu(\mathcal{E})$ is not connected. Then, for any two distinct connected components H_μ^1 and H_μ^2 of $H_\mu(\mathcal{E})$, there exists a cycle $\mathcal{C}$ in G separating H_μ^1 and H_μ^2, as otherwise H_μ^1 and H_μ^2 would be incident to a common face of $\mathcal{E}$, hence they would not be distinct connected components of $H_\mu(\mathcal{E})$. Therefore, the boundary of any region R_μ representing μ intersects (at least) one of the edges of $\mathcal{C}$. As for Condition (ii), suppose that a cycle $\mathcal{C}$ exists whose vertices belong to μ and whose interior contains in $\mathcal{E}$ a vertex not belonging to μ. Then, in any drawing of R_μ as a region homeomorphic to a closed disk containing all and only the vertices in μ, the boundary of R_μ intersects (at least) one edge of $\mathcal{C}$.

Next, we prove the sufficiency. Suppose that Conditions (i) and (ii) hold. Consider any subgraph H'_μ of $H_\mu(\mathcal{E})$ such that **(a)** $G_\mu \subseteq H'_\mu$, where G_μ is the subgraph of G induced by V_μ; **(b)** H'_μ is connected; and **(c)** for every cycle $\mathcal{C}$ in H'_μ, if any, all the edges of $\mathcal{C}$ belong to G. Observe that the fact that $H_\mu(\mathcal{E})$ satisfies Conditions (i) and (ii) implies the existence of graph H'_μ. Draw each edge of H'_μ not in G inside the corresponding face of $\mathcal{E}$. Represent μ as a region slightly surrounding the (possibly non-simple) cycle delimiting the outer face of H'_μ. Denote by $\Gamma'_\mathcal{C}$ the resulting drawing and denote by $\Gamma_\mathcal{C}$ the drawing of $\mathcal{C}(G, \mathcal{T})$ obtained from $\Gamma'_\mathcal{C}$ by removing the edges not in G. We have that $\Gamma_\mathcal{C}$ contains no ee-crossing, since $\mathcal{E}$ is

Table 12.1 Upper and lower bounds for the number of crossings in $\langle\infty, 0, 0\rangle$-, $\langle 0, \infty, 0\rangle$-, and $\langle 0, 0, \infty\rangle$-drawings of c-graphs. Flags *c-c* and *flat* mean that the c-graph is *c-connected* and *flat*, respectively. Results written in gray derive from those in black, while a "✠" means that there exist c-graphs not admitting the corresponding drawings. A "0" occurs if the c-graph is c-planar

c-c	Flat	$\langle\alpha, 0, 0\rangle$		$\langle 0, \beta, 0\rangle$		$\langle 0, 0, \gamma\rangle$	
		α UB	α LB	β UB	β LB	γ UB	γ LB
No	No	$O(n^2)$ Theorem 3	$\Omega(n^2)$	$O(n^3)$ Theorem 4	$\Omega(n^2)$	$O(n^3)$✠ Theorem 5	$\Omega(n^3)$ [4]
No	Yes	$O(n^2)$	$\Omega(n^2)$	$O(n^2)$ Theorem 4	$\Omega(n^2)$ [4]	$O(n^2)$✠ Theorem 5	$\Omega(n^2)$ [4]
Yes	No	$O(n^2)$	$\Omega(n^2)$	$O(n^2)$ Theorem 4	$\Omega(n^2)$ [4]	0✠ [31]	0✠ [31]
Yes	Yes	$O(n^2)$	$\Omega(n^2)$ [4]	$O(n)$ Theorem 4	$\Omega(n)$ [4]	0✠ [31]	0✠ [31]

a planar embedding. Also, it contains no *er*-crossing, since the only edges crossing clusters in Γ'_C are those belonging to H'_μ and not belonging to G_μ. □

A polynomial-time algorithm to test whether a c-graph $\mathcal{C}(G, \mathcal{T})$ whose underlying graph is biconnected admits a $\langle 0, 0, \infty\rangle$-drawing has been presented in [4], based on the characterization of Theorem 6. In particular, the authors exploit the SPQR-tree decomposition of G to test whether, for each node τ of the SPQR-tree of G, the pertinent graph of τ admits an embedding that can be extended to an embedding of G satisfying Conditions (i) and (ii) of Theorem 6 for each cluster $\mu \in \mathcal{T}$.

12.2.3 Lower Bounds for Crossings in $\langle\alpha, \beta, \gamma\rangle$-Drawings

The algorithms in the previous subsection provide upper bounds on the number of crossings for the three kinds of drawings. In this subsection, we show that the majority of these upper bounds are tight by providing matching lower bounds. These results are summarized in Table 12.1.

We present a lower bound on the total number of crossings in an $\langle\alpha, \beta, \gamma\rangle$-drawing of a c-graph when all the three types of crossings are admitted.

Theorem 7 ([4], Theorem 6) *There exists an infinite family $\mathcal{F}$ of n-vertex non-c-connected flat c-graphs that admit $\langle\infty, 0, 0\rangle$-, $\langle 0, \infty, 0\rangle$-, and $\langle 0, 0, \infty\rangle$-drawings, and such that $\alpha + \beta + \gamma \in \Omega(n^2)$ in every $\langle\alpha, \beta, \gamma\rangle$-drawing.*

Proof Each c-graph $\mathcal{C}(G, \mathcal{T})$ in $\mathcal{F}$ is defined as follows. Initialize G with five vertices a, b, c, d, e. For each two vertices $u, v \in \{a, b, c, d, e\}$, with $u \neq v$, and for $i = 1, \dots, m = \frac{n-5}{20}$, add to G vertices $[uv]_i$, $[vu]_i$, and edges $(u, [uv]_i)$ and $(v, [vu]_i)$, and add to $\mathcal{T}$ a cluster $\mu(u, v)_i = \{[uv]_i, [vu]_i\}$. Vertices a, b, c, d, e belong to clusters $\mu_a, \mu_b, \mu_c, \mu_d, \mu_e$, respectively. See Fig. 12.6. We denote by $M(u, v) = \{(u, [uv]_i), (v, [vu]_i), \mu(u, v)_i | i = 1, \dots, m\}$.

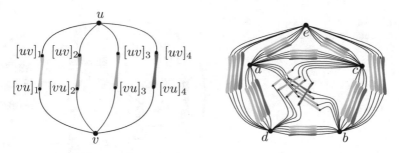

Fig. 12.6 Illustrations for the proof of Theorem 7: (left) Edges and clusters in $M(u, v)$; (right) clustered graph $C(G, T)$

Observe that $C(G, T)$ admits a $\langle \infty, 0, 0 \rangle$-drawing, by Theorem 3, and it admits a $\langle 0, \infty, 0 \rangle$-drawing, by Theorem 4 and since G is planar. To see that $C(G, T)$ also admits a $\langle 0, 0, \infty \rangle$-drawing note that, since G is a forest of stars, it satisfies the characterization of Theorem 6.

We show that $\alpha + \beta + \gamma \in \Omega(n^2)$ in every $\langle \alpha, \beta, \gamma \rangle$-drawing of $C(G, T)$. Consider any such a drawing Γ. Starting from Γ, we obtain a drawing Γ' of a subdivision of a graph obtained by replacing each edge of a K_5 with a set of m parallel edges. For each $u, v \in \{a, b, c, d, e\}$, with $u \neq v$, and for each $i = 1, \ldots, m$, insert a drawing of edge $([uv]_i, [vu]_i)$ inside $R_{\mu(u,v)_i}$ and remove region $R_{\mu(u,v)_i}$. Further, remove regions $R_{\mu_a}, R_{\mu_b}, R_{\mu_c}, R_{\mu_d}$, and R_{μ_e} to obtain Γ'. By [4, Lemma 6], Γ' has $\Omega(m^2) = \Omega(n^2)$ crossings. Moreover, each crossing in Γ' corresponds either to an ee-crossing, to an er-crossing, or to an rr-crossing in Γ, thus proving the theorem. □

As a corollary of Theorem 7, there exists a family of n-vertex c-graphs such that $\alpha \in \Omega(n^2)$ in every $\langle \alpha, 0, 0 \rangle$-drawing, such that $\beta \in \Omega(n^2)$ in every $\langle 0, \beta, 0 \rangle$-drawing, and such that $\gamma \in \Omega(n^2)$ in every $\langle 0, 0, \gamma \rangle$-drawing. As shown in Table 12.1, further lower bounds can be achieved when the instances are c-connected and/or flat.

12.3   NodeTrix Representations

NodeTrix representations have been introduced by Henry, Fekete, and McGuffin [46] as a hybrid model for the visualization of social networks, where the node-link paradigm is used to visualize the overall structure of the network, within which adjacency matrices show communities.

A *NodeTrix representation (NT-representation)* of a flat c-graph $C(G, T)$ is defined as follows; refer to Fig. 12.7. **(i)** For each cluster $\mu \in T$, the subgraph of G induced by V_μ is represented as a symmetric $|V_\mu| \times |V_\mu|$ adjacency matrix M_μ drawn in the plane so that its boundary is a square Q_μ with sides parallel to the coordinate axes. Thus, matrix M_μ conveys the information about the intra-cluster edges of μ. **(ii)** There is no overlap between a matrix and an inter-cluster edge or another

Fig. 12.7 An
NT-representation of a flat
c-graph

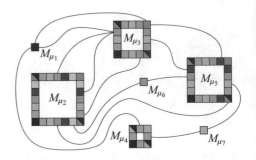

matrix. **(iii)** Each inter-cluster edge (u, v), with $u \in V_\mu$ and $v \in V_\nu$, is represented as
a Jordan arc connecting a point on Q_μ with a point on Q_ν, where the point on Q_μ
(on Q_ν) belongs to the column or to the row of M_μ (resp. of M_ν) associated with u
(resp. with v).

Several papers aimed at improving the readability of NT-representations by reducing the number of crossings between inter-cluster edges [10, 47]. Da Lozzo et
al. [24] introduced a notion of planarity for NT-representations. In a *planar NT-
representation* no two inter-cluster edges cross each other, except possibly at a common endpoint. The NODETRIX PLANARITY problem of testing whether a flat c-graph
admits planar NT-representation combines traditional graph drawing issues, like the
placement of a set of geometric objects in the plane (here the squares $Q_1, \ldots, Q_\ell$)
and the routing of Jordan arcs (here the inter-cluster edges), with a novel algorithmic challenge: To handle the degrees of freedom given by the choice of the
row-column order of each matrix and by the choice of the *sides assignment* for the
inter-cluster edges attached to it. More formally, a *row-column order* σ_μ for a clus-
ter μ is a bijection $\sigma_\mu : V_\mu \leftrightarrow \{1, \ldots, |V_\mu|\}$, while a *side assignment* s_μ for μ is a
mapping $s_\mu : \bigcup_{\nu \neq \mu} E_{\mu,\nu} \to \{\text{TOP, BOTTOM, LEFT, RIGHT}\}$, where $E_{\nu,\mu}$ is the set of
inter-cluster edges between clusters ν and μ. If a side assignment s_μ, for each cluster
$\mu \in \mathcal{T}$, is provided as part of the input, and we seek an NT-representation consistent
with such a side assignment, the variant of the problem is known as NodeTrix Pla-
narity with Fixed Sides. Also, if a row-column order σ_μ, for each cluster $\mu \in \mathcal{T}$, is
provided as part of the input, and we seek an NT-representation in which each matrix
M_μ is ordered according to σ_μ, then we have the NodeTrix Planarity with Fixed
Order problem. Finally, by fixing both a side assignment and a row-column order
for each matrix, we have the NodeTrix Planarity with Fixed Order and Fixed
Sides problem. Table 12.2 summarizes the known results about the complexity of
the NODETRIX PLANARITY problem and its variants. We overview some of them in
the following.

We start by presenting the main polynomial-time results concerning NODETRIX
PLANARITY. The first positive result derives from constraining both the row-column
orders and the side assignments for all clusters.

Theorem 8 ([24], Theorem 4) *The* NODETRIX PLANARITY WITH FIXED ORDER
AND FIXED SIDES *problem can be solved in linear time.*

Table 12.2 Complexity results for NODETRIX PLANARITY. The results marked † assume that the number of non-trivial clusters be constant. We denote by k the size of the largest cluster of the considered instances

		Free sides		Fixed sides	
		O(1) Size	O(1) clusters	O(1) Size	O(1) clusters
R/C Order	Free	NPC [11] $k \geq 2$	NPC [Theorem 11] †	P [Theorem 9] $k \leq 2$ NPC [41] $k \geq 3$	NPC [24]
	Fixed		NPC [Theorem 10] †	P [Theorem 8]	

Consider a flat c-graph $\mathcal{C}(G, \mathcal{T})$. Let H be the auxiliary graph obtained from G by collapsing each cluster $\mu \in \mathcal{T}$ into a vertex v_μ. Observe that H coincides with the clusters-adjacency graph of $\mathcal{C}(G, \mathcal{T})$ when suppressing multiple edges. Let σ_μ and s_μ be the row-column order and the side assignment, for each cluster $\mu \in \mathcal{T}$, specified in the input. Intuitively, c-graph $\mathcal{C}(G, \mathcal{T})$ admits a planar NodeTrix representation consistent with the given row-column orders and side assignments if and only if H is planar with the additional constraint that the clockwise order of the edges incident to each vertex v_μ is "compatible" with σ_μ and s_μ.

More formally, denote by E_μ the set of the inter-cluster edges incident to μ and denote by $v_\mu(e)$ the vertex of μ incident to an edge $e \in E_\mu$. The edges in E_μ can be decomposed into a circular sequence of sets $\mathcal{S} = E_{T,1}, E_{T,2}, \ldots, E_{T,|V_\mu|}, E_{R,1}, E_{R,2}, \ldots, E_{R,|V_\mu|}, E_{B,|V_\mu|}, E_{B,|V_\mu|-1|}, \ldots, E_{B,1}, E_{L,|V_\mu|}, E_{L,|V_\mu|-1|}, \ldots, E_{L,1}$, where each $E_{X,j}$, with $X \in \{T, B, L, R\}$ and $j \in \{1, \ldots, |V_\mu|\}$, contains the edges $e \in E_\mu$ such that $s_\mu(e) = X$ and $\sigma_\mu(v_\mu(e)) = j$. Let $\mathcal{E}$ be a planar embedding of H and let λ_μ denote the clockwise order of the edges incident to vertex v_μ of H in $\mathcal{E}$. The embedding $\mathcal{E}$ is *compatible* with functions σ_μ and s_μ if: (i) all the edges belonging to the same set $E_{X,j}$ appear consecutively in λ_μ and (ii) for any three edges $e' \in E_{X',j'}$, $e'' \in E_{X'',j''}$, and $e''' \in E_{X''',j'''}$, where $E_{X',j'}$, $E_{X'',j''}$, and $E_{X''',j'''}$ are all distinct, appear in this clockwise order in λ_μ if and only if $E_{X',j'}$, $E_{X'',j''}$, and $E_{X''',j'''}$ appear in this circular order in $\mathcal{S}$. By construction, the input instance of NODETRIX PLANARITY WITH FIXED ORDER AND FIXED SIDES admits a solution if and only if the corresponding auxiliary graph H admits an embedding $\mathcal{E}$ that is compatible with σ_μ and s_μ, for each cluster $\mu \in \mathcal{T}$. Since the compatibility defined above can be expressed by imposing on H a set of *hierarchical embedding constraints* [44], which define an instance of EMBEDDING CONSTRAINED PLANARITY, and since the latter problem can be solved in linear time [44], the proof of Theorem 8 immediately follows.

The second polynomial-time result concerns the case in which the side assignment of each matrix is fixed and the maximum size k of the clusters is 2.

Theorem 9 ([41], Theorem 3) *The* NODETRIX PLANARITY WITH FIXED SIDES *problem can be solved in* $O(n^3)$ *time for instances such that the maximum size of any cluster is 2.*

Di Giacomo et al. [41] have proved that the NODETRIX PLANARITY WITH FIXED SIDES problem can be solved in $O(k^{3k+\frac{3}{2}}n^3)$ time for flat c-graphs whose clusters-adjacency graph is a partial 2-tree. Further, they prove Theorem 9 by extending their polynomial-time solution to flat c-graphs with an arbitrary clusters-adjacency graph when $k = 2$. Their algorithm exploits SPQR-trees to solve a suitably defined constrained embedding problem on the graph H obtained by contracting each cluster to a single vertex, which they prove equivalent to NODETRIX PLANARITY WITH FIXED SIDES in their setting. In particular, in order to handle the rigid components of the clusters-adjacency graph, they encode the embedding problem for the pertinent graph of the corresponding R-node as an instance of the COHERENT-LABELING problem defined as follows: Given a plane graph H where each vertex w is associated with a set X_w, containing at least a label in $\{\pi_w^+, \pi_w^-\}$, and each edge (u, v) is associated with a set $Y_{(u,v)}$, containing at least one label in $\{(\pi_u^+, \pi_v^+), (\pi_u^+, \pi_v^-), (\pi_u^-, \pi_v^+), (\pi_u^-, \pi_v^-)\}$, is it possible to select a label $\pi_w^* \in X_w$, for each vertex w, in such a way that $(\pi_u^*, \pi_v^*) \in Y_{(u,v)}$, for each edge (u, v)? They provide a linear-time algorithm for the COHERENT-LABELING problem. The quadratic blow up in the running time then stems from rooting the SQPR-tree and the BC-tree of the instance in all possible ways.

Observe that the result of Theorem 9 is tight, as for any $k \geq 3$ the NODETRIX PLANARITY WITH FIXED SIDES problem is NP-complete [41]. On the other hand, this problem is also NP-complete if, instead of bounding the size of clusters, we bound their number. In fact, even instances with only two non-trivial clusters and no trivial clusters are difficult to solve [24].

It is easy to see that the NODETRIX PLANARITY (WITH FIXED ORDER) problem can be solved in linear time when each cluster is trivial, by a reduction to planarity testing. However, we will see that admitting even only one non-trivial cluster makes the problem computationally difficult.

Theorem 10 ([24], Theorem 3) NODETRIX PLANARITY WITH FIXED ORDER *is NP-complete even for instances containing only one non-trivial cluster.*

We give an outline of the proof of this result. The NP-hardness is proved via a reduction from the 4-coloring problem for a *circle graph* [61], that is, a graph $H = (N, A)$ that admits a representation $\langle \mathcal{P}, \mathcal{O} \rangle$ of H, where $\mathcal{P}$ is a linear sequence of distinct points on a circle and $\mathcal{O}$ is a set of chords between points in $\mathcal{P}$ such that (i) each chord $c \in \mathcal{O}$ corresponds to a vertex $n \in N$ and (ii) two chords $c', c'' \in \mathcal{O}$ intersect if and only if $(n', n'') \in A$, where n' and n'' are the vertices in N corresponding to c' and c'', respectively. Starting from $\langle \mathcal{P}, \mathcal{O} \rangle$, we construct an instance of NODETRIX PLANARITY WITH FIXED ORDER by defining a c-graph $\mathcal{C}(G, \mathcal{T})$ and a row-column order for the unique non-trivial cluster in $\mathcal{T}$ as follows (refer to Fig. 12.8). C-graph $\mathcal{C}(G, \mathcal{T})$ contains **(i)** a cycle D composed of vertices $v_{tl}, v'_{tr}, v''_{tr}, v_{br}, v'_{bl}$, and v''_{bl} (each

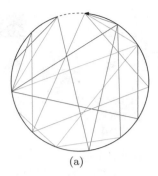 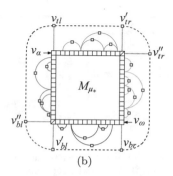

(a)   (b)

Fig. 12.8   a An intersection representation $\langle \mathcal{P}, \mathcal{O} \rangle$ of a circle graph. **b** The corresponding instance $\langle \mathcal{C}(GT), \sigma_{\mu_*} \rangle$ of NODETRIX PLANARITY WITH FIXED ORDER; bounding edges are dashed, while corner edges are dotted

in a distinct cluster of $\mathcal{T}$ containing that vertex only) connected by *bounding edges*; **(ii)** a cluster $\mu_* \in \mathcal{T}$ containing a vertex v_i for each point $p_i \in \mathcal{P}$, plus vertices v_α and v_ω; **(iii)** *corner edges* connecting $v_{tl}, v'_{tr}, v''_{tr}, v_{br}, v'_{bl}$, and v''_{bl} with either v_α or v_ω; and **(iv)** for every chord $c = (p_i, p_j) \in \mathcal{O}$, a path corresponding to c composed of a vertex v_c (cluster of $\mathcal{T}$ containing that vertex only) and of two *chord edges* (v_i, v_c) and (v_c, v_j). Finally, let the row-column order σ_{μ_*} of V_{μ_*} be $v_\alpha, \mathcal{P}, v_\omega$ where, with a slight abuse of notation, we denote by $\mathcal{P}$ not only the order of the points on the circle, but also the corresponding order of the vertices in $V_* - \{v_\alpha, v_\omega\}$. The proof of equivalence between the instances is based on the following property of any planar NodeTrix representation Γ of instance $\langle \mathcal{C}(GT), \sigma_{\mu_*} \rangle$. Namely, the bounding and corner edge together with the drawing Q_{μ_*} of M_{μ_*} define in Γ four regions, each incident to an entire side of Q_{μ_*}. Thus, for each chord $c = (p_i, p_j) \in \mathcal{O}$, if vertex v_c lies in one of these regions, then also the two chord edges (v_i, v_c) and (v_c, v_j) incident to v_c lie in the same region. Since Γ is planar, no two paths composed of chord edges and corresponding to different chords of H cross; that is, the endpoints of such paths do not alternate along the side of M_{μ_*} they are incident to. Hence, the chords corresponding to such paths do not alternate in $\mathcal{P}$, and thus can be assigned the same color.

Concerning the case in which the clusters have maximum size k, we observe that for instances with $k = 2$ the Fixed Order setting is equivalent to the general NODETRIX PLANARITY problem. In a recent work, Besa et al. [11] showed that NODETRIX PLANARITY is NP-complete even for instances in which $k \geq 2$. This strengthens a previous NP-completeness of the NODETRIX PLANARITY problem when $k \geq 5$ [41] and implies the NP-completeness of the NODETRIX PLANARITY WITH FIXED ORDER problem when $k \geq 2$.

Finally, we present a proof that the general version of the problem is NP-complete even when the number of non-trivial clusters is bounded.

Theorem 11 NODETRIX PLANARITY *is NP-complete even if at most three clusters contain more than one vertex.*

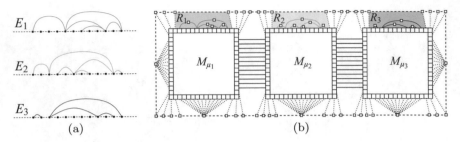

Fig. 12.9 **a** An instance of PARTITIONED 3- PAGE BOOK EMBEDDING and **b** the corresponding instance of NODETRIX PLANARITY; bounding edges are dashed, corner edges are dotted, and order transmitting edges are thick solid

To establish the NP-hardness, we show a reduction from the PARTITIONED 3-PAGE BOOK EMBEDDING problem [5], which takes as input a graph (V, E), together with a partition of E into three sets E_1, E_2, and E_3, and asks whether there exists a total ordering $\mathcal{O}$ of V such that the end-vertices of any two edges e and e' in the same set E_i do not alternate in $\mathcal{O}$; refer to Fig. 12.9. The reduction exploits three non-trivial clusters μ_1, μ_2, and μ_3, all containing a copy of vertex set V and some additional corner vertices, plus some trivial clusters which are adjacent to μ_1, μ_2, and μ_3 by means of *boundary* and *corner edges* as in the figure. For each vertex $v \in V$ the constructed c-graph contains inter-cluster *order transmitting edges* (v_1, v_2) and (v_2, v_3), where v_i is the copy of v in μ_i. Further, for $i \in \{1, 2, 3\}$ and for each edge $e = (u, v)$ in E_i, the c-graph contains a trivial cluster $\{e_i\}$ and inter-cluster *page edges* (u_i, e_i) and (e_i, v_i), where u_i and v_i are the copies of u and v in μ_i, respectively. The proof of equivalence is based on the following property of any planar NodeTrix representation Γ of the constructed c-graph. Namely, the bounding, corner, and order transmitting edges together with the drawing Q_{μ_i} of M_{μ_i}, with $i \in \{1, 2, 3\}$, defined in Γ a unique region R_i for each matrix M_{μ_i} incident to an entire side of Q_{μ_i}. Thus, each vertex v_e corresponding to an edge $e = (u, v) \in E_i$ must lie in R_i together with the two page edges incident to it. Since Γ is planar, no two paths composed of page edges cross in R_i; hence, the two corresponding edges in E_i do not alternate in $\mathcal{O}$.

We conclude the section by observing that the version of the NODETRIX PLA-NARITY problem in which the adjacency matrices representing the clusters need not be symmetric, called ROW- COLUMN INDEPENDENT NODETRIX PLANARITY problem, has been recently studied in the Fixed Sides setting [57].

12.4   Intersection-Link Representations

In this section, we consider *intersection-link representations*, which are hybrid representations introduced for the visualization of networks that are, as many real-world networks, locally dense and globally sparse. This drawing paradigm combines two

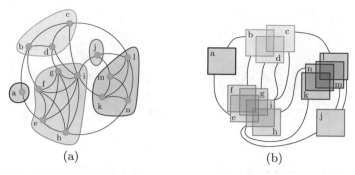

Fig. 12.10   a A flat c-graph and **b** one of its clique-planar representations

classical drawing styles as *intersection representations* and *node-link diagrams*, used to effectively convey the relational information provided by parts of the network having different density.

More formally, in an *intersection-link representation* [7] of a flat c-graph $\mathcal{C}(G, \mathcal{T})$, each vertex $v \in V$ is represented by a geometric object $R(v)$, each intra-cluster edge (u, v) is represented by an intersection between $R(u)$ and $R(v)$, and each inter-cluster edge (u, v) is represented by a Jordan arc connecting the boundaries of $R(u)$ and $R(v)$ without intersecting the interior of any object. A specific type of intersection-link representation is then defined by selecting a specific family of geometric objects for the vertices. Finally, the quality of an intersection-link representation can be measured in terms of the number of crossings between the inter-cluster edges, and its planarity can be expressed in terms of the absence of such crossings.

In the original paper introducing intersection-link representations [7], Angelini et al. consider flat c-graphs whose clusters are maximally dense, that is, each cluster defines a clique, and focus on intersection-link representations of such c-graphs in which the geometric objects are translates of the same rectangle. They study the planarity of this model, under the name of CLIQUE PLANARITY ; refer to Fig. 12.10. The main motivation behind focusing on cliques is that recognizing intersection graphs of translates of the same rectangle is NP-complete [14], while every clique trivially admits such a representation.

Similarly to the NODETRIX PLANARITY considered in Sect. 12.3, the CLIQUE PLANARITY problem requires to compute the placement of a set of geometric objects in the plane (here the rectangles representing the vertices) and the routing of Jordan arcs (here the inter-cluster edges) under the constraints imposed by the possible orders in which rectangles appear along the boundary of the arrangement of rectangles representing each cluster of the c-graph. We now show that such possible orders obey a simple pattern. Let B be the outer boundary of an arrangement of rectangles representing a clique K_n.

Lemma 1 *Traversing B clockwise, the sequence of encountered rectangles is a subsequence of $R(u_1), R(u_2), \ldots, R(u_n), R(u_{n-1}), \ldots, R(u_2)$, for some permutation $u_1, \ldots, u_n$ of the vertices of K_n.*

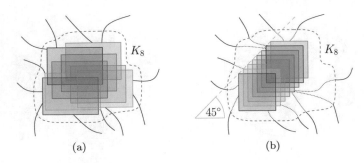

Fig. 12.11 **a** An arrangement representing a K_8 and **b** the corresponding canonical representation

The proof is based on the following claims. **(Claim A)**: Every maximal portion of B belonging to a single rectangle $R(u)$ contains (at least) one corner of $R(u)$. **(Claim B)**: If two adjacent corners of the same rectangle $R(u)$ both belong to B, then the entire side of $R(u)$ between them belongs to B. **(Claim C)**: Traversing B clockwise, the sequence of encountered rectangles is not of the form $\ldots, R(u), \ldots, R(v), \ldots, R(u), \ldots, R(v)$, for any $u, v \in K_n$. By Claims A and B, we conclude that no rectangle $R(u)$ appears three times along B. In fact, by Claim A, each maximal portion of B belonging to $R(u)$ contains a corner of $R(u)$. Thus, if $R(u)$ appears more than twice along B, there exist two adjacent corners of $R(u)$ belonging to two distinct maximal portions of B. However, by Claim B the side of $R(u)$ between those corners belongs to B, hence those corners belong to the same maximal portion of B, a contradiction. This and Claim C imply Lemma 1.

In view of Lemma 1, as shown in [7], we can focus on special clique-planar representations, called *canonical*; refer to Fig. 12.11

Definition 2 In a *canonical clique-planar representation* of a flat c-graph $\mathcal{C}(G, \mathcal{T})$ whose clusters induce cliques: 1. each vertex is represented as an axis-aligned unit square and 2. for each cluster $\mu \in \mathcal{T}$, all the squares representing vertices in V_μ have their upper-left corner along a common line with slope $45°$.

Next, we are going to show that the CLIQUE PLANARITY problem is not solvable in polynomial time, unless $P = NP$.

Theorem 12 ([7], Theorem 1) *The* CLIQUE PLANARITY *problem is NP-complete even for flat c-graphs containing exactly one non-trivial cluster.*

To prove Theorem 12 we exploit a polynomial-time reduction from a constrained clustered planarity problem, proved NP-complete even for c-graphs with a single non-trivial cluster [7].

It has long been known that a clustered graph $\mathcal{C}(G, \mathcal{T})$ is c-planar if and only if a set of edges can be added to G so that the resulting graph is c-planar and c-connected [32]. Any such a set of edges is called a *saturator*. In a *linear saturator* the edges between vertices of the same cluster form a path. The Clustered Planarity with Linear Saturators (CPLS) problem takes as input a flat c-graph $C(G, T)$ such that each cluster in T induces an independent set, and asks whether $C(G, T)$ admits a linear saturator.

The following lemma relates the CLIQUE PLANARITY and the CPLS problems, and together with the NP-hardness of CPLS [7] implies Theorem 12.

Lemma 2 ([7], Lemma 5) *Given an instance $C(G, T)$ of* CLIQUE PLANARITY, *it is possible to construct in linear time an instance $C^*(G^*, T^*)$ of* CPLS *equivalent to $C(G, T)$.*

The c-graph $C^*(G^*, T^*)$ is obtained from $C(G, T)$ by setting $T^* = T$ and by removing all intra-cluster edges. We show that $C^*(G^*, T^*)$ admits a linear saturator if and only if $C(G, T)$ is clique-planar; see Fig. 12.12a, b.

Suppose that $C^*(G^*, T^*)$ admits a linear saturator. This implies that there exists a c-planar drawing $\Gamma^\diamond$ of the c-graph $C(G^\diamond, T^\diamond)$, where $G^\diamond$ is obtained from G^* by adding the linear saturator. Consider a cluster $\mu \in T^\diamond$ represented by region R_μ, and let $u_1, \ldots, u_k$ be the vertices of μ ordered as they appear along the saturator path. Note that the order in which the inter-cluster edges cross the boundary of R_μ is a subsequence of $u_1, \ldots, u_{k-1}, u_k, u_{k-1}, \ldots, u_2$, when each edge is identified with its endpoint in μ. Thus, we can construct a clique-planar representation Γ of $C(G, T)$ starting from $\Gamma^\diamond$ as follows.

For each cluster μ, remove from $\Gamma^\diamond$ all the vertices and (part of the) edges contained in the interior of R_μ. Represent $u_1, \ldots, u_k$ by pairwise intersecting squares $S(u_1), \ldots, S(u_k)$ that are translates of each other and whose upper-left corners touch a common line in this order. Scale $\Gamma^\diamond$ such that the arrangement can be placed in the interior of R_μ. Then, complete the drawing of the inter-cluster edges from their intersection point with the boundary of R_μ with the square representing their endpoints in μ. This can be done without crossings since the order of such intersections along R_μ defines an order of their end-vertices in μ which is a subsequence of $u_1, \ldots, u_{k-1}, u_k, u_{k-1}, \ldots, u_2$, while the circular order in which the squares occur along the boundary of their arrangement is $S(u_1), \ldots, S(u_k), S(u_{k-1}), \ldots, S(u_2)$. This results in a clique-planar representation of $C(G, T)$.

Conversely, suppose that $C(G, T)$ has a clique-planar representation Γ, which we assume to be canonical by Lemma 1. We define a set $E^\diamond$ as follows. For each cluster $\mu \in T$, let $S(u_1), \ldots, S(u_k)$ be the order in which the squares representing cluster μ touch the line with slope 1 through their upper-left corners in Γ; add to $E^\diamond$ all the edges (u_i, u_{i+1}), for $i = 1, \ldots, k - 1$. We claim that $E^\diamond$ is a linear saturator for $C^*(G^*, T^*)$. Indeed, by the definition of (linear) saturator, it suffices to show that $G + E^\star$ admits a planar drawing.

Initialize $\Gamma^\diamond = \Gamma$. Place each vertex v at the center of square $S(v)$ and remove $S(v)$ from $\Gamma^\diamond$. Extend each edge (u, v) with two straight-line segments from the boundaries of $R(u)$ and $R(v)$ to u and v, respectively. This does not produce crossings, as only the segments of two vertices u and v such that $R(u)$ and $R(v)$ intersection might cross; however, these segments are separated by the line through the intersection points of the boundaries of $R(u)$ and $R(v)$ (Fig. 12.12c(top)). Draw the edges in

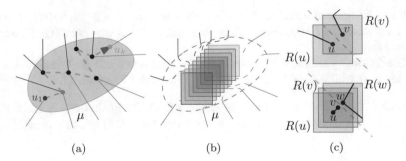

Fig. 12.12 **a** A cluster with a linear saturator and **b** the corresponding clique-planar representation. **c** Illustrations for the construction of $\Gamma^\diamond$

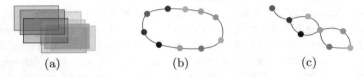

Fig. 12.13 **a** An arrangement Γ of rectangles representing a cluster μ. **b** A simple cycle with a vertex for each maximal portion of the boundary of Γ. **c** Planar drawing $\mathcal{H}'_\mu$ of graph H'_μ corresponding to Γ

$E^\diamond$ as straight-line segments without introducing crossings. In fact consider an edge (u, v) in $E^\diamond$ and any segment e_w incident to a vertex $w \neq u, v$ in the same cluster. Assume u, v, w are in this order along the line with slope 1 through them. Then (u, v) is separated from e_w by the line through the two intersection points of the boundaries of $R(v)$ and $R(w)$; see Fig. 12.12c(bottom). This concludes the proof of Lemma 2.

CLIQUE PLANARITY has also been considered in the well-studied framework of extending a partial solution to a full one [6, 15, 16, 51, 58]. Namely, given a c-graph $\mathcal{C}(G, \mathcal{T})$ and arrangements of rectangles representing the clusters in $\mathcal{T}$, the goal is to test whether the inter-cluster edges can be drawn in Γ^* to obtain a clique-planar representation Γ of $\mathcal{C}(G, \mathcal{T})$. In contrast with result of Theorem 12, the extension problem for clique-planar representations turns out to be solvable in linear time. This algorithmic result exploits a reduction to the PARTIAL EMBEDDING PLANARITY problem [6], which asks whether a planar drawing of a graph H exists extending a given drawing $\mathcal{H}'$ of a subgraph H' of H. Next, we give an overview of the reduction; refer to Fig. 12.13.

For each cluster $\mu \in \mathcal{T}$, we add to H' a connected component H'_μ corresponding to a cluster μ and we define a drawing $\mathcal{H}'_\mu$ of H'_μ. Denote by B^*_μ the boundary of the representation of μ in Γ^* (see Fig. 12.13a). If μ has one or two vertices, then H'_μ is a vertex or an edge, respectively (and $\mathcal{H}'_\mu$ is any drawing of H'_μ). Otherwise, initialize H'_μ to a simple cycle containing a vertex for each maximal portion of B^*_μ belonging to a single rectangle (see Fig. 12.13b). Let $\mathcal{H}'_\mu$ be any planar drawing of H'_μ with the same orientation as B^*_μ. Each rectangle in Γ^* may correspond to two vertices of H'_μ, but no more than two by Lemma 1. Insert an edge in H'_μ between every two vertices representing the same rectangle and draw it in the interior of $\mathcal{H}'_\mu$. Contract the inserted edges in H'_μ and $\mathcal{H}'_\mu$ (see Fig. 12.13c). This completes the construction of H'_μ, together with its planar drawing $\mathcal{H}'_\mu$. We remark that H'_μ is a *cactus graph*, that is a connected graph that admits a planar embedding in which all the edges are incident to the outer face. Graph H' is the union of graphs H'_μ, over all the clusters $\mu \in \mathcal{T}$; the drawings $\mathcal{H}'_\mu$ of H'_μ are in the outer face of each other in $\mathcal{H}'$. Finally, we define H as the graph obtained from H' by adding, for each inter-cluster edge (u, v) of G, an edge between the vertices of H' corresponding to u and v. We have the following.

Lemma 3 ([7], Lemma 6) *There exists a planar drawing of H extending $\mathcal{H}'$ if and only if there exists a clique-planar representation of $\mathcal{C}(G, \mathcal{T})$ extending Γ^*.*

Lemma 3 and the linear-time algorithm for the PARTIAL EMBEDDING PLANARITY problem [6] imply the following.

Theorem 13 CLIQUE PLANARITY *can be decided in linear time if the representation of each cluster is given as part of the input.*

In view of Lemma 2, the CLIQUE PLANARITY problem can be reformulated as the following beyond-planarity problem, called h- CLIQUE2PATH PLANARITY: Given a graph G, whose vertices are partitioned into subsets of size at most h, each inducing a clique, remove edges from each clique so that the subgraph induced by each subset is a path, in such a way that the resulting subgraph of G is planar. This problem has been studied for simple topological graphs and geometric graphs in relation with the class of k-planar graphs. In particular, the h- CLIQUE2PATH PLANARITY problem has been shown NP-complete even when $h = 4$ and G is a 3-plane simple topological graph [9] or when $h = 4$ and G is a 4-plane simple geometric graph [54], while it can be solved in linear time when $h = 3$ [54] or, for any h, when G is a 1-plane topological or geometric graph [9].

12.5 Open Problems

We conclude the chapter with a list of interesting open problems concerning visualization of clustered networks arising from the topics covered in the previous sections.

Open Problems of Sect. 12.2.

OP 2.1 Close the non-tight bounds in Table 12.1.

OP 2.2 Study classes of c-graphs that have drawings where the values of α, β, and γ are balanced in some way.

OP 2.3 Extend the testing algorithm for the existence of $\langle 0, 0, \infty \rangle$-drawings to simply-connected c-graphs.

Open Problems of Sect. 12.3.

OP 3.1 Study families of flat c-graphs for which the NodeTrix Planarity problem and its variants are polynomial-time solvable in the free sides or free-order scenario.

OP 3.2 Are the NodeTrix Planarity problem and its variants fixed-parameter tractable with respect to relevant parameters of the clusters-adjacency graph (e.g., treewidth) or of the c-graphs (e.g., number of clusters)? In this direction, we note that Liotta et al. [56] have studied FPT algorithms for the Fixed Sides setting with respect to a combination of these two types of parameters.

Open Problems of Sect. 12.4.

OP 4.1 What is the complexity of computing planar intersection-link representations when clusters induce graphs other than cliques, still belonging to families of intersection graphs that can be recognized in polynomial time?

OP 4.2 How about using geometric objects different from translates of the same rectangle for representing vertices? For instance, Besa et al. [11] have studied the case of non-convex shapes.

OP 4.3 What is the complexity of CLIQUE PLANARITY problem for flat c-graphs with a bounded number of clusters? Interestingly, when there are exactly two clusters, the problem is equivalent to a special book embedding problem, called BIPARTITE 2- PAGE BOOK EMBEDDING WITH SPINE CROSSINGS [7].

OP 4.4 What is the complexity of the CLIQUE PLANARITY problem when clusters have constant size?

OP 4.5 What is the complexity of the h- CLIQUE2PATH PLANARITY problem for simple topological or geometric 2-plane graphs?

References

1. Akitaya, H.A., Fulek, R., Tóth, C.D.: Recognizing weak embeddings of graphs. ACM Trans. Algorithms, **15**(4):50:1–50:27 (2019)
2. Angelini, P., Da Lozzo, G.: SEFE = C-Planarity? Comput. J. **59**(12), 1831–1838 (2016)
3. Angelini, P., Da Lozzo, G.: Clustered planarity with pipes. Algorithmica **81**(6), 2484–2526 (2019)
4. Angelini, P., Da Lozzo, G., Di Battista, G., Frati, F., Patrignani, M., Roselli, V.: Relaxing the constraints of clustered planarity. Comput. Geom. **48**(2), 42–75 (2015)
5. Angelini, P., Da Lozzo, G., Neuwirth, D.: Advancements on SEFE and partitioned book embedding problems. Theor. Comput. Sci. **575**, 71–89 (2015)
6. Angelini, P., Di Battista, G., Frati, F., Jelínek, V., Kratochvíl, J., Patrignani, M., Rutter, I.: Testing planarity of partially embedded graphs. ACM Trans. Algorithms **11**(4):32:1–32:42 (2015)
7. Angelini, P., Da Lozzo, G., Di Battista, G., Frati, F., Patrignani, M., Rutter, I.: Intersection-link representations of graphs. J. Graph Algorithms Appl. **21**(4), 731–755 (2017)
8. Angelini, P., Da Lozzo, G., Di Battista, G., Frati, F.: Strip planarity testing for embedded planar graphs. Algorithmica **77**(4), 1022–1059 (2017)
9. Angelini, P., Eades, P., Hong, S.-H., Klein, K., Kobourov, S.G., Liotta, G., Navarra, A., Tappini, A.: Turning cliques into paths to achieve planarity. In: Biedl, T.C., Kerren, A. (eds.) 26th

International Symposium on Graph Drawing and Network Visualization, vol. 11282 of LNCS, pp. 67–74. Springer (2018)

10. Batagelj, V., Brandenburg, F.-J., Didimo, W., Liotta, G., Palladino, P., Patrignani, M.: Visual analysis of large graphs using (x, y)-clustering and hybrid visualizations. IEEE Trans. Vis. Comput. Graph. **17**(11), 1587–1598 (2011)
11. Besa, J.J., Da Lozzo, G., Goodrich, M.T.: Computing k-modal embeddings of planar digraphs. In: Bender, M.A., Svensson, O., Herman, G. (eds.) 27th Annual European Symposium on Algorithms, vol. 144 of LIPIcs, pp. 19:1–19:16. Schloss Dagstuhl - Leibniz-Zentrum für Informatik (2019)
12. Bläsius, T., Rutter, I.: A new perspective on clustered planarity as a combinatorial embedding problem. Theor. Comput. Sci. **609**, 306–315 (2016)
13. Braß, P., Cenek, E., Duncan, C.A., Efrat, A., Erten, C., Ismailescu, D., Kobourov, S.G., Lubiw, A., Mitchell, J.S.B.: On simultaneous planar graph embeddings. Comput. Geom. **36**(2), 117–130 (2007)
14. Breu, H.: Algorithmic Aspects of Constrained Unit Disk Graphs. PhD thesis, The University of British Columbia, Canada (1996)
15. Brückner, G., Rutter, I.: Partial and constrained level planarity. In: Klein, P.N. (ed.) Twenty-Eighth Annual ACM-SIAM Symposium on Discrete Algorithms, pp. 2000–2011. SIAM (2017)
16. Chaplick, S., Guspiel, G., Gutowski, G., Krawczyk, T., Liotta, G.: The partial visibility representation extension problem. Algorithmica **80**(8), 2286–2323 (2018)
17. Chimani, M., Klein, K.: Shrinking the search space for clustered planarity. In: Didimo, W., Patrignani, M. (eds.) 20th International Symposium on Graph Drawing, vol. 7704 of LNCS, pp. 90–101. Springer (2012)
18. Chimani, M., Gutwenger, C., Jansen, M., Klein, K., Mutzel, P.: Computing maximum c-planar subgraphs. In: Tollis, I.G., Patrignani, M. (ed.) 16th International Symposium on Graph Drawing, vol. 5417 of LNCS, pp. 114–120. Springer (2008)
19. Chimani, M., Di Battista, G., Frati, F., Klein, K.: Advances on testing c-planarity of embedded flat clustered graphs. Int. J. Found. Comput. Sci. **30**(2), 197–230 (2019)
20. Cornelsen, S., Wagner, D.: Completely connected clustered graphs. J. Discrete Algorithms **4**(2), 313–323 (2006)
21. Cortese, P.F., Patrignani, M.: Clustered planarity = flat clustered planarity. In: Biedl, T.C., Kerren, A. (eds.) 26th International Symposium on Graph Drawing and Network Visualization, vol. 11282 of LNCS, pp. 23–38. Springer (2018)
22. Cortese, P.F., Di Battista, G., Frati, F., Patrignani, M., Pizzonia, M.: C-planarity of c-connected clustered graphs. J. Graph Alg. Appl. **12**(2), 225–262 (2008)
23. Cortese, P.F., Di Battista, G., Patrignani, M., Pizzonia, M.: On embedding a cycle in a plane graph. Discrete Math. **309**(7), 1856–1869 (2009)
24. Da Lozzo, G., Di Battista, G., Frati, F., Patrignani, M.: Computing NodeTrix representations of clustered graphs. J. Graph Algorithms Appl. **22**(2), 139–176 (2018)
25. Da Lozzo, G., Eppstein, D., Goodrich, M.T., Gupta, S.: Subexponential-time and FPT algorithms for embedded flat clustered planarity. In: Brandstädt, A., Köhler, E., Meer, K. (eds.) 44th International Workshop on Graph-Theoretic Concepts in Computer Science, vol. 11159 of LNCS, pp. 111–124. Springer (2018)
26. Da Lozzo, G., Eppstein, D., Goodrich, M.T., Gupta, S.: C-planarity testing of embedded clustered graphs with bounded dual carving-width. In: Jansen, B.M.P., Telle, J.A. (eds.) 14th International Symposium on Parameterized and Exact Computation, vol. 148 of LIPIcs, pp. 9:1–9:17. Schloss Dagstuhl - Leibniz-Zentrum für Informatik (2019)
27. Dahlhaus, E.: A linear time algorithm to recognize clustered graphs and its parallelization. In: Lucchesi, C.L., Moura, A.V. (eds.) Third Latin American Symposium on Theoretical Informatics, vol. 1380 of LNCS, pp. 239–248. Springer (1998)
28. Di Battista, G., Frati, F.: Efficient c-planarity testing for embedded flat clustered graphs with small faces. J. Graph Alg. Appl. **13**(3), 349–378 (2009)
29. Erten, C., Kobourov, S.G.: Simultaneous embedding of planar graphs with few bends. J. Graph Algorithms Appl. **9**(3), 347–364 (2005)

30. Feng, Q.-W., Cohen, R.F., Eades, P.: How to draw a planar clustered graph. In: Du, D.-Z., Li, M. (eds.) First Annual International Conference on Computing and Combinatorics, vol. 959 of LNCS, pp. 21–30. Springer (1995)
31. Feng, Q.-W., Cohen, R.F., Eades, P.: Planarity for clustered graphs. In: Spirakis, P.G. (ed.) Third Annual European Symposium on Algorithms, vol. 979 of LNCS, pp. 213–226. Springer (1995)
32. Feng, Q.-W., Cohen, R.F., Eades, P.: Planarity for clustered graphs. In: Spirakis, P.G. (ed.) Algorithms - ESA '95, Third Annual European Symposium, Corfu, Greece, September 25-27, 1995, Proceedings, vol. 979 of LNCS, pp. 213–226. Springer (1995)
33. Fulek, R.: Bounded embeddings of graphs in the plane. In: Mäkinen, V., Puglisi, S.J., Salmela, L. (eds.) 27th International Workshop on Combinatorial Algorithms, vol. 9843 of LNCS, pp. 31–42. Springer (2016)
34. Fulek, R.: C-planarity of embedded cyclic c-graphs. Comput. Geom. **66**, 1–13 (2017)
35. Fulek, R.: Embedding graphs into embedded graphs. In: Okamoto, Y., Tokuyama, T. (eds.) 28th International Symposium on Algorithms and Computation, vol. 92 of LIPIcs, pp. 34:1–34:12. Schloss Dagstuhl - Leibniz-Zentrum fuer Informatik (2017)
36. Fulek, R., Kyncl, J.: Hanani-Tutte for approximating maps of graphs. In: Speckmann, B., Tóth, C.D. (eds.) 34th International Symposium on Computational Geometry, vol. 99 of LIPIcs, pp. 39:1–39:15. Schloss Dagstuhl - Leibniz-Zentrum fuer Informatik (2018)
37. Fulek, R., Tóth, C.D.: Atomic embeddability, clustered planarity, and thickenability. In: Chawla, S. (ed.) 2020 ACM-SIAM Symposium on Discrete Algorithms, pp. 2876–2895. SIAM (2020)
38. Fulek, R., Kyncl, J., Malinovic, I., Pálvölgyi, D.: Efficient c-planarity testing algebraically. CoRR, abs/1305.4519 (2013)
39. Garey, M.R., Johnson, D.S.: The Rectilinear steiner tree problem is NP-complete. SIAM J. Appl. Math. **32**, 826–834 (1977)
40. Garey, M.R., Johnson, D.S.: Crossing number is NP-complete. SIAM J. Algebr. Discrete Methods **4**(3), 312–316 (1983)
41. Giacomo, E.D., Liotta, G., Patrignani, M., Rutter, I., Tappini, A.: NodeTrix planarity testing with small clusters. Algorithmica **81**(9), 3464–3493 (2019)
42. Goodrich, M.T., Lueker, G.S., Sun, J.Z.: C-planarity of extrovert clustered graphs. In: Healy, P., Nikolov, N.S. (eds.) 13th International Symposium on Graph Drawing, vol. 3843 of LNCS, pp. 211–222. Springer (2005)
43. Gutwenger, C., Jünger, M., Leipert, S., Mutzel, P., Percan, M., Weiskircher, R.: Advances in c-planarity testing of clustered graphs. In: Kobourov, S.G., Goodrich, M.T. (eds.) 10th International Symposium on Graph Drawing, vol. 2528 of LNCS, pp. 220–235. Springer (2002)
44. Gutwenger, C., Klein, K., Mutzel, P.: Planarity testing and optimal edge insertion with embedding constraints. J. Graph Algorithms Appl. **12**(1), 73–95 (2008)
45. Gutwenger, C., Mutzel, P., Schaefer, M.: Practical experience with hanani-tutte for testing c-planarity. In: McGeoch, C.C., Meyer, U. (eds.) Sixteenth Workshop on Algorithm Engineering and Experiments, pp. 86–97. SIAM (2014)
46. Henry, N., Fekete, J.-D., McGuffin, M.J.: NodeTrix: a hybrid visualization of social networks. IEEE Trans. Vis. Comput. Graph. **13**(6), 1302–1309 (2007)
47. Henry, N., Bezerianos, A., Fekete, J.-D.: Improving the readability of clustered social networks using node duplication. IEEE Trans. Vis. Comput. Graph. **14**(6), 1317–1324 (2008)
48. Hong, S.-H., Nagamochi, H.: Simpler algorithms for testing two-page book embedding of partitioned graphs. Theor. Comput. Sci. **725**, 79–98 (2018)
49. Jelínek, V., Jelínková, E., Kratochvíl, J., Lidický, B.: Clustered planarity: Embedded clustered graphs with two-component clusters. In: Tollis, I.G., Patrignani, M. (eds.) 16th International Symposium on Graph Drawing, vol. 5417 of LNCS, pp. 121–132. Springer (2008)
50. Jelínek, V., Suchý, O., Tesar, M., Vyskocil, T.: Clustered planarity: clusters with few outgoing edges. In: Tollis, I.G., Patrignani, M. (eds.) 16th International Symposium on Graph Drawing, vol. 5417 of LNCS, pp. 102–113. Springer (2008)
51. Jelínek, V., Kratochvíl, J., Rutter, I.: A Kuratowski-type theorem for planarity of partially embedded graphs. Comput. Geom. **46**(4), 466–492 (2013)

52. Jelínková, E., Kára, J., Kratochvíl, J., Pergel, M., Suchý, O., Vyskocil, T.: Clustered planarity: Small clusters in cycles and Eulerian graphs. J. Graph Algorithms Appls **13**(3), 379–422 (2009)
53. Kammer, F.: Simultaneous embedding with two bends per edge in polynomial area. In: Arge, L., Freivalds, R. (eds.) 10th Scandinavian Workshop on Algorithm Theory, vol. 4059 of LNCS, pp. 255–267. Springer (2006)
54. Kindermann, P., Klemz, B., Rutter, I., Schnider, P., Schulz, A.: The partition spanning forest problem. CoRR, abs/1809.02710 (2018)
55. Lengauer, T.: Hierarchical planarity testing algorithms. J. ACM **36**(3), 474–509 (1989)
56. Liotta, G., Rutter, I., Tappini, A.: Graph planarity testing with hierarchical embedding constraints. CoRR, abs/1904.12596 (2019)
57. Liotta, G., Rutter, I., Tappini, A.: Simultaneous FPQ-ordering and hybrid planarity testing. In: Chatzigeorgiou, A., Dondi, R., Herodotou, H., Kapoutsis, C.A., Manolopoulos, Y., Papadopoulos, G.A., Sikora, F. (eds.) 46th International Conference on Current Trends in Theory and Practice of Informatics, vol. 12011 of LNCS, pp. 617–626. Springer (2020)
58. Mchedlidze, T., Nöllenburg, M., Rutter, I.: Extending convex partial drawings of graphs. Algorithmica **76**(1), 47–67 (2016)
59. Purchase, H.C.: Which aesthetic has the greatest effect on human understanding? In: Battista, G.D. (ed.) 5th International Symposium on Graph Drawing, vol. 1353 of LNCS, pp. 248–261. Springer (1997)
60. Schaefer, M.: Toward a theory of planarity: Hanani-tutte and planarity variants. J. Graph Algorithms Appl. **17**(4), 367–440 (2013)
61. Unger, W.: On the k-colouring of circle-graphs. In: Cori, R., Wirsing, M. (eds.) 5th Annual Symposium on Theoretical Aspects of Computer Science, vol. 294 of LNCS, pp. 61–72. Springer (1988)

Chapter 13
Simultaneous Embedding

Ignaz Rutter

Abstract Given two planar graphs G_1 and G_2 that share some vertices and edges, a *simultaneous embedding with fixed edges* (SEFE) is a pair of planar topological drawings Γ_i of G_i, for $i = 1, 2$, that coincide on the shared graph $G_1 \cap G_2$. Despite much progress in the last years, the complexity of the corresponding decision problem is still open. This chapter surveys the developments in this area from the last decade. We first describe the recently discovered relations between the SEFE problem (which asks to decide whether a given pair of graphs admits a SEFE) and several other graph drawing problems, which show that SEFE is one of the most general problems in the context of planarity. Afterward, we survey algorithmic approaches to the SEFE problem, give an overview of recent results, and discuss their limitations. We close with a brief discussion of some recent variations of the simultaneous embedding problem.

13.1 Introduction

Let $G_1 = (V_1, E_1)$ and $G_2 = (V_2, E_2)$ be two graphs that potentially share some vertices and edges. We call the graph $G = G_1 \cap G_2 = (V_1 \cap V_2, E_1 \cap E_2)$ the *shared graph* of G_1 and G_2, and we typically denote its vertex set by V and its edge set by E. Given the graphs G_1 and G_2, the problem SIMULTANEOUS EMBEDDING WITH FIXED EDGES (SEFE for short) asks whether there exists a planar topological drawing Γ_1 of G_1 and Γ_2 of G_2 such that the drawings Γ_1 and Γ_2 coincide on the shared graph G. Such a pair of drawings is called a *simultaneous planar drawing* or *simultaneous embedding with fixed edges* (SEFE for short).[1] The SEFE problem can be naturally generalized to $k > 2$ input graphs, where one requires that the topological planar

---

[1]Note that both the drawing one seeks and the problem of deciding whether given input graphs admit such a drawing are called SEFE. This is a somewhat unfortunate double-meaning. On the other hand, the meaning is typically clear, and we follow this convention from the literature.

I. Rutter (✉)
Faculty of Computer Science and Mathematics, University of Passau, Passau, Germany
e-mail: rutter@fim.uni-passau.de

© Springer Nature Singapore Pte Ltd. 2020
S.-H. Hong and T. Tokuyama (eds.), *Beyond Planar Graphs*,
https://doi.org/10.1007/978-981-15-6533-5_13

drawings Γ_i and Γ_j coincide on the graph $G_i \cap G_j$. An important special case of this is the *sunflower case*, where it is assumed that the shared graph is the same for any two different input graphs, i.e., $G_i \cap G_j$ is the same graph for each pair $i, j \in \{1, \ldots, k\}, i \neq j$.

There are several other variants of the simultaneous embedding problem that take the same input, but place different requirements on the drawings. The problem SIMULTANEOUS EMBEDDING only requires that the shared vertices are mapped to the same points, whereas the same edge may be represented by different curves in the different drawings. For the problem SIMULTANEOUS GEOMETRIC EMBEDDING, the planar drawings are required to be straight-line drawings. It is well known that every pair of planar graphs admits a simultaneous embedding[2] [42]; so in that case the focus is mostly on the number of bends per edge that such a drawing requires. The problem SIMULTANEOUS GEOMETRIC EMBEDDING on the other hand is known to be NP-hard even for two input graphs [23]. This holds even if the combinatorial embedding of the input graphs is provided as part of the input, though in that case it is only known for $k \geq 14$ graphs [6]. We are not aware of any new results in this area over the last couple of years, and therefore refer to the survey of Bläsius et al. [14]. Further variants, such as simultaneous RAC drawings, where crossings have to occur at right angles [10, 12], and simultaneous orthogonal drawings [1] have been studied. The main focus of this chapter is the SEFE problem, and for the rest of the chapter, the terms simultaneous planar drawing or simultaneous embedding refer to a SEFE, unless explicitly stated otherwise.

The survey [14] provides a thorough overview of the results up to 2012. The purpose of this chapter is to review the recent progress on this topic and also to discuss the status of the open questions from [14].

The SEFE problem has long been motivated by applications in dynamic graph drawing, where it models the requirements that a visualization of an evolving network needs to provide both an aesthetic visualization of each individual snapshot (planarity of the individual drawings) and stability over time (coinciding with the other drawing on the shared graph). In the last years, however, SEFE has taken on an even more central role. Namely, besides the usual concept of planarity of a graph, there are several variants of it such as *level planarity* [38] (where edges have fixed y-coordinates, and edges have to be drawn as y-monotone curves), a radial variant of it [11], *partially embedded planarity*, where part of the drawing of a graph is already prescribed and the question is whether it can be completed without crossings [7], and clustered planarity [24] (where one seeks a drawing that respects a given planar clustering). As it turns out, each of these problems can be reduced to SEFE in polynomial time. Thus, a polynomial-time algorithm for SEFE would not only solve the original problem itself, but at the same time it would provide a unified planarity testing algorithm that encompasses almost all known efficiently testable planarity variants. The

---

[2]Pach and Wenger [42] show that any planar graph with a fixed combinatorial embedding can be drawn with fixed vertex positions and a linear number of bends per edge. Fixing the positions of all vertices arbitrarily at distinct points in the plane and applying the result by Pach and Wenger independently for both graphs yields the desired drawing.

relation to planarity variants along with the implications for complexity questions will be discussed in Sect. 13.2.

Afterward, we will turn toward the algorithmic problem of testing whether two graphs admit a simultaneous planar drawing. In the literature there exist essentially three approaches for showing that two graphs admit a simultaneous planar drawing. The oldest, and arguably most natural approach, is simply based on constructing the corresponding drawings. From here on two independent directions have been developed.

The first is based on so-called Hanani–Tutte characterizations, and tries to relax the properties required on simultaneous planar drawings by weakening them to finding a drawing of the union of both graphs such that the only pairs of edges that may cross an odd number of times are exclusive edges from distinct graphs. As it turns out, the existence of such a drawing can be tested algebraically; the crux is that the Hanani–Tutte characterization, which claims that if a pair of graphs admit such a drawing, then it admits a simultaneous planar drawing, has only been conjectured, but so far not proven in the general case. We will review the approach and the current state in Sect. 13.3.1.

The second approach is based on the work by Jünger and Schulz [39], who show that the existence of a SEFE is equivalent to the existence of *compatible embeddings* of the input graphs, i.e., combinatorial embeddings that induce the same combinatorial embedding on the shared graph. Here a combinatorial embedding refers to the information about the *rotation scheme*, i.e., the circular ordering of the edges around the vertices, and the *relative positions*, which describe the relative positioning of the connected components to each other by specifying which faces of the connected components are identified. Two combinatorial embeddings of the shared graph are considered to be the same if both these information coincide. This has enabled what we refer to as the *constraint-based approach*. The idea is to consider all the possible planar embeddings of the shared graph and to study the constraints that the two input graphs impose on these embeddings. A compatible embedding is precisely one that satisfies both sets of constraints. The main issue is that while the structure of all the planar embeddings of a graph is sufficiently well understood, also in the presence of constraints, there is no known data structure that allows to intersect two such sets of constraints efficiently. However, solutions have been found for increasingly general cases over the course of the last years, and we will review these and the open questions deriving from them in Sect. 13.3.2.

Finally, it is worth noting that both the constraint-based approach and the Hanani–Tutte approach will only construct compatible combinatorial embeddings of the input graphs. While, technically, this answers the question of the existence of a simultaneous planar drawing, these algorithms only provide a certificate of existence for such a drawing. Thus, the shift away from the classical drawing-based approach toward the Hanani–Tutte and the constraint-based approach has given rise to a new type of problem; constructing a simultaneous planar drawing from a pair of compatible embeddings. Here the two most prominent questions are how many bends per edge such drawings require and the number of times a pair of edges may need to cross. We survey results of this type in Sect. 13.4 along with some recent developments in the area of dynamic graph drawing.

13.2 Relations to Planarity Variants and Complexity

The work of Schaefer [43] shows that a large part of the known variants for graph planarity for which efficient algorithms exist can (in most cases quite easily) be reduced to SEFE for two graphs. As mentioned before, this puts SEFE in the position that an efficient algorithm would not only solve a long-standing question in graph drawing, but would also provide a unified planarity testing algorithm for almost all planarity variants. The position of SEFE in this setting has been further emphasized by the connections to the well-known problem of *clustered planarity* (*c-planarity* for short), which was uncovered by Schaefer [43] and by Angelini and Da Lozzo [2]. Finally, while the complexity of SEFE for two input graphs remains open, there has been success in showing NP-hardness for some quite restricted generalizations of the problem. Together with the reductions mentioned before, over the last few years we have seen a border of tractability emerge, with several planarity variants that are efficiently testable on the one side and NP-hard problems on the other side. In the following, we will outline this border of tractability in more detail by explaining the connections to several planarity variants, discussing various recent complexity results related to SEFE, and finally sketching the reductions that establish the connection to the c-planarity problem.

13.2.1 Planarity Variants

In the following we show examples of reductions from different planarity variants to SEFE. The fact that most of these reductions are not difficult highlights the expressive power of the SEFE problem and underlines the importance of determining its complexity. Most of the following reductions are due to Schaefer [43].

Partially Embedded Planarity

Given a graph G, and a subgraph $H \subseteq G$ with a fixed planar embedding $\mathcal{E}_H$, the problem PARTIALLY EMBEDDED PLANARITY (PEP) is to determine whether there exists a planar embedding $\mathcal{E}_G$ of G whose restriction to H coincides with $\mathcal{E}_H$. The embedding $\mathcal{E}_G$ is then called an *embedding extension*. Figure 13.1a, b show an instance of PEP and an embedding extension.

An instance $(G, H, \mathcal{E}_H)$ of this problem can be transformed into an equivalent instance of SEFE as follows [43]. Choose $G_1 := G$, and choose G_2 as a triangulation

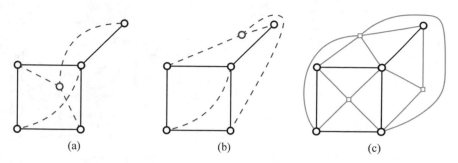

Fig. 13.1 Example instance $(G, H, \mathscr{E}_H)$ of PARTIALLY EMBEDDED PLANARITY (**a**) and an embedding extension $\mathscr{E}_G$ (**b**). The graph H is drawn black, the vertices of $V(G) - V(H)$ are red, and the edges of $E(G) - E(H)$ are red and dashed. **c** Shows the triangulation G_2 constructed from H by adding vertices (green squares) and edges (green)

of H with its embedding $\mathscr{E}_H$ that is obtained by adding vertices into the faces of H. Since we do not add any edges between vertices of H, it follows that $G_1 \cap G_2 = H$. Moreover, since G_2 is a triangulated planar graph, its combinatorial embedding on the sphere is uniquely determined up to reflection, and therefore the combinatorial embedding of H induced by a planar embedding of G_2 is either $\mathscr{E}_H$ or its reflection. It follows that G admits an embedding that extends $\mathscr{E}_H$ if and only if G_1 and G_2 admit a SEFE. Figure 13.1c shows the graph G_2 for the instance from Fig. 13.1a.

Theorem 1 ([43, Theorem 6.2]) *There is a linear-time reduction from* PARTIALLY EMBEDDED PLANARITY *to* SEFE.

Similar constructions can be made for more general types of planarity constraints. A *PQ-constraint* for a vertex v is a PQ-tree T_v that constrains the circular order of a subset of the edges incident to v to be one of the orders represented by v. Such a constraint is *total* if it constrains all the edges incident to v. The problem PARTIALLY PQ-CONSTRAINED PLANARITY asks whether a given graph G with a PQ-constraint T_v for each vertex v of G admits a planar embedding that satisfies all the constraints. Gutwenger et al. (who call such constraints *ec-constraints*) show that the problem PQ-CONSTRAINED PLANARITY, which requires that all constraints are total, can be solved in linear time [36]. Using gadgets similar to the ones by Gutwenger et al. [36] and the triangulation construction above, Schaefer [43] shows the following.

Theorem 2 ([43, Theorem 6.16] *There is a linear-time reduction from* PARTIALLY PQ-CONSTRAINED PLANARITY *to* SEFE.

Bläsius and Rutter showed that PARTIALLY PQ-CONSTRAINED PLANARITY can be solved in linear time if the input graph is biconnected [16]. The general case remains open.

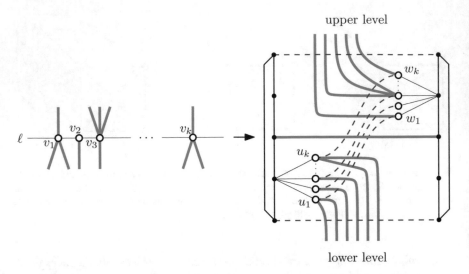

Fig. 13.2 Illustration for the reduction of level planarity to SEFE. A level ℓ of a level graph with vertices $v_1, \ldots, v_k$ and edges coming from the previous and going to the next level (left) together with the corresponding gadget (right). Edges of the shared graph are solid black, exclusive edges are dashed blue and bold green

(Radial) Level Planarity

It is well known that level planarity reduces to radial level planarity (e.g., by adding a single edge that connects a new vertex on the first level to a new vertex on the last level). Schaefer [43] describes transformations that reduce level planarity and radial level planarity to SEFE. The key idea is to model each level ℓ of the input graph that contains vertices $v_1, \ldots, v_k$ as a box-shaped gadget as in Fig. 13.2, where for each vertex v_i there are two copies u_i and w_i that are used for attaching the incoming and outgoing edges of v_i, respectively. The fact that the dashed blue edges $u_i w_i$ may not cross each other, enforces that the order of the vertices $u_1, \ldots, u_k$ in the lower half of the box is transferred to the order of the vertices $w_1, \ldots, w_k$ in the upper half of the box. As mentioned before, the incoming edges of v_i are attached to u_i and the outgoing edges to w_i. The reduction from level planarity to SEFE is formed by creating one gadget per level of the input graph and identifying the upper horizontal dashed blue edge of each gadget with the lower horizontal dashed blue edge of the gadget representing the next level. Note that the fact that the thick green edges may not cross each other, models exactly the fact that the drawing should be level planar.

Given a SEFE of the resulting instance, a level planar drawing of the original graph is obtained by taking for each level the order of the vertices $u_1, \ldots, u_k$ from top to bottom. By extending each gadget with an additional ring structure as in Fig. 13.3, one can also allow the edges leaving the gadget to "wrap around" the gadget, which models the behavior of radial levels.

Fig. 13.3 Modified gadget
for the reduction of radial
level planarity to SEFE

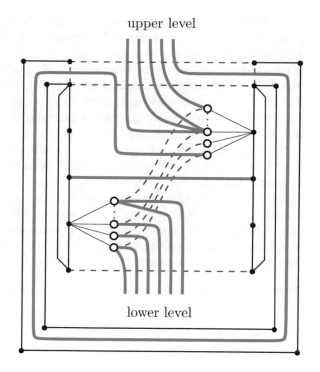

upper level

lower level

Theorem 3 ([43, Lemma 6.12, Corollary 6.13]) *There are linear-time reductions from level planarity and radial level planarity to* SEFE, *respectively.*

By replacing the star used to attach the vertices $u_1, \ldots, u_k$ to the rest of the frame of the gadget by a suitable construction, it is also possible to further constrain the vertex ordering on each level by an arbitrary PQ-tree.

Note that in the reduction from radial level planarity, for each level ℓ the gadgets representing the levels above and below ℓ lie in different faces of the gadget representing the level ℓ. It is thus not possible to link the last level to the first one to model level planarity on the torus. Note that this problem can be solved in polynomial time [4]. It is not entirely clear whether this problem can also be reduced to SEFE, though it seems that a construction similar to [3, arXiv version 1] can be used to modify the gadgets so that the ordering on the last level is propagated through all gadgets without interference.

13.2.2 Relation to Clustered Planarity

Similarly as above, Schaefer also gives a reduction from c-planarity to SEFE. Due to interesting and recent developments in this area, we devote its own section to the relation between SEFE and c-planarity.

An instance of c-planarity consists of a pair $(G, \mathcal{T})$, where $G = (V, E)$ is a planar graph and $\mathcal{T}$ is a rooted tree with leaf set V. The tree $\mathcal{T}$ defines a hierarchical clustering of G by considering for each node v of $\mathcal{T}$ the leaves of the subtree rooted at v as a cluster in G. A c-planar drawing of $(G, \mathcal{T})$ consists of a planar drawing of G together with a closed curve for each cluster C defined by $\mathcal{T}$ such that (i) the curves are pairwise non-crossing, (ii) each curve encloses exactly the vertices contained in the corresponding cluster, and (iii) each edge crosses each curve at most once. Note that the curves enclosing the clusters may only cross edges but are not allowed to contain vertices. The problem c-planarity asks whether a given instance of c-planarity admits a c-planar drawing. The problem was first studied by Lengauer [40]; the name c-planarity was coined by Feng et al. [24]. The computational complexity of the c-planarity problem is open both if the combinatorial embedding of G is fixed and if it is variable. Though recently, Cortese and Patrignani [21] have shown that the general case with hierarchical clusters can be reduced to the case where $\mathcal{T}$ has height 2, i.e., besides the trivial clusters containing only one vertex or the whole graph, there is only one partition into non-trivial clusters. Such a clustering is called *flat clustering*.

Theorem 4 [43] *There is a polynomial-time reduction from c-planarity to* SEFE.

Proof Sketch

Let $(G, \mathcal{T})$ be an instance of c-planarity. The key idea is to replace each cluster by a gadget as shown in Fig. 13.4. Each such gadget has two special *attachment vertices* I and O. The gadget for a cluster C consists of a *frame*, which is composed from the shared black edges and the bold exclusive edges (dashed blue and solid green). This frame is a subdivision of a 3-connected graph and therefore has a unique planar embedding. The gadget consists of three regions, the triangular *inner region* on the left, the shaded *transmission structure*, and the *outer region*, which lies outside the gadget.

The interior of the cluster is modeled by the content of the inner region. Each edge that leaves the cluster C is split into an *inner edge*, which starts in the inner region and ends in the transmission structure, and an *outer edge*, which starts in the transmission structure and ends in the outer region. In Fig. 13.4, these edges are bold solid green. The transmission structure, which is similar to the gadget used for the reduction of level planarity, synchronizes the order in which the inner and outer edges leave their regions. If the cluster consists of a single vertex, we place a single vertex inside the triangular region, connect all the inner edges to it and attach it to the inner attachment vertex I via a dashed blue edge. Otherwise, the cluster consists of several child clusters (observe that according to our definition each vertex also forms a cluster). We then position the corresponding gadgets inside the inner region and connect their outer attachments to the inner attachment I of cluster C. We further identify the inner edges of C with the outer edges of the children of C; see Fig. 13.4, where the gadgets of the child clusters are illustrated by a scaled outline of the gadget for C. Note that the dashed blue attachment edges ensure that the gadgets

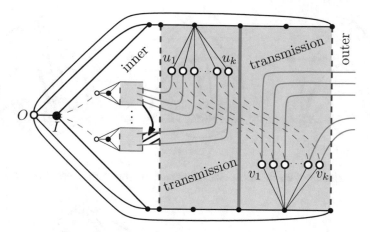

Fig. 13.4 Reduction of c-planarity to SEFE

representing the child clusters remain inside the inner region. Observe further that the order in which the inner edges enter the transmission structure is highly flexible; for example, in Fig. 13.4 by moving the upper child cluster into the tiled region as indicated by the arrow, the edges leaving these clusters can be nested. Since the inner edges may not cross each other, it is, however, not possible that the edges of two child clusters alternate.

Thus, in a SEFE of the resulting instance, we can essentially use the inner region of each cluster gadget as the cluster boundary and draw the corresponding clusters in the inside. Since the inner and outer edges are ordered consistently, they can then be joined back to form the original edges. Conversely, it is not difficult to see that, given a c-planar drawing, the gadgets can be drawn in such a way that a SEFE is obtained.   □

Interestingly, a variant of the converse also holds, though slightly stronger requirements are needed; namely the shared graph needs to be connected. We call the variant of SEFE with this input restriction CONNECTED SEFE. The following reduction is due to Angelini and Da Lozzo [2].

Theorem 5 *There is a polynomial-time reduction from* CONNECTED SEFE *to c-planarity.*

Proof Sketch

The construction works in several steps. First, the instance of CONNECTED SEFE is modified such that (1) the shared graph is a tree T and (2) only the leaves of T are incident to exclusive edges. To achieve property (1), an edge of the shared graph that lies on a cycle of the shared graph is subdivided twice, and afterward the middle segment is replaced by two parallel paths of length 3, of which the outer two

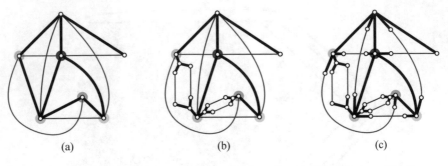

Fig. 13.5 Illustration of the first step of the reduction from CONNECTED SEFE to c-planarity

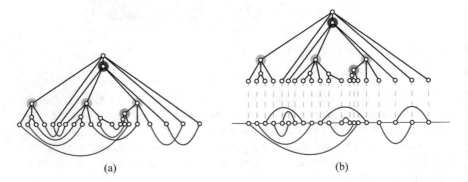

Fig. 13.6 Illustration of the second step of the reduction

edges are shared and the middle edges are exclusive and belong to different graphs; Fig. 13.5b shows the result of iteratively applying this transformation to the instance from Fig. 13.5a. Afterward, as long as there still is an exclusive edge uv such that v is not a leaf of T, subdivide uv into two edges uw, vw and declare uw a shared edge; see Fig. 13.5c. The fact that this transformation preserves the existence of a SEFE is due to Angelini et al. [8].

Observe that any SEFE of the resulting instance can be modified such that the leaves of T lie on the x-axis, the tree is drawn crossing-free in the halfplane above the x-axis, and the exclusive edges are drawn in the lower halfplane; see Fig. 13.6a. By detaching the tree from the remaining edges and flipping the red exclusive edges above the x-axis, we find that this is equivalent to a planar 2-page book embedding of the graph consisting only of the exclusive edges such that (1) the edge partition given by the book embedding coincides with the partition of the exclusive edges to the two graphs and (2) the vertex ordering along the spine is *coherent* with T, i.e., for each node v of T the vertices that correspond to the leaves of the subtree rooted at v are (circularly) consecutive along the spine. This problem is called T-COHERENT PARTITIONED 2-PAGE BOOK EMBEDDING. Again this transformation works both ways [8], and thus shows that this is equivalent to CONNECTED SEFE.

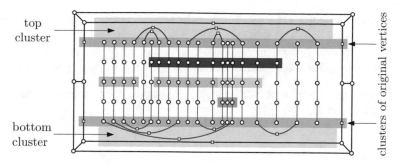

Fig. 13.7 Illustration of the third step of the reduction

Finally, and this is the novel step by Angelini and Da Lozzo [2], an instance of *T*-COHERENT PARTITIONED 2-PAGE BOOK EMBEDDING can be transformed into an equivalent instance of c-planarity as illustrated in Fig. 13.7.

The construction starts with a frame (thick black edges) that forms a subdivision of a 3-connected planar graph, and hence has a unique (up to reversal) combinatorial embedding on the sphere. Then, for each vertex of the spine, we create an upper and a lower copy and two clusters, one that contains all upper copies and one that contains all lower copies (dark shaded cluster). The inclusion of two subdivision vertices of the frame ensures that both clusters are attached to the frame. We further connect each pair of copies by a path that contains as many subdivision vertices as the tree *T* has levels (with respect to some arbitrarily chosen root). Altogether, this enforces that the clusters and the paths are embedded as shown in Fig. 13.7. In particular, the copies of the spine vertices must be embedded in the inner face of the frame, and the order of the upper and the lower copies given by the order in which their incident edges cross the boundary of their cluster is the same. The consecutivity constraints imposed by the inner nodes of *T* can then be modeled as clusters on the subdivision vertices of the paths between the copies of the spine vertices. Finally, a top and a bottom cluster is added (light gray). For each red exclusive edge, we create a subdivision vertex and assign it to the top cluster, and for each blue exclusive edge, we create a subdivision vertex and assign it to the bottom cluster. Since the top and bottom clusters also contain a subdivision vertex of the top and bottom of the frame, respectively, this ensures that the red and blue edges must be embedded crossing-free above and below the clusters containing the copies, respectively. Again, it is readily seen that the transformation is correct.   □

13.2.3   Complexity

The above reductions show that SEFE is one of the most general planarity variants. Despite much research, the complexity question is still open. While Gassner et al. [33] have shown that SEFE is NP-hard for $k \geq 3$ input graphs, their reduction yields graphs that have little structure. One may thus wonder whether structural assumptions allow to solve SEFE for $k \geq 3$ graphs, e.g., in the sunflower case or if the intersection graphs are particularly simple. The following recent result by Angelini, Da Lozzo and Neuwirth [5] shows hardness of SEFE for $k \geq 3$ even for highly restricted cases.

Theorem 6 ([5]) *Sunflower* SEFE *for $k \geq 3$ graphs is NP-complete even if the shared graph is a star.*

Proof Sketch

The reduction is from the well-known NP-hard problem BETWEENNESS [41]. Its input consists of a ground set X and a set T of triplets of distinct elements of X. The question is whether X admits a linear ordering $<$ such that for each triplet $(x, y, z) \in T$ the element y is between x and z, that is, it is $x < y < z$ or $z < y < x$.

The construction of the hardness proof is illustrated in Fig. 13.8. It consists of a frame (thick solid black and blue edges), which is a wheel graph with center vertex o and a special subdivision vertex t on one non-spine edge. Observe that the embedding of the frame is unique. For each triplet $(x, y, z) \in T$, the wheel contains three consecutive spines with end vertices a, b, c. The vertices a and c are connected to t by dotted red edges, and the vertex b is connected to t by a dashed green edge. To model the ordering constraint expressed by (x, y, z), we create for each $w \in X$ a vertex w_ℓ and a vertex w_r, and connect all of them to o by shared edges. Moreover, we pick some fixed element $w' \in X$ and connect w'_ℓ to a and b, as well as to all vertices w_ℓ with $w \in X \setminus \{w'\}$ by blue edges (thin). This enforces that all vertices w_ℓ are embedded in the face of the frame enclosed by the cycle oab. Further, we add the five blue edges $x_r y_r$, $y_r z_r$, cx_r, cy_r, cz_r, which enforces that x_r, y_r and z_r are embedded inside the face of the frame bounded by obc. Also, their ordering

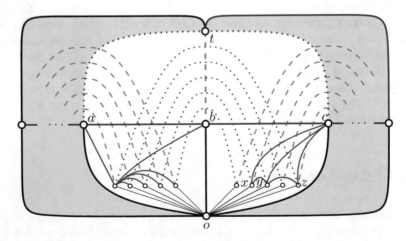

Fig. 13.8 Illustration of the hardness proof of SEFE for three graphs that share a star

around o must be such that y lies between x and z. We connect each pair of vertices w_ℓ, w_r by a red edge (dotted). This ensures that, if the vertices w_r are all embedded in the interior of the face formed by bco, then their linear orderings indeed coincide. Finally, we attach to each vertex w_ℓ a *left connector edge*, and to each vertex w_r a *right connector edge*, which belong to the green graph (green, dashed). If (x, y, z) is the first or the last triplet, we simply attach the left or the right edges to the frame. To place multiple triplets, we simply put copies of the above construction next to each other, identifying them along the boundary and also identifying their connector edges. This not only ensures that indeed all vertices w_r have to reside in the face of the frame bounded by obc, but also ensures that for each face of the frame, the ordering of the elements of X determined by the ordering of the edges ow_ℓ (or ow_r) around o is the same.

Since the construction can be performed in polynomial time, the theorem follows. $\square$

Angelini et al. [5] show further interesting hardness results. For example, two of the graphs in the construction above can be made 2-connected. They also show that the following natural optimization version MAX SEFE of SEFE is NP-hard, where one asks for planar drawings Γ_1, Γ_2 of G_1, G_2, respectively, such that as many shared edges as possible are drawn identically in G_1 and in G_2. Recently, Grilli [34] has shown that testing the existence of a SEFE, where any edge may receive at most k crossings is NP-complete for $k \geq 1$.

13.3   Algorithmic Approaches to Simultaneous Planarity Testing

In this section, we survey the two main directions that have been pursued for obtaining polynomial-time algorithms for testing simultaneous planarity for restricted inputs. Here, the two main contenders are the Hanani–Tutte approach, pioneered by Schaefer [43], and an approach that is based on modeling the SEFE problem as a planar embedding problem with constraints. The first instances of the latter seem to be the works by Angelini et al. [8] and Jampani et al. [37].

13.3.1   The Hanani–Tutte Approach

The work in this area is based on the Hanani–Tutte characterization of planarity, which states that a graph is planar if and only if it has a drawing where any two edges cross an even number of times [20]. The proof is based on a redrawing procedure that shows how to transform a drawing where any pair of edges crosses an even number of times into a planar drawing by redrawing edges one by one. In this setting, it can even be shown that the rotation system of the initial drawing can be preserved.

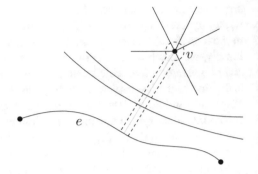

This is known as the *weak Hanani–Tutte theorem*. There is also a strong version of
the theorem, which requires only that any pair of independent edges (i.e., edges not
sharing an endpoint) crosses evenly. More formally, for a drawing D of a graph G,
iocr(D) denotes the number of pairs of independent edges that cross an odd number
of times. The independent odd crossing number iocr(G) is the minimum of iocr(D)
over all drawings D of G. The strong Hanani–Tutte theorem states that a graph G is
planar if and only if iocr$(G) = 0$. When transforming a drawing D with iocr$(D) = 0$
into an embedding, it is generally not possible to preserve the rotation system [43].
In fact, Fulek, Kynčl, and Pálvölgyi [28] show a common strengthening of the strong
and the weak Hanani–Tutte theorem by proving that a drawing where any pair of
independent edges crosses evenly can be made planar such that the rotation system
is preserved at vertices whose incident edges cross evenly.

At first sight, this looks like a purely combinatorial tool, and it is unclear how to
use it for planarity testing. However, the strong Hanani–Tutte theorem can be used
to design an algebraic planarity test. The idea is to start with an arbitrary drawing of
the input graph $G = (V, E)$. Now consider an edge $e \in E$ and a vertex v that is not
an endpoint of e. An (e, v)-*move* is a modification of the drawing, which redraws
the edge e around v. This is achieved by taking an arbitrary curve c from a point
on the curve representing e to v and then rerouting e around c; see Fig. 13.9. The
key insight is that the only additional crossings are with edges incident to v and
with edges crossed by the curve c, where the latter crossings come in pairs and do
not change the crossing parity. In total, it is exactly the edges incident to v whose
crossing parities with e change.

In particular, if $e = \{u, v\}$ and $f = \{x, y\}$ are two edges, then the crossing parity
of e and f is changed only by moves of e around endpoints of f and vice versa.
We use a variable $x_{e,v}$ over the field $\mathbb{F}_2$ with two elements to encode whether an
(e, v)-move should be performed. The requirement that edges e and f should cross
evenly after performing the moves can then be expressed by the equation

$$x_{e,x} + x_{e,y} + x_{f,u} + x_{f,v} = i_{e,f}, \tag{13.1}$$

where $i_{e,f}$ is the number of crossings between e and f modulo 2 in the initial drawing. Taking this equation for each pair of independent edges e, f yields a linear system of equations, which has a solution if and only if there is a set of (e, v)-moves that results in a drawing where any two independent edges cross evenly. By the strong Hanani–Tutte theorem this is equivalent to the existence of a planar drawing.

Schaefer [43] attempts to generalize this approach to simultaneous planar drawings by considering the *simultaneous independent odd crossing number*. For a drawing D of $G_1 \cup G_2$, we denote by $D[G_1]$ and $D[G_2]$ the subdrawings induced by G_1 and G_2, respectively. For a pair of graphs (G_1, G_2), Schaefer defines the *simultaneous independent odd crossing number* $\mathrm{siocr}(D) = \mathrm{iocr}(D[G_1]) + \mathrm{iocr}(D[G_2])$. The *simultaneous independent odd crossing number* $\mathrm{siocr}(G_1, G_2)$ is the minimum of $\mathrm{siocr}(D)$ over all drawings D of $G_1 \cup G_2$. Schaefer conjectures the following.

Conjecture 1   Two graphs (G_1, G_2) admit a SEFE if and only if $\mathrm{siocr}(G_1, G_2) = 0$.

If the conjecture holds, then an equation system similar to the one described above yields an efficient algorithm for SEFE: One starts with an arbitrary drawing and then seeks to remove odd crossings between edges of the same graph by (e, v)-moves.

So far a proof has been found only in restricted cases.

Theorem 7 ([43])   *Let G_1, G_2 be two planar graphs such that each connected component of $G_1 \cap G_2$ is biconnected or subcubic. Then G_1 and G_2 admit a SEFE if and only if $\mathrm{siocr}(G_1, G_2) = 0$.*

The advantage of the Hanani–Tutte approach is that it comes with a simple, though somewhat inefficient algorithm (solving a system of $O(n^2)$ linear equations over $\mathbb{F}_2$), and all the complexity is hidden in the redrawing steps inside the proof. In fact, Schaefer shows how to make the algorithm constructive so that it actually produces a SEFE, provided the conjecture holds. Thus, we do actually have an algorithm that either produces for each input instance a solution, or it gets stuck in a counterexample to the conjecture.

Unfortunately, even though the Hanani–Tutte approach has been extended to various drawing styles [29–31], there has not been further progress on the simultaneous version. This may partially be due to a counterexample of Fulek et al. [27]; it disproves a Hanani–Tutte variant for clustered planarity, which by means of the reduction from Sect. 13.2 from clustered planarity to simultaneous planarity might yield a counterexample to Conjecture 1.

13.3.2   Constraint-Based Simultaneous Planarity Testing

The second quite successful approach to solving restricted cases of SEFE is the constraint-based approach, where one considers the restrictions that the input graphs G_1 and G_2 impose on the planar embedding of the shared graph G as constraints. The question is then, does G admit a planar embedding that satisfies all the constraints? The key ingredient is the following theorem due to Jünger and Schulz [39].

Theorem 8 ([39]) *Let G_1 and G_2 be two planar graphs with shared graph G. Then G_1 and G_2 admit a SEFE if and only if they admit compatible embeddings.*

One might argue that this is but a small step akin to passing from planar drawings to embeddings, which are equivalence classes of drawings. To illustrate the power of this theorem, we show a simple derivation of a known result.

Theorem 9 ([25]) *A planar graph and a tree always admit a SEFE.*

Proof Let $G_1 = (V_1, E_1)$ be a planar graph, let $G_2 = (V_2, E_2)$ be a tree, and let $G = (V_1 \cap V_2, E_1 \cap E_2)$ be their shared graph. Let $\mathscr{E}_1$ be a planar embedding of G_1, which exists since G_1 is planar. Let $\mathscr{E}$ denote the restriction of $\mathscr{E}_1$ to G, i.e., it specifies for each vertex $v \in V$ the circular ordering of the edges in $E_1 \cap E_2$ around it. We arbitrarily extend this information to a rotation system of G_2. Since every rotation system of a tree is an embedding, this defines a planar embedding $\mathscr{E}_2$ of G_2 whose restriction to G is $\mathscr{E}$. That is, $\mathscr{E}_1$ and $\mathscr{E}_2$ are compatible embeddings of G_1 and G_2, and therefore, a SEFE exists by Theorem 8. □

We note that this proof is significantly shorter than the original one [25], which highlights the power of the underlying Theorem 8 that essentially does all of the heavy lifting. On the other hand, in contrast to the original construction, it is only existential and does not actually provide a SEFE in the form of a drawing. For the purpose of testing the existence of a SEFE, the increased simplicity seems key, and throughout the remainder of this section we will concern ourselves only with the (equivalent) problem of determining the existence of compatible embeddings.

The outline of a generic SEFE algorithm for (G_1, G_2) with shared graph G then becomes as follows.

1. Compute the set Ω_i of all planar embeddings of G that are *compatible with G_i*, for $i = 1, 2$, in the sense that they can be extended to a planar embedding of G_i.
2. Check whether $\Omega_1 \cap \Omega_2 \neq \emptyset$.

Indeed the same algorithm works also for $k \geq 2$ input graphs in the sunflower case. The crux is that, usually, the size of Ω_i is not polynomially bounded in the size of the input, and it is thus not feasible to compute these sets by listing all their elements. To implement the algorithm efficiently, it is thus necessary to design a data structure that can implicitly represent all the planar embeddings of G that are compatible with some planar supergraph G_i. The requirements on such a data structure are twofold. First, it needs to be computable for input graphs (G_1, G_2) in polynomial time and, second, given the data structures representing Ω_1 and Ω_2, it must be efficiently testable whether $\Omega_1 \cap \Omega_2 \neq \emptyset$. Note that the second property is essential; the pair (G_i, G) can be seen as an implicit representation of all the planar embeddings of G that are compatible with G_i. But then the intersection test is equivalent to the original SEFE problem making this a somewhat useless choice. To date, the existence of a general data structure for answering these questions is open; however, there have been fruitful developments over the past years, of which we will sketch the most

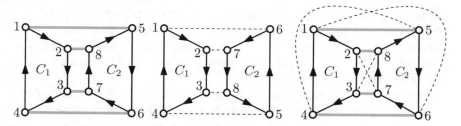

Fig. 13.10 Two graphs G_1 and G_2 with embeddings that induce the same rotation scheme on the shared graph G. The right figure shows that G_1 and G_2 do not admit a SEFE. The reason for this is that the embeddings of G_1 and G_2 induce different relative positions on the connected components of G. For G_1, cycle C_2 lies to the left of C_1 and cycle C_1 lies to the left of C_2, whereas for G_2, cycle C_2 also lies to the left of C_1, but C_1 lies to the right of C_2

important ones, as they will also nicely highlight the general idea of developing a constraint-based SEFE algorithm.

The first algorithms following this type of paradigm adressed the restriction of SEFE to inputs where the shared graph is biconnected. Interestingly, the simultaneous and independent papers [8, 37] followed the same basic approach but differed in their choice of representing all the possible planar embeddings of the shared graph. While Haupler, Jampani and Lubiw [37] based their algorithm on PQ-trees, a well-known data structure used in, e.g., planarity testing, Angelini et al. [8] used SPQR-trees, a data structure specifically designed for representing all planar embeddings of a biconnected planar graph [22]. However, the way these algorithms are constructed inherently limit them to the respective class of instances. In the following, we will survey some further algorithmic approaches, which together overcome this limitation and form the state of the art of SEFE testing algorithms to this date.

13.3.3   Relative Positions

Recall that two embeddings of a planar graph are the same if they have both the same rotation system and the same relative positions. Up to 2012, all SEFE testing algorithms focused entirely on the rotation system of the shared graph, and were thus unable to solve instances of SEFE with a disconnected shared graph; see Fig. 13.10 for a pair of graphs that do not admit a SEFE but do admit embeddings that induce the same rotation scheme on the shared graph.

In the following, we sketch the work of Bläsius and Rutter [15], who first studied this problem in isolation by designing a constraint-based algorithm for testing whether two graphs whose shared graph is a collection of disjoint cycles admit a SEFE. Observe that, in this case, the shared graph $G = (V, E)$ has a unique rotation system, so the only relevant information about a planar embedding of G are the relative positions of the connected components of G with respect to each other.

The key is of course the representation of the planar embeddings of G. To this end, they adopt the following simple idea. Fix an arbitrary orientation for each connected component of G as a directed cycle. For each orderd pair of cycles $C, C', C \neq C'$ create a Boolean variable $\text{pos}_C(C')$ that is true if and only if C' lies on the right side of C in the planar embedding of G.

Note that an arbitrary assignment of true/false values to these variables does not necessarily correspond to a planar embedding of G. As an example consider three directed cycles C_0, C_1, and C_2. Then the assignment $\text{pos}_{C_i}(C_{i-1}) = \text{true}$ and $\text{pos}_{C_i}(C_{i+1}) = \text{false}$ for $i \in \{0, 1, 2\}$, where indices are taken modulo 3, does not correspond to a planar embedding. Namely, by the values of $\text{pos}_{C_2}(\cdot)$, it follows that C_1 is drawn as a cycle in the plane that has C_0 and C_2 on different sides. But then, for each of the remaining cycles C_0 and C_2 the other two cycles have to lie on the same side, which is not the case in the given assignment.

Bläsius and Rutter [15] characterize the truth assignments of the variables $\text{pos}(\cdot)$ that correspond to a planar embedding of a connected planar graph G_i containing the cycles and show that they can be expressed as the satisfying assignments of a set of 2-SAT clauses on these variables. Therefore, Ω_i can be formulated as the set of satisfying assignments of a 2-SAT formula φ_i for $i = 1, 2$. Hence $\Omega_1 \cap \Omega_2$ is described by the satisfying assignments of the formula $\varphi = \varphi_1 \wedge \varphi_2$, which is again an instance of 2-SAT. Thus it can easily be checked whether $\Omega_1 \cap \Omega_2 \neq \emptyset$. The above naive construction yields a quadratic-size formula (as it has a quadratic number of variables). Bläsius and Rutter proceed to show that the number of variables and constraints can be reduced to a linear number. Overall, they obtain the following result.

Theorem 10 ([15]) *Given two planar graphs G_1 and G_2 whose shared graph is a collection of disjoint cycles, it can be tested in linear time whether G_1 and G_2 admit a SEFE.*

Moreover, the result can be generalized to the case where the shared graph consists of several connected components $C_1, \ldots, C_k$, each of which has a fixed rotation system, and it only remains to choose consistent relative positions. In this case, the relative position $\text{pos}_{C_i}(C_j)$ is described by giving a face of C_i whose interior contains C_j. The key insight is that, in most cases, this is a binary choice, and therefore, can be encoded as a Boolean variable. In the other cases, a linear number of different choices is possible, but these choices do not affect each other or the final outcome of the algorithm.

Theorem 11 ([15]) *SEFE can be solved in quadratic time, if the embedding of each connected component of the common graph is fixed.*

13.3.4 Simultaneous PQ-Ordering

We sketch a second constraint-based approach to the SEFE problem. The key here is a novel type of embedding representation that describes all planar embeddings of a

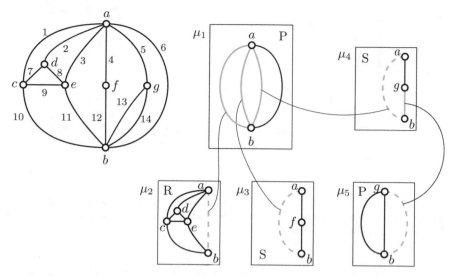

Fig. 13.11 A graph G and its SPQR-tree, where each node of the tree is annotated with its type (P for parallel, S for series, or R for rigid) and its skeleton graph. For clarity, Q-nodes are omitted. Instead, skeletons contain real edges (black) and virtual edges (gray). Observe that the embedding choice for G amounts precisely to choosing the permutations of the edges of μ_1 and μ_5 as well as choosing the flip of the 3-connected skeleton of μ_5

connected graph, the shared graph G, that can be induced by a planar embedding of a biconnected supergraph of G, e.g., the input graph G_i. Since two of these representations can be intersected efficiently, we obtain a constraint-based algorithm for SEFE when the input graphs are biconnected and the shared graph is connected [16]. The algorithm can be further generalized to admit cutvertices with at most two non-trivial blocks. In particular, this includes all cases where the input graphs have maximum degree 5 and the shared graph is connected.

The idea for the construction of this embedding representation is the following. Consider a biconnected graph G. One possibility to describe all planar embeddings of G is to consider the SPQR-tree of G and the embedding choices presented there; see Fig. 13.11.

Another possibility is to consider the embedding from the perspective of a single vertex v by making use of PQ-trees [17]. A *PQ-tree T* is a tree with a fixed rotation system whose internal vertices are either P- or Q-nodes. We consider two PQ-trees as *equivalent* if one is obtained from the other by arbitrarily changing the rotations at P-nodes and by possibly reversing the rotations at Q-nodes. Traversing a PQ-tree along an Euler tour that respects the rotation system defines a circular ordering of its leaves. A PQ-tree T represents a set of circular orders $\mathcal{O}(T)$ of its leaves $L(T)$, namely the circular orders defined by all PQ-trees that are equivalent to T.

It is well known that, for a biconnected graph G, the possible rotations around a vertex v over all planar embeddings of G can be represented by a PQ-tree $T(v)$ (see e.g., [16]), which we call the *embedding tree* of v. In fact, the embedding tree for

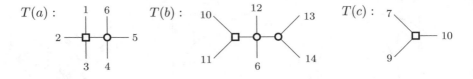

Fig. 13.12 Embedding trees of the graph G from Fig. 13.11. P-nodes are represented by disks, their neighbors may be permuted arbitrarly. Q-nodes are represented as squares and the ordering of their neighbors may be either kept or fully reversed

a vertex v can easily be read from the SPQR-tree of G. Essentially, P-nodes of the SPQR-tree become P-nodes in the PQ-tree, whereas R-nodes of the SPQR-tree become Q-nodes of the PQ-tree. Figure 13.12 shows the embedding trees of the vertices of the graph G from Fig. 13.11.

The advantage of this consideration is the following. The SPQR-tree gives a very global perspective on the embedding, where determining the ordering around v for a specific set of embedding choices requires checking the various embedding choices of the SPQR-tree that together determine the rotation of v. The PQ-tree perspective, on the other hand, is far more local as $T(v)$ directly represents the ordering of the edges around v. Since in the setting of a connected shared graph all that matters are the rotations of vertices, it seems that a more local embedding representation should be advantageous as it allows to express constraints in a more local fashion.

The main difficulty is, however, that the local embedding choices, where for each vertex $v \in V$, we have to choose one of the orders represented by the embedding tree $T(v)$, are not independent. Not every choice of orientations of the trees in Fig. 13.12 actually yields an embedding for the graph G from Fig. 13.11. Namely, the edge orderings of P- and Q-nodes of the PQ-trees that stem from the same P- and R-nodes of the SPQR-tree, respectively, need to be oriented consistently. We thus need to be able to express constraints on these choices. To model these constraints, Bläsius and Rutter [16] introduce networks of PQ-trees and the corresponding problem SIMULTANEOUS PQ-ORDERING.

Recall that each PQ-tree T with leaf set $L(T)$ represents a set $\mathcal{O}(T)$ of circular orders of $L(T)$. Let now T and T' be two PQ-trees. An *arc* e from T to T' is a triplet $e = (T, T', \varphi)$ such that φ is an injective mapping $\varphi \colon L(T') \to L(T)$. A choice of orderings $O \in \mathcal{O}(T)$ and $O' \in \mathcal{O}'(T)$ *satisfies* the arc e if $\varphi(O')$ coincides with the restriction of O to $\varphi(L(T'))$. That is, arc e expresses the condition that the ordering for T has to be chosen in such a way that the subordering of the elements whose ordering is restricted by T' is compatible with T'. Similarly, Bläsius and Rutter also allow *reversing arcs*, where one insists that the ordering $\varphi(O')$ is the reverse of the restriction of O to $\varphi(L(T'))$.

An instance of the SIMULTANEOUS PQ-ORDERING problem is a DAG N where each node v is equipped with a PQ-tree T_v, and moreover, each arc (u, v) comes with an injective mapping $\varphi_{u,v} : L(T_v) \to L(T_u)$ and is either marked as a normal or as a reversing arc. The problem SIMULTANEOUS PQ-ORDERING asks whether it is possible to choose for each node $v \in V(N)$ an ordering $O_v \in \mathcal{O}(T_v)$ such that all arcs of N are satisfied, This is a very powerful problem, and in fact it is easily shown that this problem is NP-complete. To build some familiarity with the problem, we repeat the proof from [16].

Theorem 12 SIMULTANEOUS PQ-ORDERING *is NP-complete.*

Proof The problem is clearly in NP, since we can guess a circular ordering for each node of an instance of SIMULTANEOUS PQ-ORDERING and then check in polynomial time, whether it satisfies all arcs.

We show NP-hardness by giving a reduction from the NP-hard problem CYCLIC ORDERING [32], which given a ground set X and a set of triplets $C = \{(x_1^i, x_2^i, x_3^i)\}_{i=1}^k$ asks whether there exists a circular ordering O of X such that for each triplet $(x_1^i, x_2^i, x_3^i) \in C$ the circular order in O is (x_1^i, x_2^i, x_3^i).

The network N we construct has $k + 2$ nodes. A source s whose tree T_s consists of a *universal* PQ-tree with leaf set X, i.e., it consists of a single P-node whose neighbors are the elements of X. In this way, T_s represents all circular orderings of X. For each triplet (x_1^i, x_2^i, x_3^i) of C we create a corresponding node i whose tree is a single Q-node whose leaves are x_1^i, r_2^i, x_3^i, and we create an arc (s, i, φ_i) where $\varphi_i : \{x_1^i, r_2^i, x_3^i\} \to X$ is the identity. Finally, we create a sink node t whose PQ-tree consists of a single Q-node with leaves $1, 2, 3$. For $i = 1, \ldots, k$, we add the arc (i, t, φ_i') with $\varphi_i' : \{1, 2, 3\} \to \{x_1^i, x_2^i, x_3^i\}$ given by $\varphi_i'(j) = x_j^i$ for $j = 1, 2, 3$. We claim that the constructed instance of SIMULTANEOUS PQ-ORDERING is satisfiable if and only if the original instance of CYCLIC ORDERING admits a valid solution.

Given an ordering O of X that is compatible with all triplets in C, we choose the ordering of s as O, and we choose the circular ordering for node i as (x_1^i, x_2^i, x_3^i) and the ordering for node t as $(1, 2, 3)$. This choice clearly satisfies all arcs of the instance of SIMULTANEOUS PQ-ORDERING.

Conversely, assume that we find an ordering for each node of N such that all arcs are satisfied. Without loss of generality, we may assume that the order chosen for t is $(1, 2, 3)$; otherwise we reverse all chosen orders. Let O be the order chosen for node s. The fact that the arc (i, t) is satisfied implies that the order chosen for node i is (x_1^i, x_2^i, x_3^i). And since the arcs (s, i) are all satisfied it follows that O is compatible with all triplets (x_1^i, x_2^i, x_3^i) for $i = 1, \ldots, k$, i.e., it forms a solution of the instance of CYCLIC ORDERING.   □

We now return to the construction of an embedding representation of a biconnected planar graph $G = (V, E)$ by means of an instance of SIMULTANEOUS PQ-ORDERING. As mentioned before, the task of deciding an embedding is equivalent to choosing an ordering for each of the embedding trees $T(v), v \in V$, such that

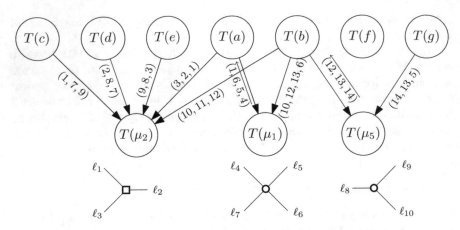

Fig. 13.13 SIMULTANEOUS PQ-ORDERING embedding representation of the graph G from Fig. 13.11. Each arc is annotated with a vector that encodes the mapping from the leaves of the child tree to the leaves of the parent by listing the images of the leaves ℓ_i in the order of their subscripts. For example, the vector $(1, 7, 9)$ for the arc from $T(c)$ to $T(\mu_2)$ encodes that ℓ_1 maps to 1, ℓ_2 to 7, and ℓ_3 to 9. Moreover, the vectors of reversing arcs are marked with a backward arrow

the edge orderings of P- and Q-nodes of the PQ-trees that stem from the same P- and R-nodes of the SPQR-tree, respectively, are oriented consistently. We ensure this by creating additional constraint trees; one for each P- and R-node of the SPQR-tree. The resulting instance, which is called the SIMULTANEOUS PQ-ORDERING *embedding representation*, for the graph from Fig. 13.11 is shown in Fig. 13.13.

For each R-node μ of the SPQR-tree of G, we create a PQ-tree $T(\mu)$, called the *constraint tree* of μ, that consists of one Q-node and three leaves, say ℓ_1, ℓ_2, ℓ_3. For each Q-node in an embedding tree $T(v)$ that stems from μ, we create an arc from $T(v)$ to $T(\mu)$, where ℓ_1, ℓ_2, ℓ_3 are mapped to edges incident to v that are in different virtual edges of the skeleton of μ and that are all ordered clockwise in one planar embedding of that skeleton. As an example, consider the tree $T(\mu_2)$ in Fig. 13.13.

For each P-node μ of the SPQR-tree, we create a constraint tree $T(\mu)$ that consists of a single P-node and whose leaves correspond to the parallel edges in the skeleton of μ. For each P-node in an embedding tree $T(v)$ that stems from μ, we create an arc from $T(v)$ to $T(\mu)$, where each leaf ℓ of $T(\mu)$ is mapped to an edge of G that is contained in the virtual edge of the skeleton of μ the leaf ℓ corresponds to. Note that for each P-node μ there are precisely two trees $T(v)$ for which we create such arcs (the two poles), and we mark exactly one of the arcs as reversing. This models precisely that the clockwise orders of the virtual edges around each of the poles of a P-node skeleton must be the reverse of each other. Examples of this are the trees $T(\mu_1)$ and $T(\mu_5)$ in Fig. 13.13.

Bläsius and Rutter [16] show that the solutions to the SIMULTANEOUS PQ-ORDERING embedding representation of a biconnected graph G are exactly the rotation systems of planar embeddings of G. Moreover, if $H = (V', E')$ is a subgraph of G, then the rotation system of H induced by restricting the rotation system of G can be obtained by creating for each $v \in V'$ a *subgraph tree* $T'(v)$ consisting of a single P-node whose leaves are the edges of H incident to v and creating an arc from the embedding tree $T(v)$ to $T'(v)$, whose associated mapping is the identity.

Now, given two biconnected planar graphs (G_1, G_2) with shared graph G, we can decide the existence of a SEFE as follows. We form an instance of SIMULTANE-OUS PQ-ORDERING that is obtained by taking the SIMULTANEOUS PQ-ORDERING embedding representations I_1 and I_2 of G_1 and G_2, respectively, including the sub-graph trees for the graph G. The instance for (G_1, G_2) is obtained by joining I_1 and I_2 into a single instance, by identifying the subgraph trees that correspond to the same vertex v of G. We denote the resulting instance by $I(G_1, G_2)$. Observe that the solutions for the individual instances I_1 and I_2 correspond bijectively to planar rotation systems of G_1 and G_2. Hence, by construction, the solutions of instance I bijectively correspond to pairs of planar rotation systems of G_1 and G_2 that induce the same rotation system on the shared graph. If the shared graph is connected, then the relative positions do not matter, and the existence of a solution is equivalent to the existence of a SEFE. By showing that the instance $I(G)$ can be solved in quadratic time, the following result is obtained.

Theorem 13 ([16]) SEFE *can be decided in* $O(n^2)$ *time for two biconnected input graphs whose shared graph is connected.*

Moreover, the result can be slightly generalized. For example, if each cutvertex of the input graphs is incident to at most *two non-trivial block*, i.e., blocks that do not consist of a single edge. This includes, for example, all graphs with maximum degree 5. The reason is that, in this case, the rotations also at cutvertices can be represented by a PQ-tree. Moreover, the result also applies if the shared graph is disconnected but acyclic, since also in this case the relative positions do not matter. The last result has found recent applications in clustered planarity [9]. The follow-ing theorem summarizes these generalizations. To the best of our knowledge, these general forms can currently not be handled by the Hanani–Tutte approach.

Theorem 14 SEFE *can be decided in* $O(n^2)$ *time for two input graphs* G_1 *and* G_2 *with shared graph* G *if both the following conditions hold:*

 (i) *each cutvertex of the input graphs is incident to at most one non-trivial block,*
(ii) *the shared graph* G *is connected or acyclic.*

We note that the results from this section crucially depend on the fact that there are only two input graphs. They do not apply to three or more input graphs even in the sunflower case.

13.3.5   Combination of Rotation System and Relative Positions

The results of Sects. 13.3.3 and 13.3.4 are in a sense complementary. The former only deals with relative positions and ignores all information concerning the rotation system, whereas the latter ignores relative positions entirely and deals only with the rotation system. It is a natural question whether it is possible to combine these two results to obtain the best of both worlds.

Note that the Hanani–Tutte approach does not make this distinction and thus can handle both aspects of embeddings at the same time. Indeed, the work of Schaefer [43], which appeared almost simultaneously with the results from the previous two subsections, was the first algorithmic approach that could handle both the rotation system and relative positions together. It was only a year later when it was found out how to combine the two approaches from the previous two subsections.

Bläsius et al. [13] proceed in two ways. First, they give a reduction rule that allows to replace in a SEFE instance each connected component of the shared graph that is biconnected by a cycle [13, Lemma 10]. If all connected components of G are biconnected, then iteratively applying this procedure yields an instance whose shared graph is a collection of cycles, which can hence be solved using Theorem 10.

To deal with non-biconnected components, Bläsius et al. restrict their attention to graphs of maximum degree 3 (they can also handle a slightly more general case that allows higher degrees within biconnected components, as long as they do not produce large parallel structures). A good starting point is of course the result from Theorem 10. However, the approach there encodes the relative positions of two distinct connected components C_1 and C_2 of the shared graph by a variable $pos_{C_1}(C_2)$, which can take as value any of the faces of C_1. But if the embedding of C_1 is no longer fixed, then its faces can change and the possible values one may choose for $pos_{C_1}(C_2)$ depend on the embedding of C_1, making it difficult to implement this idea.

To enable this approach, Bläsius et al. [13] choose a cycle cover of the connected component C_1 of G and express the relative positions of C_2 with respect to each cycle in the cycle cover of C_1. They show that these relative positions uniquely determine the face f of C_1 that contains C_2, if it exists. However, such a cycle basis usually contains a linear number of different cycles, and only a small fraction of the possible relative position assignments corresponds to a face in any embedding. Here Bläsius et al. [13] make use of the fact that, due to the low degrees of the shared graph, the embedding choice of G can essentially be described by independent Boolean decisions, and the relations between these embedding choices and which relative positions correspond to faces of the embedding can be encoded in an equation system over $\mathbb{F}_2$.

Theorem 15 ([13, Corollary 1]) SEFE *can be solved in* $O(n^3)$ *time if each connected component of the shared graph is biconnected or has maximum degree 3.*

We note that the instances covered by this result coincide with the one from the Hanani–Tutte approach from Theorem 7. The approach, here, can be generalized to also handle cases where the shared graph is not biconnected but has a degree more than 3, provided that the interaction with the two input graphs is sufficiently well behaved. Bläsius et al. [13] formalize this with the notion of the common P-node degree. However, the details are somewhat technical, and we omit them here.

13.3.6 Obstacles to Further Progress

As of today, it remains open how to choose consistent relative positions in the presence of rotation system choices of the shared graph, where more than three edges may be permuted arbitrarily, which appears both for (high-degree) cutvertices of the shared graph, and for large parallel structures of the shared graph that are not grouped into binary choices by the individual graphs.

It seems that to overcome the current stagnation, a better understanding of the interplay of the rotation at cutvertices on the one hand and the relative positions of connected components on the other hand is necessary. In fact, understanding the cutvertices alone is a formidable task. Even the case where the shared graph G is a tree (i.e., all non-leaves are cutvertices) is still open and, by Theorem 5, intimately related to the complexity of the infamous c-planarity problem.

13.4 Realizability Problems

As mentioned in the introduction to Sect. 13.3.2, the characterization of SEFE in terms of the existence of compatible embeddings from Theorem 8 has provided a new perspective on the development of SEFE algorithms, which now focus entirely on the existence of compatible embeddings, rather than on actual simultaneous planar drawings. While this has certainly been advantageous from the algorithmic point of view, it comes with the disadvantage that these algorithms only test the existence of a drawing but do not compute one. By contrast, many previous results have indeed given quality guarantees on the obtained drawings. This opens the field for the study of so-called *realizability problems*, which ask, for a given pair of compatible embeddings, how drawings realizing these embeddings can look like. We note that, though the combinatorial embeddings of G_1 and G_2 are both fixed, the topology of $G_1 \cup G_2$ is not, as we may still choose which of the exclusive edges of G_1 and G_2 cross each other. It thus makes sense to try and minimize the number of crossings or to minimize the number of bends in the drawing.

From the perspective of crossing minimization, Chimani et al. [19] have shown that in a SEFE it may be necessary that a pair of edges crosses multiple times, even linearly often in non-sunflower instances.

The question of minimizing the bends in a SEFE that realizes a given pair of compatible embedding was first introduced by Haeupler et al. [37]. They showed that if the shared graph is connected, then one can draw G_1 straight-line with the given embedding, and one can then draw the graph G_2 with at most $O(|V(G)|)$ bends per edge. Later it was shown that this can in fact be made to work even if the shared graph is disconnected.

Theorem 16 ([18]) *Let G be a graph with a fixed embedding and let H be a subgraph of G with a fixed straight-line drawing. Then there exists a planar drawing of G that extends the drawing of H and has at most $72|V(H)|$ bends per edge.*

It is an open problem whether the factor of 72 can be significantly decreased. It is not hard to see that drawing one of the graphs straight-line in an arbitrary fashion may impose a linear number of bends per edge on the other one. It is thus natural to ask whether it is possible to draw all edges with a constant number of bends. Grilli et al. [35] define a (k_1, k_2)-drawing as one where the shared edges have at most k_1 bends and the exclusive edges have at most k_2 bends. They answer the above question in the affirmative by showing that every pair of graphs with given compatible embeddings admits a $(0, 9)$ drawing, i.e., the shared graph is drawn with straight-line segments, and the exclusive edges have at most nine bends each. Grilli et al. give better results if the shared graph is connected or even biconnected. In particular, if the shared graph is biconnected and an induced subgraph of both input instances, then there exists a $(0, 0)$-drawing, i.e., a straight-line simultaneous drawing.

Theorem 17 ([35, Theorem 1, Corollaries 4,5]) *Let $G_1, \ldots, G_k$ be a sunflower instance of* SEFE *with shared graph G that admits compatible embeddings.*

(i) *If G is biconnected and an induced subgraph of each G_i, then there exists a $(0, 0)$-drawing.*

(ii) *If G is biconnected, then there exists a $(0, 1)$-drawing.*

(iii) *If G is connected and an induced subgraph of each G_i, then there exists a $(0, 1)$-drawing.*

(iv) *If G is connected, then there exists a $(0, 3)$-drawing.*

(v) *If G is an induced subgraph of each G_i, then there exists a $(0, 4)$-drawing.*

(vi) *In all cases there exists a $(0, 9)$-drawing.*

Frati et al. [26] consider the same problem and give drawing algorithms for compatible embeddings of two trees, a planar graph and a tree, and two planar graphs, respectively. For two trees they use only one bend per edge, for the other cases they use six bends per edge, while shared edges are drawn without bends.

Theorem 18 *Any pair of planar graphs that admits compatible embeddings has a $(0, 6)$-drawing such that any pair of exclusive edges crosses at most 16 times.*

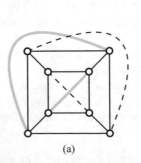 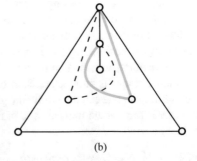

(a)	(b)

Fig. 13.14 Compatible embeddings that do not admit a straight-line simultaneous drawing even when the shared graph is **a** biconnected or **b** connected and an induced subgraph of each instance

It remains an open question whether these results can be improved. Grilli et al. [35] show that there exist graphs with compatible embeddings that do not admit a $(0, 0)$-drawing; see Fig. 13.14. But it is unclear whether a $(0, 1)$-drawing is always possible. This, however, seems quite unexpected and rather indicates a distinctive lack of lower bounds for such drawings. For SEFE of k graphs in the sunflower model, Grilli et al. prove the following asymptotic lower bound.

Theorem 19 *There exist compatible embeddings of k graphs with sunflower intersection such that any $(0, c)$-drawing has $c \in \Omega((\sqrt{2})^k / k)$.*

13.5   Conclusion

We have surveyed the state of the art in the area of simultaneous embeddings with a focus on the findings that appeared after the previous survey [14] from 2012. Since then, it has turned out that simultaneous embedding with fixed edges is indeed one of the most general variants of planarity. It encompasses almost all planarity variants for which efficient testing algorithms are known and also some whose complexity is unknown, such as the problem of clustered planarity.

We have further described the current algorithmic approaches and their limitations. In particular, it seems that further progress in the constrained-embedding approach hinges on a better understanding of the interplay between the relative positions of the connected components of the shared graph and the embedding choices offered by cutvertices of the input graphs.

Finally, we have discussed realization problems, which ask to further optimize aesthetic criteria of simultaneous planar drawings, whose existence is certified by compatible embeddings, such as the number of crossings per edge pair and the number of bends per edge.

References

1. Angelini, P., Chaplick, S., Cornelsen, S., Da Lozzo, G., Di Battista, G., Eades, P., Kindermann, P., Kratochvíl, J., Lipp, F., Rutter, I.: Simultaneous orthogonal planarity. In: 24th International Symposium on Graph Drawing and Network Visualization (GD'16), vol. 9801 of Lecture Notes in Computer Science, pp. 532–545. Springer (2016)
2. Angelini, P., Da Lozzo, G.: SEFE = c-planarity? Comput. J. **59**(12), 1831–1838 (2016)
3. Angelini, P., Da Lozzo, G., Di Battista, G., Frati, F., Patrignani, M., Rutter, I.: Testing cyclic level and simultaneous level planarity (2015). CoRR, arXiv:1510.08274
4. Angelini, P., Da Lozzo, G., Di Battista, G., Frati, F., Patrignani, M., Rutter, I.: Beyond level planarity. In: Hu, Y., Nöllenburg, M. (eds.) Proceedings of the 24th International Symposium on Graph Drawing and Network Visualization (GD'16), vol. 9801 of Lecture Notes in Computer Science, pp. 482–495. Springer (2016)
5. Angelini, P., Da Lozzo, G., Neuwirth, D.: Advancements on SEFE and partitioned book embedding problems. Theor. Comput. Sci. **575**, 71–89 (2015)

6. Angelini, P., Di Battista, G., Frati, F.: Simultaneous embedding of embedded planar graphs. Int. J. Comput. Geometry Appl. **23**(2), 93–126 (2013)
7. Angelini, P., Di Battista, G., Frati, F., Jelínek, V., Kratochvíl, J., Patrignani, M., Rutter, I.: Testing planarity of partially embedded graphs. ACM Trans. Algorithms **11**(4), 32:1–32:42 (2015)
8. Angelini, P., Di Battista, G., Frati, F., Patrignani, M., Rutter, I.: Testing the simultaneous embeddability of two graphs whose intersection is a biconnected or a connected graph. J. Discrete Algorithms **14**, 150–172 (2012)
9. Angelini, P., Lozzo, G.D.: Clustered planarity with pipes. Algorithmica **81**(6), 2484–2526 (2019)
10. Argyriou, E.N., Bekos, M.A., Kaufmann, M., Symvonis, A.: Geometric RAC simultaneous drawings of graphs. J. Graph Algorithms Appl. **17**(1), 11–34 (2013)
11. Bachmaier, C., Brandenburg, F.J., Forster, M.: Radial level planarity testing and embedding in linear time. J. Graph Algorithms Appl. **9**(1), 53–97 (2005)
12. Bekos, M.A., van Dijk, T.C., Kindermann, P., Wolff, A.: Simultaneous drawing of planar graphs with right-angle crossings and few bends. J. Graph Algorithms Appl. **20**(1), 133–158 (2016)
13. Bläsius, T., Karrer, A., Rutter, I.: Simultaneous embedding: edge orderings, relative positions, cutvertices. Algorithmica **80**(4), 1214–1277 (2018)
14. Bläsius, T., Kobourov, S.G., Rutter, I.: Simultaneous embedding of planar graphs. In: Tamassia, R. (ed.) Handbook of Graph Drawing and Visualization, Discrete Mathematics and its Applications, pp. 349–373. CRC Press (2014)
15. Bläsius, T., Rutter, I.: Disconnectivity and relative positions in simultaneous embeddings. Comput. Geom. **48**(6), 459–478 (2015)
16. Bläsius, T., Rutter, I.: Simultaneous PQ-ordering with applications to constrained embedding problems. ACM Trans. Algorithms **12**(2), 16:1–16:46 (2016)
17. Booth, K.S., Lueker, G.S.: Testing for the consecutive ones property, interval graphs, and graph planarity using PQ-tree algorithms. J. Comput. Syst. Sci. **13**(3), 335–379 (1976)
18. Chan, T.M., Frati, F., Gutwenger, C., Lubiw, A., Mutzel, P., Schaefer, M.: Drawing partially embedded and simultaneously planar graphs. J. Graph Algorithms Appl. **19**(2), 681–706 (2015)
19. Chimani, M., Jünger, M., Schulz, M.: Crossing minimization meets simultaneous drawing. In: Proceedings of the IEEE Pacific Visualization Symposium (PacificVis '08), pp. 33–40. IEEE (2008)
20. Chojnacki, C.: Über wesentlich unplättbare Kurven im dreidimensionalen Raume. Fundamenta Mathematicae **23**(1), 135–142 (1934)
21. Cortese, P.F., Patrignani, M.: Clustered planarity = flat clustered planarity (2018). CoRR arXiv:1808.07437
22. Di Battista, G., Tamassia, R.: On-line maintenance of triconnected components with SPQR-trees. Algorithmica **15**(4), 302–318 (1996)
23. Estrella-Balderrama, A., Gassner, E., Jünger, M., Percan, M., Schaefer, M., Schulz, M.: Simultaneous geometric graph embeddings. In: Proceedings of the 15th International Symposium on Graph Drawing, (GD'07), pp. 280–290 (2007)
24. Feng, Q., Cohen, R.F., Eades, P.: Planarity for clustered graphs. In: Spirakis, P.G. (ed.) Proceedings of the 3rd Annual European Symposium on Algorithms (ESA'95), vol. 979 of Lecture Notes in Computer Science, pp. 213–226. Springer (1995)
25. Frati, F.: Embedding graphs simultaneously with fixed edges. In: Kaufmann, M., Wagner, D. (eds.) Proceedings of the 14th International Symposium on Graph Drawing (GD'06), vol. 4372 of Lecture Notes in Computer Science, pp. 108–113. Springer (2006)
26. Frati, F., Hoffmann, M., Kusters, V.: Simultaneous embeddings with few bends and crossings. In: Giacomo, E.D., Lubiw, A. (eds) 23rd International Symposium on Graph Drawing and Network Visualization (GD'15), vol. 9411 of Lecture Notes in Computer Science, pp. 166–179. Springer (2015)
27. Fulek, R., Kyncl, J., Malinovic, I., Pálvölgyi, D.: Clustered planarity testing revisited. Electr. J. Comb. **22**(4), P4.24 (2015)

28. Fulek, R., Kyncl, J., Pálvölgyi, D.: Unified Hanani-Tutte theorem. Electr. J. Comb. **24**(3), P3.18 (2017)
29. Fulek, R., Pelsmajer, M., Schaefer, M.: Hanani-Tutte for radial planarity. In: Di Giacomo, E., Lubiw, A. (eds.) Graph Drawing and Network Visualization, pp. 99–110. Springer International Publishing (2015)
30. Fulek, R., Pelsmajer, M., Schaefer, M.: Hanani-Tutte for radial planarity II. In: Hu, Y., Nöllenburg, M. (eds.) Graph Drawing and Network Visualization, pp. 468–481. Springer International Publishing (2016)
31. Fulek, R., Pelsmajer, M.J., Schaefer, M., Štefankovič, D.: Hanani-Tutte, monotone drawings, and level-planarity. In: Pach, J. (ed.) Thirty Essays on Geometric Graph Theory, pp. 263–287. Springer, New York (2013)
32. Galil, Z., Megiddo, N.: Cyclic ordering is NP-complete. Theor. Comput. Sci. **5**, (1977)
33. Gassner, E., Jünger, M., Percan, M., Schaefer, M., Schulz, M.: Simultaneous graph embeddings with fixed edges. In: Proceedings of the 32nd International Workshop on Graph-Theoretic Concepts in Computer Science (WG'06), pp. 325–335 (2006)
34. Grilli, L.: On the NP-hardness of GRacSim drawing and k-SEFE problems. J. Graph Algorithms Appl. **22**(1), 101–116 (2018)
35. Grilli, L., Hong, S., Kratochvíl, J., Rutter, I.: Drawing simultaneously embedded graphs with few bends. In: Proceedings of the 22nd International Symposium on Graph Drawing (GD'04)
36. Gutwenger, C., Mutzel, P., Weiskircher, R.: Inserting an edge into a planar graph. Algorithmica **41**(4), 289–308 (2005)
37. Haeupler, B., Jampani, K.R., Lubiw, A.: Testing simultaneous planarity when the common graph is 2-connected. J. Graph Algorithms Appl. **17**(3), 147–171 (2013)
38. Jünger, M., Leipert, S.: Level planar embedding in linear time. In: Kratochvíl, J. (ed.) Graph Drawing and Network Visualization, pp. 72–81. Springer, Berlin Heidelberg (1999)
39. Jünger, M., Schulz, M.: Intersection graphs in simultaneous embedding with fixed edges. J. Graph Algorithms Appl. **13**(2), 205–218 (2009)
40. Lengauer, T.: Hierarchical planarity testing algorithms. J. ACM **36**(3), 474–509 (1989)
41. Opatrny, J.: Total ordering problem. SIAM J. Comput. **8**(1), 111–114 (1979)
42. Pach, J., Wenger, R.: Embedding planar graphs at fixed vertex locations. Graphs Comb. **17**(4), 717–728 (2001)
43. Schaefer, M.: Toward a theory of planarity: Hanani-tutte and planarity variants. J. Graph Algorithms Appl. **17**(4), 367–440 (2013)

Index

© Springer Nature Singapore Pte Ltd. 2020
S.-H. Hong and T. Tokuyama (eds.), *Beyond Planar Graphs*,
https://doi.org/10.1007/978-981-15-6533-5